NEUROMETHODS

Series Editor
Wolfgang Walz
University of Saskatchewan
Saskatoon, SK, Canada

For further volumes:
http://www.springer.com/series/7657

Genomic Mosaicism in Neurons and Other Cell Types

Editors

José María Frade

*Department of Molecular, Cellular and Developmental Neurobiology,
Cajal Institute (IC–CSIC), Madrid, Spain*

Fred H. Gage

Biological Studies and Laboratory of Genetics, The Salk Institute, La Jolla, CA, USA

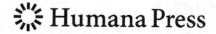 Humana Press

Editors
José María Frade
Department of Molecular, Cellular
    and Developmental Neurobiology
Cajal Institute (IC-CSIC)
Madrid, Spain

Fred H. Gage
Biological Studies and Laboratory of Genetics
The Salk Institute
La Jolla, CA, USA

ISSN 0893-2336   ISSN 1940-6045   (electronic)
Neuromethods
ISBN 978-1-4939-8440-4   ISBN 978-1-4939-7280-7   (eBook)
DOI 10.1007/978-1-4939-7280-7

This Humana Press imprint is published by Springer Nature
The registered company is Springer Science+Business Media LLC
The registered company address is: 233 Spring Street, New York, NY 10013, U.S.A.

Preface to the Series

Experimental life sciences have two basic foundations: concepts and tools. The *Neuromethods* series focuses on the tools and techniques unique to the investigation of the nervous system and excitable cells. It will not, however, shortchange the concept side of things as care has been taken to integrate these tools within the context of the concepts and questions under investigation. In this way, the series is unique in that it not only collects protocols but also includes theoretical background information and critiques which led to the methods and their development. Thus it gives the reader a better understanding of the origin of the techniques and their potential future development. The *Neuromethods* publishing program strikes a balance between recent and exciting developments like those concerning new animal models of disease, imaging, in vivo methods, and more established techniques, including, for example, immunocytochemistry and electrophysiological technologies. New trainees in neurosciences still need a sound footing in these older methods in order to apply a critical approach to their results.

Under the guidance of its founders, Alan Boulton and Glen Baker, the *Neuromethods* series has been a success since its first volume published through Humana Press in 1985. The series continues to flourish through many changes over the years. It is now published under the umbrella of Springer Protocols. While methods involving brain research have changed a lot since the series started, the publishing environment and technology have changed even more radically. *Neuromethods* has the distinct layout and style of the Springer Protocols program, designed specifically for readability and ease of reference in a laboratory setting.

The careful application of methods is potentially the most important step in the process of scientific inquiry. In the past, new methodologies led the way in developing new disciplines in the biological and medical sciences. For example, Physiology emerged out of Anatomy in the nineteenth century by harnessing new methods based on the newly discovered phenomenon of electricity. Nowadays, the relationships between disciplines and methods are more complex. Methods are now widely shared between disciplines and research areas. New developments in electronic publishing make it possible for scientists that encounter new methods to quickly find sources of information electronically. The design of individual volumes and chapters in this series takes this new access technology into account. Springer Protocols makes it possible to download single protocols separately. In addition, Springer makes its print-on-demand technology available globally. A print copy can therefore be acquired quickly and for a competitive price anywhere in the world.

Saskatoon, Canada   *Wolfgang Walz*

Preface

A variety of techniques developed over many years of research have made it possible to demonstrate that vertebrate neurons are not genomically homogeneous and that this genomic mosaicism contributes to the cellular diversity that characterizes the vertebrate nervous system. This volume summarizes the currently available methods for the analysis of genomic variability in vertebrate neurons. These methods are continuously evolving to face this challenging problem, and we foresee that future methodologies will surely improve our knowledge about the actual genomic composition of vertebrate neurons. Another aim of this volume has been to attract the attention of readers to a novel field of research that opens new avenues in the way we understand the brain and its basic constituents: the neurons. A deep knowledge of the mechanisms that trigger the enormous amount of variability in the normal and pathological brain will surely facilitate the design in the future of previously unimaginable therapies against brain disease.

The first method that unequivocally demonstrated the existence of genomic variability in neurons was fluorescence in situ hybridization (FISH). By using chromosome-specific, composite probe pools, Jerold Chun's laboratory showed in 2001 the existence of aneuploid neurons in the adult brain [1]. This finding was confirmed by alternative FISH methods that used probes against whole chromosomes [2] or multiple probes recognizing different regions from the same chromosome, such as those described by Montagna and coworkers in this volume.

In addition to the FISH technology, other new techniques have improved our knowledge about the variability in the genomic composition of vertebrate neurons. Next generation sequencing (NGS) applied to single-cell genomes, a technology initially developed for the analysis of cancer cells [3], has been a major advance in the analysis of genomic variability in cells. This methodology has been rapidly adopted by neuroscientists, transforming the way in which genomic mosaicism in neurons can be studied. Examples of two specific methods for whole genome amplification of single cells followed by deep sequencing are shown in Chaps. 6 and 7. Single-cell DNA sequencing has shown not only the existence of gains or losses of full chromosomes from a euploid complement [4], thus confirming previous FISH analyses, but also the presence of subchromosomal copy number variation consisting of relatively small deletions and duplications of genomic DNA [5, 6] as well as variations in the distribution throughout the genome of LINE-1 sequences [7].

Single-cell NGS applied to single neurons requires a previous genome amplification step that might introduce artifacts in the analysis. In addition, single probe FISH might misinterpret specific deletions or duplications as losses or gains of full chromosomes. Therefore, a main goal in the field has been to develop improved protocols for unbiased analysis of neuronal genomes; some ideas for this improvement are discussed in Chap. 1. Further development of the protocols of single-cell sequencing will surely clarify current discrepancies about the actual proportion of aneuploid neurons in the mammalian brain, which seems to differ depending on whether FISH or single-cell sequencing is used for the analysis [4, 8].

Neither FISH nor single-cell genomic sequencing can identify the existence of full genome duplications in neurons. Indeed, the presence of four FISH spots in a tetraploid neuron could be misinterpreted as an event of tetrasomy, and single-cell sequencing cannot identify the ploidy level. Flow cytometry and slide-based cytometry have emerged as highly useful methods that unambiguously detect hyperploidy in neurons [9–11]. These procedures are often complemented with FISH to unequivocally demonstrate the existence of tetraploidy. In this regard, Chap. 3 describes protocols for the use of cell nuclei sorting followed by FISH, which would be useful for this kind of analyses.

The molecular mechanisms involved in retrotransposition are essentially known [12–14], and the regulation of cell cycle progression in differentiating neurons leading to neuronal tetraploidy is beginning to be understood [10, 11, 15]. In contrast, the mechanisms leading to deletions and duplication in specific genomic regions, as well as the way in which chromosomal gains or losses appear in the developing and adult vertebrate nervous system, remain obscure. The possibility that random mutagenesis is involved in these latter processes cannot be ruled out. Indeed, a number of studies have reported massive cell death during neural development in mouse models defective in DNA double-strand break (DSB) repair [16–21], suggesting that stochastic DNA breaks occurring during development could result in genomic mosaicism. Similarly, aneuploidy could also derive from randomly occurring aberrant mitosis [1]. Nevertheless, both aneuploidies and other genomic alterations occurring during the development of the vertebrate nervous system could also derive, at least partially, from a developmental program aimed at generating genomic variability in the nervous system. This view is supported by the observation that, during their last division, S-phase is shortened in neural precursors committed to neuronal differentiation [22, 23]. This observation led [22] to suggest that terminally differentiating neuronal precursors invest less time during S-phase in controlling the quality of replicated DNA, thus facilitating the existence of replicative stress and DSBs in their DNA. The known concatenation checkpoint deficiency in neural progenitor cells [24] could facilitate mitosis entry prior to chromosome disentangling, thus triggering genomic and chromosomal aberrations in vertebrate neurons. This possibility is consistent with the observation that a substantial proportion of terminally differentiating neuronal precursors contains altered DNA profiles (both hyperploidy and hypoploidy) when studied by flow cytometry [23]. Different degrees of genomic variability are expected to result from this process, and those neurons with elevated levels of aneuploidy are then removed by apoptosis, as described by [25]. This Darwinian-like process would be reminiscent of the apoptotic removal of neurons incorrectly innervating their targets during the neurotrophic phase [26].

In the adult brain, DSBs can be generated by oxidative stress and other genotoxic agents [27], and deficiencies in DSB repair associated with neurodegeneration [28] or viral infections [29] may be the basis of DNA copy variation in the adult brain. In addition, aneuploidy [30], which may be derived from aberrant cell cycle events in neurons [31], is increased in the adult brain. It is crucial, therefore, to develop methods for the evaluation of cell cycle progression in the aging brain to decipher the events involved in the creation of this type of mosaicism associated with aging and neurodegeneration.

We hope this volume has allowed readers to learn the most prominent techniques currently available for the analysis of genome and genetic mosaicism in vertebrate neurons and other cell types. We also urge our readers to join us in this still developing enterprise to describe the mechanisms and effects that genomic variability triggers in both normal and pathological neurons.

Madrid, Spain   *José María Frade*
La Jolla, CA, USA   *Fred H. Gage*

Acknowledgments

We thank Ministerio de Economía y Competitividad (grant numbers SAF2015-68488-R) and Tetraneuron S.L. for the support to our research (J.M.F.). We also thank Mary Lynn Gage for editing.

References

1. Rehen SK, McConnell MJ, Kaushal D, Kingsbury MA, Yang AH, Chun J (2001) Chromosomal variation in neurons of the developing and adult mammalian nervous system. Proc Natl Acad Sci U S A 98:13361–13366

2. Iourov IY, Vorsanova SG, Liehr T, Yurov YB (2009) Aneuploidy in the normal, Alzheimer's disease and ataxia-telangiectasia brain: differential expression and pathological meaning. Neurobiol Dis 34:212–220

3. Navin N, Kendall J, Troge J, Andrews P, Rodgers L, McIndoo J, Cook K, Stepansky A, Levy D, Esposito D, Muthuswamy L, Krasnitz A, McCombie WR, Hicks J, Wigler M (2011) Tumour evolution inferred by single-cell sequencing. Nature 472:90–94

4. Knouse KA, Wu J, Whittaker CA, Amon A (2014) Single cell sequencing reveals low levels of aneuploidy across mammalian tissues. Proc Natl Acad Sci U S A 111:13409–13414

5. McConnell MJ, Lindberg MR, Brennand KJ, Piper JC, Voet T, Cowing-Zitron C, Shumilina S, Lasken RS, Vermeesch JR, Hall IM, Gage FH (2013) Mosaic copy number variation in human neurons. Science 342:632–637

6. Cai X, Evrony GD, Lehmann HS, Elhosary PC, Mehta BK, Poduri A, Walsh CA. (2014) Single-cell, genome-wide sequencing identifies clonal somatic copy-number variation in the human brain. Cell Rep 8:1280–1289

7. Erwin JA, Paquola AC, Singer T, Gallina I, Novotny M, Quayle C, Bedrosian TA, Alves FI, Butcher CR, Herdy JR, Sarkar A, Lasken RS, Muotri AR, Gage FH (2016) L1-associated genomic regions are deleted in somatic cells of the healthy human brain. Nat Neurosci 19:1583–1591

8. van den Bos H, Spierings DC, Taudt AS, Bakker B, Porubský D, Falconer E, Novoa C, Halsema N, Kazemier HG, Hoekstra-Wakker K, Guryev V, den Dunnen WF, Foijer F, Tatché MC, Boddeke HW, Lansdorp PM (2016) Single-cell whole genome sequencing reveals no evidence for common aneuploidy in normal and Alzheimer's disease neurons. Genome Biol 17:116

9. Mosch B, Morawski M, Mittag A, Lenz D, Tarnok A, Arendt T (2007) Aneuploidy and DNA replication in the normal human brain and Alzheimer's disease. J Neurosci 27:6859–6867

10. Morillo SM, Escoll P, de la Hera A, Frade JM (2010) Somatic tetraploidy in specific chick retinal ganglion cells induced by nerve growth factor. Proc Natl Acad Sci U S A 107:109–114

11. López-Sánchez N, Frade JM (2013) Genetic evidence for p75^{NTR}-dependent tetraploidy in cortical projection neurons from adult mice. J Neurosci 33:7488–7500

12. Ostertag EM, Kazazian HH Jr (2001) Biology of mammalian L1 retrotransposons. Annu Rev Genet 35:501–538

13. Muotri AR, Chu VT, Marchetto MC, Deng W, Moran JV, Gage FH (2005) Somatic mosaicism in neuronal precursor cells mediated by L1 retrotransposition. Nature 435:903–910

14. Thomas CA, Paquola AC, Muotri AR (2012) LINE-1 retrotransposition in the nervous system. Annu Rev Cell Dev Biol 28:555–573

15. Morillo SM, Abanto EP, Román MJ, Frade JM (2012) Nerve growth factor-induced cell cycle reentry in newborn neurons is triggered by p38^{MAPK}-dependent E2F4 phosphorylation. Mol Cell Biol 32:2722–2737

16. Orii KE, Lee Y, Kondo N, McKinnon PJ (2006) Selective utilization of nonhomologous end-joining and homologous recombination DNA repair pathways during nervous system development. Proc Natl Acad Sci U S A 103:10017–10022

17. Gao Y, Sun Y, Frank KM, Dikkes P, Fujiwara Y, Seidl KJ, Sekiguchi JM, Rathbun GA, Swat W, Wang J, Bronson RT, Malynn BA, Bryans M, Zhu C, Chaudhuri J, Davidson L, Ferrini R, Stamato T, Orkin SH, Greenberg ME, Alt FW (1998) A critical role for DNA end-joining proteins in both lymphogenesis and neurogenesis. Cell 95:891–902

18. Deans B, Griffin CS, Maconochie M, Thacker J (2000) Xrcc2 is required for genetic stability, embryonic neurogenesis and viability in mice. EMBO J 19:6675–6685

19. Frank KM, Sharpless NE, Gao Y, Sekiguchi JM, Ferguson DO, Zhu C, Manis JP, Horner J, DePinho RA, Alt FW (2000) DNA ligase IV deficiency in mice leads to defective neurogenesis and embryonic lethality via the p53 pathway. Mol Cell 5:993–1002

20. McKinnon PJ (2009) DNA repair deficiency and neurological disease. Nat Rev Neurosci 10:100–112

21. Baleriola J, Álvarez-Lindo N, de la Villa P, Bernad A, Blanco L, Suárez T, de la Rosa EJ (2016) Increased neuronal death and disturbed axonal growth in the Polμ-deficient mouse embryonic retina. Sci Rep 6:25928

22. Arai Y, Pulvers JN, Haffner C, Schilling B, Nüsslein I, Calegari F, Huttner WB (2011) Neural stem and progenitor cells shorten S-phase on commitment to neuron production. Nat Commun 2:154

23. Saade M, Gutiérrez-Vallejo I, Le Dréau G, Rabadán MA, Miguez DG, Buceta J, Martí E (2013) Sonic hedgehog signaling switches the mode of division in the developing nervous system. Cell Rep 4:492–503

24. Damelin M, Sun YE, Sodja VB, Bestor TH (2005) Decatenation checkpoint deficiency in stem and progenitor cells. Cancer Cell 8:479–484

25. Peterson SE, Yang AH, Bushman DM, Westra JW, Yung YC, Barral S, Mutoh T, Rehen SK, Chun J (2012) Aneuploid cells are differentially susceptible to caspase-mediated death during embryonic cerebral cortical development. J Neurosci 32:16213–16222

26. de la Rosa EJ, de Pablo F (2000) Cell death in early neural development: beyond the neurotrophic theory. Trends Neurosci 23:454–458

27. De Zio D, Bordi M, Cecconi F (2012) Oxidative DNA damage in neurons: implication of ku in neuronal homeostasis and survival. Int J Cell Biol 2012:752420

28. Kanungo J (2016) DNA-PK Deficiency in Alzheimer's Disease. J Neurol Neuromed 1:17–22

29. De Chiara G, Racaniello M, Mollinari C, Marcocci ME, Aversa G, Cardinale A, Giovanetti A, Garaci E, Palamara AT, Merlo D (2016) Herpes simplex virus-type 1 (HSV-1) impairs DNA repair in cortical neurons. Front Aging Neurosci 8:242

30. Faggioli F, Wang T, Vijg J, Montagna C (2012) Chromosome-specific accumulation of aneuploidy in the aging mouse brain. Hum Mol Genet 21:5246–5253

31. Frade JM, Ovejero-Benito MC (2015) Neuronal cell cycle: the neuron itself and its circumstances. Cell Cycle 14:712–720

Contents

Preface to the Series.. *v*

Preface.. *vii*

Contributors.. *xiii*

PART I INTRODUCTION

1   Principles and Approaches for Discovery and Validation of Somatic
    Mosaicism in the Human Brain... 3
    Alexej Abyzov, Alexander E. Urban, and Flora M. Vaccarino

PART II ANEUPLOIDY AND PLOIDY VARIATION

2   FISH-Based Assays for Detecting Genomic (Chromosomal)
    Mosaicism in Human Brain Cells.. 27
    Yuri B. Yurov, Svetlana G. Vorsanova, Ilia V. Soloviev,
    Alexei M. Ratnikov, and Ivan Y. Iourov

3   Flow Cytometric and Sorting Analyses for Nuclear DNA Content,
    Nucleotide Sequencing, and Interphase FISH.................................. 43
    Gwendolyn E. Kaeser and Jerold Chun

4   Flow Cytometric Quantification, Isolation, and Subsequent
    Epigenetic Analysis of Tetraploid Neurons.................................... 57
    Noelia López-Sánchez, Iris Patiño-Parrado, and José María Frade

5   A Cytomic Approach Towards Genomic Individuality of Neurons............... 81
    Thomas Arendt, Birgit Belter, Martina K. Brückner,
    Uwe Ueberham, Markus Morawski, and Attila Tarnok

PART III DNA COPY NUMBER VARIATION

6   Single-Cell CNV Detection in Human Neuronal Nuclei........................ 109
    Margaret B. Wierman, Ian E. Burbulis, William D. Chronister,
    Stefan Bekiranov, and Michael J. McConnell

7   Multiple Annealing and Looping-Based Amplification Cycles
    (MALBAC) for the Analysis of DNA Copy Number Variation................. 133
    Chenghang Zong

8   Competitive PCR for Copy Number Assessment by Restricting dNTPs........ 143
    Luming Zhou, Robert A. Palais, Yotam Ardon, and Carl T. Wittwer

9   Using Cloning to Amplify Neuronal Genomes for Whole-Genome
    Sequencing and Comprehensive Mutation Detection and Validation......... 163
    Jennifer L. Hazen, Michael A. Duran, Ryan P. Smith,
    Alberto R. Rodriguez, Greg S. Martin, Sergey Kupriyanov, Ira M. Hall,
    and Kristin K. Baldwin

PART IV   LINE-1 RETROTRANSPOSITION

10   Analysis of LINE-1 Retrotransposition
     in Neural Progenitor Cells and Neurons   189
     Angela Macia and Alysson R. Muotri

11   Estimation of LINE-1 Copy Number in the Brain
     Tissue and Isolated Neuronal Nuclei.   209
     Miki Bundo, Tadafumi Kato, and Kazuya Iwamoto

12   Analysis of Somatic LINE-1 Insertions in Neurons   219
     Francisco J. Sanchez-Luque, Sandra R. Richardson,
     and Geoffrey J. Faulkner

13   Single-Cell Whole Genome Amplification and Sequencing
     to Study Neuronal Mosaicism and Diversity   253
     Patrick J. Reed, Meiyan Wang, Jennifer A. Erwin,
     Apuã C. M. Paquola, and Fred H. Gage

PART V   GENETIC AND GENOMIC MOSAICISM IN AGING AND DISEASE

14   FISH Analysis of Aging-Associated Aneuploidy in Neurons
     and Nonneuronal Brain Cells   271
     Grasiella A. Andriani and Cristina Montagna

15   Genomic Analysis and In Vivo Functional Validation
     of Brain Somatic Mutations Leading to Focal Cortical Malformations   299
     Jae Seok Lim and Jeong Ho Lee

16   Using Fluorescence In Situ Hybridization (FISH) Analysis
     to Measure Chromosome Instability and Mosaic Aneuploidy
     in Neurodegenerative Diseases   329
     Julbert Caneus, Antoneta Granic, Heidi J. Chial,
     and Huntington Potter

17   Identification of Low Allele Frequency Mosaic Mutations
     in Alzheimer Disease.   361
     Carlo Sala Frigerio, Mark Fiers, Thierry Voet,
     and Bart De Strooper

Index *379*

Contributors

ALEXEJ ABYZOV • *Department of Health Sciences Research, Center for Individualized Medicine, Mayo Clinic, Rochester, MN, USA*

GRASIELLA A. ANDRIANI • *Department of Genetics, Albert Einstein College of Medicine, Yeshiva University, Bronx, NY, USA*

YOTAM ARDON • *Department of Pathology, University of Utah Medical School, Salt Lake City, UT, USA*

THOMAS ARENDT • *Paul Flechsig Institute of Brain Research, Universitat Leipzig, Leipzig, Germany*

KRISTIN K. BALDWIN • *Department of Neuroscience, Dorris Neuroscience Center, Scripps Research Institute, San Diego, CA, USA*

STEFAN BEKIRANOV • *Department of Biochemistry and Molecular Genetics, School of Medicine, University of Virginia, Charlottesville, VA, USA*

BIRGIT BELTER • *Department of Radiopharmaceutical Biology, Institute of Radiopharmacy, Forschungszentrum Dresden-Rossendorf, Dresden, Germany*

MARTINA K. BRÜCKNER • *Paul Flechsig Institute of Brain Research, Universitat Leipzig, Leipzig, Germany*

MIKI BUNDO • *Department of Molecular Brain Science, Graduate School of Medical Sciences, Kumamoto University, Chuo-ku, Kumamoto, Japan; PRESTO, Japan Science and Technology Agency, Chiyodaku, Tokyo, Japan*

IAN E. BURBULIS • *Department of Biochemistry and Molecular Genetics, School of Medicine, University of Virginia, Charlottesville, VA, USA; Escuela de Medicina, Universidad San Sebastian, Puerto Montt, Chile*

JULBERT CANEUS • *Department of Neurology, Rocky Mountain Alzheimer's Disease Center, and Linda Crnic Institute for Down Syndrome, University of Colorado School of Medicine, Aurora, CA, USA; Neuroscience Program, University of Colorado, Anschutz Medical Campus, Aurora, CO, USA*

HEIDI J. CHIAL • *Department of Neurology, Rocky Mountain Alzheimer's Disease Center, and Linda Crnic Institute for Down Syndrome, University of Colorado School of Medicine, Aurora, CA, USA*

WILLIAM D. CHRONISTER • *Department of Biochemistry and Molecular Genetics, School of Medicine, University of Virginia, Charlottesville, VA, USA*

JEROLD CHUN • *Sanford Burnham Prebys Medical Discovery Institute, La Jolla, CA, USA*

MICHAEL A. DURAN • *Dorris Neuroscience Center, Scripps Research Institute, San Diego, CA, USA*

JENNIFER A. ERWIN • *Salk Institute for Biological Studies, La Jolla, CA, USA; Leiber Institute for Brain Development, Baltimore, MD, USA*

GEOFFREY J. FAULKNER • *Mater Research Institute–University of Queensland, Woolloongabba, QLD, Australia; Queensland Brain Institute, University of Queensland, Brisbane, QLD, Australia*

MARK FIERS • *VIB Center for Brain & Disease Research, Leuven, Belgium; Center for Human Genetics, Universitaire ziekenhuizen and LIND, KU Leuven, Leuven, Belgium*

JOSÉ MARÍA FRADE • *Department of Molecular, Cellular and Developmental Neurobiology, Cajal Institute (IC-CSIC), Madrid, Spain*

CARLO SALA FRIGERIO • *VIB Center for Brain & Disease Research, Leuven, Belgium; Center for Human Genetics, Universitaire ziekenhuizen and LIND, KU Leuven, Leuven, Belgium*

FRED H. GAGE • *Salk Institute for Biological Studies, La Jolla, CA, USA*

ANTONETA GRANIC • *Department of Neurology, Rocky Mountain Alzheimer's Disease Center, and Linda Crnic Institute for Down Syndrome, University of Colorado School of Medicine, Aurora, CA, USA; Institute of Neuroscience, Newcastle Institute for Ageing, Newcastle University, Newcastle upon Tyne, UK; NIHR Newcastle Biomedical Research Centre in Ageing and Chronic Disease, Newcastle University, Newcastle upon Tyne, UK; Newcastle upon Tyne NHS Foundation Trust, Campus for Ageing and Vitality, Newcastle University, Newcastle upon Tyne, UK*

IRA M. HALL • *Dorris Neuroscience Center, Scripps Research Institute, San Diego, CA, USA*

JENNIFER L. HAZEN • *Dorris Neuroscience Center, Scripps Research Institute, San Diego, CA, USA*

IVAN Y. IOUROV • *Mental Health Research Center, Moscow, Russia; N.I. Pirogov Russian National Research Medical University, Academician Yu.E. Veltishchev Research Clinical Institute of Pediatrics, Ministry of Health of the Russian Federation, Moscow, Russia; Department of Medical Genetics, Russian Medical Academy of Postgraduate Education, Moscow, Russia*

KAZUYA IWAMOTO • *Department of Molecular Brain Science, Graduate School of Medical Sciences, Kumamoto University, Chuo-ku, Kumamoto, Japan*

GWENDOLYN E. KAESER • *Sanford Burnham Prebys Medical Discovery Institute, La Jolla, CA, USA; Biomedical Sciences Graduate Program, University of California San Diego, San Diego, CA, USA*

TADAFUMI KATO • *Laboratory for Molecular Dynamics of Mental Disorders, RIKEN Brain Science Institute, Wako, Saitama, Japan*

SERGEY KUPRIYANOV • *Dorris Neuroscience Center, Scripps Research Institute, San Diego, CA, USA*

JEONG HO LEE • *Graduate School of Medical Science and Engineering, Brain Korea 21 Plus Project, KAIST, Daejeon, South Korea*

JAE SEOK LIM • *Graduate School of Medical Science and Engineering, Brain Korea 21 Plus Project, KAIST, Daejeon, South Korea*

NOELIA LÓPEZ-SÁNCHEZ • *Department of Molecular, Cellular and Developmental Neurobiology, Cajal Institute (IC-CSIC), Madrid, Spain; Tetraneuron S.L., Valencia, Spain*

ANGELA MACIA • *Department of Pediatrics, Rady Children's Hospital San Diego, University of California San Diego, La Jolla, CA, USA*

GREG S. MARTIN • *Dorris Neuroscience Center, Scripps Research Institute, San Diego, CA, USA*

MICHAEL J. MCCONNELL • *Department of Biochemistry and Molecular Genetics, School of Medicine, University of Virginia, Charlottesville, VA, USA*

CRISTINA MONTAGNA • *Department of Genetics, Albert Einstein College of Medicine, Bronx, NY, USA; Department of Pathology, Albert Einstein College of Medicine, Bronx, NY, USA*

MARKUS MORAWSKI • *Paul Flechsig Institute of Brain Research, Universitat Leipzig, Leipzig, Germany*

ALYSSON R. MUOTRI • *Department of Pediatrics, Rady Children's Hospital San Diego, University of California San Diego, La Jolla, CA, USA*

ROBERT A. PALAIS • *Department of Pathology, University of Utah Medical School, Salt Lake City, UT, USA; Department of Mathematics, Utah Valley University, Orem, UT, USA*

APUÃ C.M. PAQUOLA • *Salk Institute for Biological Studies, La Jolla, CA, USA; Lieber Institute for Brain Development, Baltimore, MD, USA*

IRIS PATIÑO-PARRADO • *Department of Molecular, Cellular and Developmental Neurobiology, Cajal Institute (IC-CSIC), Madrid, Spain*

HUNTINGTON POTTER • *Department of Neurology, Rocky Mountain Alzheimer's Disease Center, and Linda Crnic Institute for Down Syndrome, University of Colorado School of Medicine, Aurora, CO, USA; Neuroscience program, University of Colorado, Anschutz Medical Campus, Aurora, CO, USA*

ALEXEI M. RATNIKOV • *Mental Health Research Center, Moscow, Russia; Moscow State University of Psychology and Education, Moscow, Russia*

PATRICK J. REED • *Salk Institute for Biological Studies, La Jolla, CA, USA*

SANDRA R. RICHARDSON • *Mater Research Institute–University of Queensland, Woolloongabba, QLD, Australia*

ALBERTO R. RODRIGUEZ • *Dorris Neuroscience Center, Scripps Research Institute, San Diego, CA, USA*

FRANCISCO J. SANCHEZ-LUQUE • *Mater Research Institute–University of Queensland, Woolloongabba, QLD, Australia; Pfizer-Andalusian Government-University of Granada Centre for Genomics and Oncologic Research (Genyo), Granada, Spain*

RYAN P. SMITH • *Dorris Neuroscience Center, Scripps Research Institute, San Diego, CA, USA*

ILIA V. SOLOVIEV • *Mental Health Research Center, Moscow, Russia*

BART DE STROOPER • *VIB Center for Brain & Disease Research, Leuven, Belgium; Center for Human Genetics, Universitaire ziekenhuizen and LIND, KU Leuven, Leuven, Belgium; Dementia Research Institute (UK-DRI), University College London, London, UK*

ATTILA TARNOK • *Faculty of Medicine, Institute for Medical Informatics, Statistics and Epidemiology (IMISE), Universitat Leipzig, Leipzig, Germany*

UWE UEBERHAM • *Paul Flechsig Institute of Brain Research, Universitat Leipzig, Leipzig, Germany*

ALEXANDER E. URBAN • *Department of Psychiatry and Behavioral Sciences, Program on the Genetics of Brain Function, Stanford University School of Medicine, Palo Alto, CA, USA; Department of Genetics, Stanford Center for Genomics and personalized Medicine, Stanford University School of Medicine, Palo Alto, CA, USA*

FLORA M. VACCARINO • *Child Study Center and Department of Neuroscience, Yale Kavli Institute for Neuroscience, Yale University School of Medicine, New Haven, CT, USA*

THIERRY VOET • *Department of Human Genetics, University of Leuven, KU Leuven, Leuven, Belgium; Wellcome Trust Sanger Institute, Hinxton, UK*

SVETLANA G. VORSANOVA • *Mental Health Research Center, Moscow, Russia; N.I. Pirogov Russian National Research Medical University, Academician Yu.E. Veltishchev Research Clinical Institute of Pediatrics, Ministry of Health of the Russian Federation, Moscow, Russia; Moscow State University of Psychology and Education, Moscow, Russia*

MEIYAN WANG • *Salk Institute for Biological Studies, La Jolla, CA, USA*

MARGARET B. WIERMAN • *Department of Biochemistry and Molecular Genetics, School of Medicine, University of Virginia, Charlottesville, VA, USA*

CARL T. WITTWER • *Department of Pathology, University of Utah Medical School, Salt Lake City, UT, USA*

YURI B. YUROV • *Mental Health Research Center, Moscow, Russia; N.I. Pirogov Russian National Research Medical University, Academician Yu.E. Veltishchev Research Clinical Institute of Pediatrics, Ministry of Health of the Russian Federation, Moscow, Russia; Moscow State University of Psychology and Education, Moscow, Russia*

LUMING ZHOU • *Department of Pathology, University of Utah Medical School, Salt Lake City, UT, USA*

CHENGHANG ZONG • *Baylor College of Medicine, Houston, TX, USA*

Part I

Introduction

Chapter 1

Principles and Approaches for Discovery and Validation of Somatic Mosaicism in the Human Brain

Alexej Abyzov, Alexander E. Urban, and Flora M. Vaccarino

Abstract

Mosaic variants are by definition present in just some of the cells that make up a given tissue. The frequency of such mosaic variants in that cell population depends on many factors, including when they originated during development, and whether they affect rates or patterns of cellular proliferation or are subject to selection of the cells carrying them. Their confident detection depends on combinations of the following four factors: (1) frequency, type, and functional effect of a mosaic variant; (2) strategy utilized for the discovery (single cell or bulk analyses); (3) applied experimental and analytical method (e.g., sequencing, droplet digital PCR); and (4) funds and effort that can be invested into each experiment. Furthermore, none of the existing strategies and techniques are universally applicable, nor cost effective, to find variants of all types. Studies aimed at discovering mosaic variants should carefully balance strategy, experimental and computational techniques, funds, and effort to carry out experiments and analyses that will allow the aims to be achieved.

Key words Whole-genome amplification (WGA), Fluorescence in situ hybridization (FISH), Array comparative genome hybridization (aCGH), Flow cytometry, Single-nucleotide polymorphism (SNP) array, Whole-genome sequencing, DNA fragment capture, L1-enrichment, Amplicon-Seq

1 Spectrum of Mosaic Variants

Mosaic variants can differ in type, frequency, and functional effect. They encompass variations in DNA sequence, such as single-nucleotide variants (SNV), small insertions and deletions (indel), and genome structural variations (SV), including mobile element insertions (MEI), copy number alterations (CNA), losses of heterozygosity (LOH), inversions, translocations, chromosomal aneuploidies, and multiploidies (Box 1). Not all experimental techniques are equally well suited to discover variants of all types. In fact, most of the techniques are capable of finding variants of just one type. Whole-genome sequencing (WGS) could be seen as a notable exception as it can potentially detect variants of all types. However, as will be discussed below, only when combined with the

José María Frade and Fred H. Gage (eds.), *Genomic Mosaicism in Neurons and Other Cell Types*, Neuromethods, vol. 131, DOI 10.1007/978-1-4939-7280-7_1, © Springer Science+Business Media LLC 2017

Box 1 Mosaic Variants and Alterations

SNV—Single-nucleotide variant, a change in a nucleotide at a defined position, e.g., C to T.

Indel—Short insertion (in-) or deletion (-del) in the genome. There is no common consensus about maximum indel size. Typically, an indel is defined as an insertion or a deletion smaller than 50 or 100 bp. Several studies, however, consider insertions and deletions of kbp size also as indels.

MEI—Mobile element insertion, an insertion of retrotransposon elements (endogenous retroviruses) into the genome. In humans four elements are present and can retrotranspose: ALU, LINE1, HERV, and SVA.

CNA—Copy number alteration, a region that has a higher or lower number of copies compared to other regions in the same genome, e.g., a deletion or duplication. This term is often used for large somatic alterations in the genome but is applicable to any alterations larger than indels.

LOH—Loss of heterozygosity, a normally diploid region where germline SNPs become homozygous either as a result of loss of one haplotype or replacement of one haplotype with the copy of another one. Regions of LOH are typically large, as, in order to be detected, they must contain at least a few heterozygous SNPs. The loss of one haplotype, i.e., a heterozygous deletion, is a CNA as well.

Inversion—Replacement of a sequence with its reverse complement. This term is applicable to alterations larger than indels.

Translocation—Rearrangement leading to covalent connection of DNA from two different chromosomes.

Chromosomal Aneuploidy—Change in the number of copies for a entire chromosome or arm(s) of a chromosome.

Multiploidy—Global change in the genome ploidy.

SV—Structural variant, a general term to denote variations, alterations, and rearrangements in the genome. SVs include CNAs, MEIs, inversions, translocations, chromosomal aneuploidies, multiploidies, and complex rearrangements that bear signatures of multiple aforementioned types. Complex rearrangements are not very frequent but can be observed in appreciable numbers as inherited variants, somatic variants in cancers, and mosaic variants in normal cells. Highly complex and clustered rearrangements that may also involve multiple chromosomes are called chromothripsis.

appropriate experimental design and analytical components can WGS be utilized to its full benefit.

All types of variants have common characteristics that are relevant regarding options for their detection. Rare variants are harder to find regardless of the utilized strategy and experimental technique. This follows from the general logic that the more rare a variant is, the smaller a contribution it will make to the measurable experimental signal, regardless of what that signal is. Variants that may give selective advantage to a cell, for instance faster proliferation or better viability, will have a higher frequency in a given tissue [1]. Conversely mosaic variants can be disadvantageous for the cells that carry them, for instance by lengthening the cell cycle or reducing cellular viability, and could then be expected to have a lower frequency in the tissue.

Some variants could reduce cell fitness under the experimental approaches that may be used to detect them, for example cell transformation and/or clonal culturing. This would preclude discovering these mutations by using such approaches. And, obviously, mosaic mutations leading to cell death by, for example, knocking out an essential gene, will not be discovered by any approach.

2   Strategies for Mosaic Variant Discovery

Two major strategies for mosaic variant detection are bulk analysis and single-cell analysis (Fig. 1). When a tissue sample or cell type is analyzed in bulk, the genomes of many cells or nuclei are assayed together in a single experiment. In single-cell analysis the genome of only an individual cell or nucleus is assayed in a single experiment. Each strategy can be combined with multiple experimental techniques for data generation and analysis such as sequencing and FISH. And, in fact, almost all techniques for detection can be combined with either of the strategies. Often, the DNA of analyzed cells or tissue is prepared in a strategy- and technique-specific way prior to analysis. A typical example is whole-genome amplification (WGA) or clonal amplification of single cells. Because such manipulations can introduce artifacts into the DNA, the discovery of variants in such material does not necessarily imply their presence in the original tissue. Experimental validation of the likely existence of a given variant call or set of variant calls is required prior to reporting it, and it is essential that such validation is done in a way that excludes being confounded by artifacts of DNA preparation. Therefore, we differentiate between technical validation, by which we mean validation of variants present in manipulated (e.g., WGA treated) DNA, and biological validation, by which we mean validation of variants in cells from the original tissue sample (Fig. 1).

2.1   Bulk Analysis

Bulk samples, i.e., populations of cells to be analyzed together, can be directly obtained from primary brain tissue, e.g., from postmortem tissue or from surgically resected material, in which case they represent a mix of different cell types. DNA can be extracted directly from this mix of cells. Alternatively, the bulk tissue sample can be subjected to sorting for particular cell fractions, such as neuronal nuclei positive for the NeuN marker, leading to a relatively more homogeneous cell type of sample [2]; see also Chaps. 3 and 4. Experimental techniques typically applied to bulk tissue are WGS, SNP arrays, array comparative genome hybridization (aCGH), oligomer-capture or PCR-based targeted high-depth sequencing, regular PCR followed by Sanger sequencing, and digital PCR (the latter two approaches are of low throughput and typically used for validation rather than discovery).

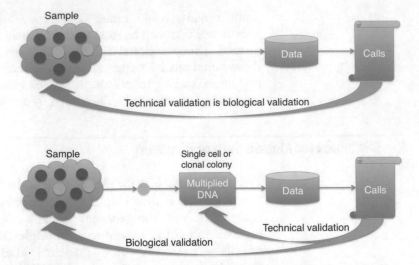

Fig. 1 Conceptual strategies for mosaic variants discovery. In the first strategy a sample is analyzed in bulk. At a standard sequencing depth of 30×, variants below 20% allele frequency (AF) are not detectable and thus very rare variants are unlikely to be found. In this strategy technical validation of calls in the original sample is equivalent to biological validation. In the second strategy individual cells are analyzed. Most experimental techniques will not work with DNA from just a single cell and amplification of a cell's DNA is required. This is achieved through either clonal expansion or in vitro whole-genome amplification (WGA). Because of this, technical validation of a call on amplified DNA is not the same as biological validation in the original sample. Extremely rare variants that can be called correctly are challenging to validate in the original sample. Therefore, the technique chosen for single-cell analysis must be one that gives high confidence in variant detection

Analysis of bulk material is widely used, primarily because most experimental techniques require large quantities of DNA (at least a few thousand cells). Its main disadvantage is that it mostly discovers only common variants (in modern studies typically only variants with AF above 10–20% can be detected). Rare variants, such as those present in only the two green cells in Fig. 1, are not likely to be discovered. While an argument could be made that such rare variants are unlikely to have a strong phenotypic effect on the studied tissue, this has not been proven. In fact, evidence exists that variants with a frequency of just a few percent can result in profound phenotypic effects [3–6]. Moreover, the definition of "rare" differs between studies. Oftentimes, authors consider as "rare" those variants that are beyond their detection limits; for WGS, standard 30× average genomic coverage typically will not detect variants at AF below 20% [7]. Hence, when performing WGS on bulk tissue, the sensitivity is a function of the amount of generated sequencing data and limited by the depth of coverage by the DNA sequencing reads—the higher the coverage, the higher the sensitivity will be.

Table 1
Comparative characteristics of bulk and single-cell strategies for mosaic variant discovery

	Bulk analysis	Single-cell analyses
Discovery of variants with different frequencies within tissue	Common variants in analyzed cells (typically above 10–30%)	Variants of any frequency if present in analyzed cell(s)
Comparison to other samples from the same individual	Beneficial but not essential	Essential
Determining the presence of multiple variants in a single cell	Can be inferred in some cases	Can be confidently determined
Data-driven evaluation of analytics	Not possible	Possible
Sample preparation and handling	Straightforward, relatively affordable	More complex, relatively expensive
Technical validation	Possible for all variants	Possible for all variants
Biological validation	Possible for all variants	Possible for a subset of variants, with frequency above detection threshold in primary tissue

For targeted deep-coverage sequencing of certain regions in the human genome, sensitivity can be high enough to find variants at 0.1% AF [8]. However, carrying out high-depth analysis of the entire genome is very expensive, which represents the second essential limitation of bulk analysis. For other experimental techniques, such as SNP arrays, the detection sensitivity is further limited by inherent noise, and variants below a frequency of 10% cannot be detected [9, 10]. Improvement of discovery sensitivity towards variants at lower AF and towards smaller sized CNVs can be made by comparing tissues from the same individual or from monozygotic twins, as such comparison allows for better control of technical variability and noise [11, 12].

Advantages of bulk tissue as starting point are that the steps of sample preparation and handling are typically easier, less time consuming, and cheaper than for single-cell analysis. Finally, technical validation of calls in the original sample is equivalent to biological validation (Fig. 1), relatively straightforward at least for higher frequency variants, and can be done with techniques that are orthogonal and much more sensitive than those used for discovery (Table 1).

2.2 Single-Cell Analysis

In single-cell analysis the genome of an individual cell or nucleus is analyzed in each experiment (Fig. 1), and multiple individual cells or nuclei can be analyzed in parallel to allow for statistically significant findings. The fundamental strength of this strategy is that it can discover variants present in a given cell, regardless of their frequency in a tissue. However, this also presents two fundamental challenges. Namely that for rare variants the analysis of a large number of individual cells or nuclei will be necessary, and that most of the discovered variants may be so rare that validating them in the primary tissue or detecting them again in additional cells from the same source tissue can be very unlikely [13–17]. Consequently, when conducting such experiments, one has to rely on experimental techniques for the preparation of DNA and for calling variants from such DNA that are robust and produce a high-confidence call set. Unfortunately, for some of the currently existing techniques (e.g., amplification of DNA from single cells) there will have to be the development of extensive improvements before one can say that they reached the necessary level of maturation.

Another strength of single-cell analyses is that the majority of mosaic variants are likely to be on only one allele out of the two typically present in a cell (only sex chromosomes in male cells are present in only one copy). This provides means for filtering out false-positive calls that could be the result of DNA preparation or data generation. For example, mosaic SNVs should be present at roughly ~50% AF in single-cell sequencing experiments. Additionally this feature makes mosaic variants similar to the heterozygous germline variants present in a cell. And consequently, germline variants detected from the same single-cell-based sequencing data can be used to test, optimize, refine, validate, and estimate the sensitivity of the analytical methodology used for discovery of mosaic variants.

Because mosaic variants are indistinguishable in AF from germline variants, single-cell analyses must compare the genome of a single cell to a genome of some reference tissue (e.g., genome of bulk DNA from polyclonal tissue such as blood) from the same person in order to exclude germline variants. When analyzing cells in bulk, reference tissue is not necessarily required, as mosaic variants can be distinguished from germline variants based on their AF. Still, reference tissue can be analyzed in all cases to increase sensitivity and specificity of discovery and confidence in discovered variants.

Finally, an advantage of the single-cell strategy over bulk sequencing is that the presence of multiple variants in one and the same cell, and their sharing across cells, can be readily determined (Table 1).

2.2.1   Direct Observation of Mosaic Variants in Single Cells

Direct observation of chromosomal aneuploidies in cells is possible under the microscope but only during cell division. Therefore, such cytogenetic analyses are only applicable to astrocytes or glia cells in adult brain, or to fetal brain. Direct observation of whole-chromosome aneuploidies or large CNAs is also possible with fluorescent in situ hybridization (FISH) [18]. These classical although low-resolution techniques are very reliable and often used as validations for newer and/or higher resolution techniques (see Chaps. 2, 3, 14, and 16).

2.2.2 Single-Cell Whole-Genome Amplification (WGA)

For all experimental techniques besides microscopic examination, the amount of DNA extracted from a single cell is too small to yield any observable signal. Therefore, the DNA of a cell has to be amplified. Such whole-genome amplification is the key step in single-cell analyses, since the quality of the amplified DNA is the major determinant for finding mosaic variants in single cells (Fig. 1). There are several enzymatic processes available for the amplification of whole single-cell genomes. However, various errors and biases are inherent to all of them, with the exact types of these errors being different for different WGA processes.

The oldest WGA method is to degenerate oligonucleotide-primed PCR (DOP-PCR) [19]. This PCR-based technique yields the most uniform coverage across the genome and is therefore recommended when aiming to discover mosaic CNAs and other larger chromosomal aberrations [20–22]. More details are given in Chap. 6. Multiple displacement amplification (MDA) on the other hand [23] is an isothermal WGA technique which, while suffering from drastically less uniform coverage and a relatively high rate of allelic dropouts (i.e., regions of the genome where none or only one out of two possible haplotypes is being amplified, Fig. 2), is advantageous in that it has a much lower rate of error in the amplified DNA and also produces very long DNA fragments (up to several kbp long), as opposed to the relatively short amplicons produced by PCR-based WGA. Because of these properties, MDA is often suggested as the WGA method of choice for the discovery of certain types of mosaic variants, particularly SNVs and MEIs [16, 24] (see Chap. 13). However, there is currently no universal agreement in the field about the utility of WGA

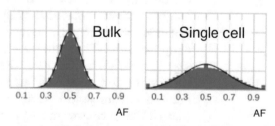

Fig. 2 Allele frequency (AF) distribution of heterozygous SNPs when sequencing bulk samples (*left*) and MDA-amplified samples from a single cell to the same coverage of 30×. The shape of the distributions can be described by Gaussian function (*black bell-shaped curve*). While in both samples AF is centered on 0.5, the distribution is much wider for the single-cell sample, reflecting uneven amplification of one allele vs. the other. Regions with allelic dropout in the single-cell sample contribute to bars at 0 and 1 AF

methods for the discovery of somatic mosaicism [13, 14, 17]. Other techniques for single-cell WGA, more recently developed and currently not widely adopted, are based on DOP-PCR or MDA and attempt to mitigate the earlier method's weaknesses: MALBAC [25], which is also described in details in Chap. 7; PicoPLEX [26]; MIDAS [27]; TruePrime [28]; and cold MDA {please insert reference to PMID:28319112}. A principally new linear WGA approach has also been suggested {please insert reference to PMID:28408603}.

2.2.3 *Clonal Expansion*

Single-cell whole-genome amplification is at the present state of the field far less faithful than in vivo genome duplication in mitotic cells. The reason is obvious: dividing cells use much more sophisticated molecular machinery to precisely duplicate their entire DNA and minimize errors by thorough proofreading and error correction: machinery that is not recapitulated by in vitro enzymatic WGA procedures. To leverage the advantage of the high-precision DNA amplification in dividing cells, the single-cell clonal expansion strategy can be used to study mosaic variants (Fig. 1) [13–15, 17, 29]. In this strategy single-cell colonies are cultured until the number of cells in culture is large enough to extract the amount of DNA necessary to apply an experimental technique for discovery without enzymatic WGA.

Although the clonal expansion strategy bypasses some technical challenges related to discovery of mosaic variants in the context of single-cell WGA, the clonal expansion strategy has its own limitations. Depending on which methodology is used, ensuring clonality of cell colonies from tissue culture can be challenging. In many cases clonality is deemed very likely [15] and verified post hoc using the generated DNA sequencing data [14].

Next, one has to filter out variants generated during culturing. In the ideal scenario, where all cells in a clonal colony survive and divide at exactly the same pace, variants created during culturing will have a small AF in the colony. For diploid chromosomes, variants created during the first division of the founder cell will be present on only one haplotype out of four in the daughter cells. Thus, their AF in the colony will be 25%, which is significantly lower than the AF of 50% for heterozygous germline variants or for mosaic variants in the founder cell. Similarly, for haploid chromosomes such variants will have an AF that is half that of the germline and of the mosaic variants. Variants created during later cell divisions will be at even lower AF. Therefore, separation of discovered variants by AF is an efficient way of removing culturing artifacts. However, in the real world cells may divide at different pace, experience positive or negative selection and senescence, or die, all of which may increase AF of variants created during culture, making them less distinguishable from mosaic variants in the founder cell. Variants that arose during the first few cell divisions are the ones most likely to be indistinguishable from mosaic variants; therefore it is desirable to monitor early stages of clonal expansions to ensure that there was no disparity in cell proliferation.

The most fundamental limitation of the clonal expansion is that only culturable cells (such as neuronal progenitors and glia) can be analyzed, while terminally differentiated neurons are inaccessible to this strategy. Ability of a cell to proliferate may not only be determined by its differentiation state but also by its mosaic mutations. Thus, mosaic mutations that lead to a cell being unculturable will not be discovered, and this represents a fundamental and essential bias inherent to the clonal expansion strategy.

2.3 Hybrid Strategies

There are a few strategies that have been suggested as options that would diminish the disadvantages of single-cell amplification and of clonal expansion. The most prominent is an adaptation of clonal expansion for somatic cells, called somatic cell nuclear transfer (SCNT) [30], discussed in details in Chap. 9. In this strategy, the nucleus of a somatic cell is transferred to an enucleated oocyte, which then, at low but appreciable rates, can proliferate and result in a cell colony or, if transferred into the uterus of the matching model organism, even lead to a living animal, as it has been done in mouse. Another strategy utilizes cell cycle S-phase arrest with in vitro amplification of the duplicated genome from the still undivided cell [31]. This strategy can be useful for finding mosaic variants in cells with limited potential for proliferation and has the advantage of starting DNA amplification from a larger DNA amount than that in a single diploid cell. Finally, a somewhat different strategy involves sorting cells into different cellular subtypes by using antibodies as lineage markers [32]. This could be particularly advantageous for studying mosaicism in brain, a very complex organ consisting of multiple types of neurons, microglial cells, astrocytes, and oligodendrocytes. Since, theoretically, cells of each type arise from only a few ancestor-founder cells in a local region of the developing CNS, this avenue potentially provides better sensitivity for resolving mosaic mutations present in the founder cells, as such mutations will be at higher AF in the relevant sorted cell fraction. However, the advantageousness of this strategy has not yet been demonstrated.

3   Experimental Techniques for Mosaic Variant Discovery and Validation

3.1 FISH (Fluorescent In Situ Hybridization)

In classical FISH, fluorescent probes made from BAC clone sequences, of typically around 100 kbp length, bind to the complementary regions of a chromosome [18]. The fluorescent signal is then observed within individual cells under a microscope. Given the large size of the probes, technical limitations on the number of probes that can be used in each given experiment, and the necessity to limit each experiment to probing a relatively small region of the genome, FISH can typically detect only very large CNAs or SVs in general. However, hybridization of probes is very strong and in general FISH is widely considered a robust technique [33]. Application of FISH to brain cells is discussed in Chaps. 2, 3, 14, 15, and 16. There are also many methods derived from the original BAC-based FISH, such as the use of a series of fluorescently labeled oligomers as probes. For the most part, in the context of somatic variation, FISH should be seen as a method to consider as a tool for validation in some narrowly defined scenarios, but not so much an option for genome-wide discovery of variants.

3.2 Flow Cytometry

In the context of somatic variation, flow cytometry has been reported as a tool for the discovery of highly aneuploid and polyploid cells in brain, by measuring the total amount of DNA per cell or nucleus. This method is being addressed in Chaps. 3 and 4.

3.3 Array Comparative Genome Hybridization

Array comparative genome hybridization (aCGH) relies on hybridization of DNA from two samples (the test sample and a reference sample) to a set of several hundred thousand, 50–100 bp long, oligonucleotide probes that represent the genome sequence and are arrayed on a glass slide [34, 35] {Haraksingh et al., this paper is now out in print, please insert it as reference number 36: PMID 28438122, Haraksingh, Abyzov, Urban, BMC Genomics 2017 (and move all the other references down one in number accordingly, i.e. the current 36 is now 37 and so on). DNA from the two samples is each labeled with a different fluorescent dye. By comparing differential fluorescence intensities for a given probe or a set of probes, one can infer a relative change in the number of copies in the two analyzed samples. The number of probes in aCGH experiment can be very large. While typically arrays contain a few million probes or less, custom arrays can have up to 42 million probes [36]. Probes are designed based on the reference genome and may cover the entire genome or some portion. The studied DNA can be either from a bulk of cells or whole-genome amplified from a single cell. aCGH is in general a very robust method for detecting genomic copy number changes from large chromosomal aneuploidies to relatively small CNAs (i.e., millions bp to tens of thousands bp for standard whole-genome oligomer-chip-based aCGH). It has been used successfully in the context of somatic genome variation [11, 37, 38]. One limitation is that for the analysis of DNA extracted from bulk tissue this technique is not very sensitive, and one should not expect to be able to detect variants that are present in less than a considerable fraction of the cells of this tissue.

3.4 SNP Arrays

SNP arrays are glass chips carrying oligonucleotide probes that represent hundreds of thousands to millions of known single-nucleotide polymorphisms (SNPs) [39] (Haraksingh et al., this paper is now out in print, please insert it as reference number 36: PMID 28438122, Haraksingh, Abyzov, Urban, BMC Genomics 2017 (and move all the other references down one in number accordingly, i.e. the current 36 is now 37 and so on)). For each targeted SNP there are oligomers representing the two alleles, commonly referred as A and B. Fluorescently labeled DNA from the test sample is then hybridized to the array and differential signal strength from the oligomers representing the various alleles makes it possible to determine which of the SNPs included in a given array design are present in the sample. At each SNP locus the sample's genotype is determined, i.e., heterozygous or homozygous for a certain SNP. Reanalyzing the data by integrating the signal intensity of all oligos for a given SNP also allows detecting copy number changes [35, 40]. This technique does not require comparative hybridization of DNA from a reference sample. Furthermore, careful analysis of A and B allele frequencies can be used to detect mosaic copy number changes from bulk tissue DNA with greater sensitivity than when using aCGH. Other than that, similar considerations apply as for aCGH—the method is robust for the detection of medium- to large-sized copy number changes and can be applied to DNA from a bulk of cells or to whole-genome-amplified DNA from a single cell [9, 10, 21, 41].

3.5 DNA Sequencing

DNA sequencing methods, in particular the now-available and revolutionary "next-generation" technologies, hold the promise of being able to discover all types of mosaic variants across the entire genome. From a sequencing library prepared from a sample of genomic DNA, DNA sequencing instruments generate outputs in the form of millions and even billions of shorter stretches of nucleotide sequence, called sequencing reads. Those reads are then computationally mapped to the reference genome and by using algorithms that detect inconsistencies or imperfections in the mapping, like mismatches for SNVs and gaps in reads for indels, one can discover variants present in the analyzed sample and absent from the human reference genome. Comparing two samples to each other allows discovering variants that are present in one sample and absent from the other. A current standard read length is 100-150 bp and the read output is typically generated as pairs of reads that are separated by a stretch of sequence of about 250 bp that is not read out. Sequencing reads in pairs improves the mapping to the reference genome and contains additional information (i.e., the expected distance and orientation of the reads) that can be used to discover structural variation in the genome under analysis (i.e., when observing discordance in pairs of reads from the expectation). DNA sequencing can be applied to single cells, clonally expanded colonies, as well as bulk samples.

3.5.1   Whole-Genome
Sequencing (WGS)

WGS is the most comprehensive and least biased way to analyze genomes. Its efficiency for single-cell analyses is currently dependent on evenness and errors during whole-genome DNA amplification, while the efficiency of bulk cell analyses by WGS depends on the depth of coverage after mapping the sequencing reads onto the reference genome. At coverages of 30×–40×, which is currently considered the adequate standard to find germline SNPs by WGS, only mosaic variants at an AF of 15–20% or higher can be discovered with confidence. Therefore, much deeper sequencing coverage is necessary for finding low-frequency variants. WGS is still relatively costly and obtaining a coverage of, for example, 500× to analyze mosaic SNVs and indels is for the most part still prohibitively expensive. However, for an analysis of the genome of single cells or clonal colonies such high coverage is not necessary, since 30×–40× coverage already allows discovering a majority of germline and mosaic variants [16, 42, 43]. The efficiency of applying WGS for finding SVs depends on physical coverage, i.e., when counting unsequenced bases between paired reads [44, 45]. Therefore, custom libraries with long spans of DNA sequence (2–20 kbp between reads) could be an efficient way of finding mosaic SVs, as physical coverage will be several fold larger than sequencing coverage.

Another advantage of WGS is that generated reads cover the genome proportionally to its copy number; that is, an increase or a decrease in copy number of a particular region of the genome will be reflected, respectively, in an increase or a decrease in read coverage. Biases in this type of read-depth analysis do exist but their sources are for the most part known and can be corrected for, so that depth of coverage by sequencing reads can be used to find large aneuploidies and CNA [46]. Even at moderate coverage, the depth of coverage method is sensitive enough to find CNA of few kbp in size and larger in clonal colonies and single cells, where the CNA will most likely be present at 50% of AF [21, 46]. However, in bulk tissue, read-depth analysis will only work if a mosaic CNA is either present in a large fraction of the cells in the tissue under analysis. Combined with a special library preparation that separates reads by DNA strands one can observe genomic rearrangements at extremely shallow coverage [47].

3.5.2   DNA Fragment
Capture and Sequencing

Enrichment for particular regions of the genome during the production of a sequencing library can be used to increase sequencing coverage of those targeted regions and consequently boost sensitivity for finding low-frequency variants. DNA capture library preparation begins with hybridizing DNA fragments from a sample to a pool of oligomer baits that have been designed to be complementary to the targeted regions in the genome. The oligomer baits carry a chemical label, typically a biotin moiety, which is used to extract the baits from the hybridization reaction and with it the captured target DNA, which is then prepared as library and sequenced. This approach has been used to perform high coverage sequencing (>200×) of a panel of candidate genes implicated in intellectual disability in blood samples, allowing several somatic mutations to be detected and validated by subcloning [48]. The targeted capture approach is most commonly used to sequence only the coding portion of the genome, so-called exome sequencing. As the coding portion of the genome is only about 1–2% of the entire genomic sequence and generally regarded as being the place where most variants with strong and direct phenotypic effects can be expected to reside, even an off-the-shelf exome sequencing may provide a relatively cost-effective option for discovering mosaic variants and not just germline variants, for which exome-sequencing was originally developed [49]. However, the capture step introduces bias into the coverage across the captured regions. Particularly regions with indels are not well captured, and, generally, coverage across the genome is not uniform. While read-depth analysis is still possible, it is only sensitive to large CNAs that span multiple exons or genes. And lastly, this approach would miss variants that occurred in any regions of the genome that are not exonic in nature but may well be functionally relevant, such as gene-distal regulatory or enhancer sequence elements.

Custom capture libraries can be made to target certain elements in the genome, as was described for retrotransposon families in the human genome [50, 51]. Sequencing of the captured DNA library will yield reads that mostly map on or around retroelements across the entire genome, independent of their location, allowing for discovery of germline and mosaic retroelements. Further details are described in Chap. 12. However, because of the biases in capture referred to above, this method cannot give a reliable quantitative estimate of mosaic retroelements, and should be validated with an orthogonal technique.

3.5.3 Amplicon-Seq and L1 Enrichment by PCR Amplification

As an alternative to targeted sequence capture, PCR-amplified DNA from multiple target regions can be pooled and sequenced in the same experiment, an approach called amplicon-seq. Because of the need to conduct separate amplification reactions for each target region this technique can only be used for a relatively limited number of regions at a time, for example for the analysis of the exons of a handful of genes. On the other hand, given that the targeted regions represent only a tiny fraction of the entire genome, even producing only a few million sequence reads from these pooled amplicons can result in extremely deep coverage of the regions to be analyzed, which often will allow detecting or confirming even low-frequency mosaic variants [52]. One major caveat is that the DNA polymerase can introduce errors into the amplified DNA that would look like mosaic variants and thereby cancel out the advantage of having deep sequence coverage. Therefore the use of high-fidelity polymerase and a careful estimation of its background error rate are necessary to reliably find mosaic variants with this technique.

In addition to using amplicon-seq for the analysis of specific target regions, PCR amplification can also be employed as a tool for genome-wide discovery to generate amplified DNA from, in theory, all loci where a particular genomic sequence element resides. For example, one way to study mosaic retrotransposition of L1 elements was by semi-targeted PCR where one of the primers contains the 3′-ends of active L1 elements [53–55]. Sequencing of such PCR amplicons will yield reads that mostly map on or around both germline and mosaic L1 elements. Comparison with the reference tissue from the same individual then allows distinguishing mosaic L1s from germline L1s and from background noise. Just as for the targeted capture sequencing approach, this approach is advantageous in that with significantly fewer sequencing reads than for WGS the targeted genomic loci are covered significantly deeper. The disadvantage is that PCR amplification can result in chimeric sequences, particularly for loci that are rich in repeat sequences common in the human genome, such as L1, which can lead to false discoveries. Additionally, the absence of a global view of variations in genomes can lead to misinterpretation of SVs as mosaic L1 insertions [55]. Various aspects of studying L1 retrotransposition activity in brain cells are discussed in Chaps. 10 through 13.

4   The Concept of Validation

Validation, in the context here meaning confirming the existence of a discovered somatic variant in the original tissue or source DNA (Fig. 1) by using methods that are orthogonal to those used for discovery, is crucial for establishing the validity of mosaic calls. Methods for validation can differ drastically in their sensitivity, throughput, required labor, and cost (Table 2). It is also important to understand that the entire concept of validation and, in particular, the interpretation of its results depend on the strategy chosen for the discovery of mosaic variants. When conducting discovery in bulk tissue one only needs to conduct validation in the original

Table 2
Comparative characteristics of validation methods

	Capture	Amplicon-seq	ddPCR
Throughput	>1000	Hundreds	Dozens
Sensitivity	~0.5%	~0.1%	~0.01%
Turn around	Month	Weeks	Weeks
Labor	Little	Moderate	High
Price per site	Low	Moderate	High

tissue with, preferably, an orthogonal experimental technique. For a straightforward interpretation of validation results one should also use a technique that is more sensitive than the one used for discovery (typically, targeted capture and sequencing or digital droplet PCR). In such a case only two outcomes for a tested variant call are possible: validated or not validated.

However, additional considerations regarding validation apply when analyzing mosaic variants from discovery in single cells or clonally expanded colonies. One can conduct validation in the original tissue; however, none of the available techniques is more sensitive than single-cell or clonal expansion analysis itself. In the most extreme scenario, one can have a variant present in a single cell and only in this cell, found during the discovery phase; but this variant could never be validated in the tissue, as it is not present there anymore: the cell in which it was present having been removed and destroyed for the purpose of discovery. Therefore, biological validation in the original tissue may provide limited results and not validating a call could both mean that the call is incorrect or that the variant is too rare in the tissue to be validated. Nevertheless, carrying out biological validation is essential as the validated calls provide a solid "lower boundary" for counts of mosaic variants.

Technical validation in the DNA sample used for discovery, i.e., in the DNA resulting from WGA of a single cell or from the clonal expansion culture as opposed to from the original tissue, is also necessary and crucial. For this, one can use the same techniques used for biological validation, i.e., targeted capture and sequencing or digital droplet PCR. The limitation of technical validation is that errors introduced in the DNA during preparation (e.g., during single-cell WGA or clonal expansion culture) may result in a false call that would then be validated. Thus, in such cases special actions need to be taken in order not only to validate a call but also to demonstrate that it is not likely an introduced artifact. For example, the AF of true mosaic variants on a diploid chromosome should be around 50% in the amplified DNA from a single cell, with strong or systematic deviation from this value being indicative of a given call or call set not representing true mosaic variants in the original tissue.

4.1 Validation by PCR/qPCR

PCR is a classical technique that, especially when combined with Sanger sequencing, can be applied for validating SNVs, indels, and SVs (if their breakpoints are predicted with near-bp resolution). Validation of SNVs can only be done on a coarse-grained scale, i.e., validating presence or absence. Determining the AF or even defining whether a given variant is present at high or low AF is either impossible or at best very subjective. However, when using PCR for the validation of indels there can be a good degree of sensitivity and objectivity. The amplified sequence containing an indel will contain a set of secondary peaks in the Sanger sequencing trace. Even if peaks are small and can be compatible with just random noise, consistency of the secondary peaks with the expected sequence downstream from the indel will validate it (Fig. 3). As such, PCR can validate indels with an AF in analyzed tissue as low as 1%. For validation of SVs, PCR may produce even greater sensitivity. PCR primers flanking SV breakpoints (in case of a deletion) will yield amplicons for haplotypes harboring an SV and yield no amplicon or a different amplicon for the haplotype without the SV. Thus, presence of a dual band or just one band with the expected size for the SV haplotype would validate the SV down to an AF of less than 1% [15]. Sequencing amplicons will provide ultimate proof and also resolve the SV breakpoints at base pair resolution [15].

Quantitative PCR (qPCR) is a laboratory technique that monitors in real time the amplification of a targeted DNA sequence during the reaction. For mosaic variant analysis it can be applied to estimate the number of copies of an amplified region in a given sample. The quantification is rather crude and its application for validation of CNAs has the same limitations as the application of PCR to the validation of mosaic SNVs.

4.2 Validation by ddPCR

Droplet digital PCR (ddPCR) is a recently developed technique that allows for the precise quantification of a target allele in a given DNA sample. In brief, the DNA sample is diluted into nanoliter-scale droplets in such a way that it is unlikely to have more than one molecule of the target allele in a given droplet. Then a PCR reaction with fluorescent tags marking the targeted region is run in each droplet. The readout for each droplet is whether it does or does not color. The reaction can be run simultaneously on multiple regions so regions with known copy number can be used as baseline to quantify the frequency of other regions. With optimization and large input DNA amounts the technique is sensitive to validate mosaic variants

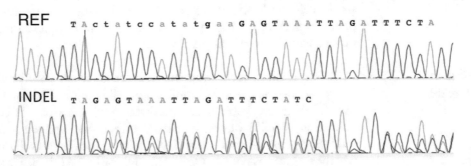

Fig. 3 Validation of indels with Sanger sequencing. The upper trace shows the reference sequence. An indel deletion is show in small letters, starting after the vertical line in the Sanger trace. The lower part shows the effect of the indel upon the sequence of the chromatogram, which could be inferred from a set of double peaks. PCR can validate indels with a frequency in analyzed tissue as low as 1%

down to an allele frequency of 0.01% in the analyzed sample. It can be applied to validation of SNVs, indel, and SVs, provided that the latter are known with breakpoint resolution. Arguably, ddPCR is considered as the gold standard validation technique for mosaic variants. However, ddPCR is a laborious technique that is hard to apply to large numbers of predicted variants.

4.3 Validation by Re-sequencing

Re-sequencing after DNA capture or amplification can be used for both biological and technical validations of large numbers of predicted variant loci. In each case candidate mosaic variants are subjected to sequencing at a depth that is much greater than the depth used for variant discovery. For biological validation in the original sample, deep coverage is necessary to get support for low-frequency variants. At a coverage of 1000× and above, variants with AF of a fraction of percent could be validated and their frequency can be precisely quantified. Such coverage is relatively easy to obtain even for many thousand sites. For example, 1000 sites with mosaic variants captured and sequenced to a depth of about 1000× are equivalent in terms of total reads to whole-genome coverage at a sequencing depth of less than 0.5×. For technical validation in DNA from a clonal colony or from DNA amplified from a single cell, high coverage is necessary to precisely establish the frequency of mosaic candidates because, as was discussed above, AF can be used as indirect but very strong evidence for the mosaic nature of candidate variants. For example, at 1000× coverage, the uncertainty in calculated AF of mosaic variant is only about 5%, meaning that for variants on diploid chromosomes their AF should typically be within a range of 50 ± 2.5%. Note that although whole-genome amplification will alter AF of mosaic variants, the distribution for many variants should still be centered around 50% (Fig. 2).

Re-sequencing can be conducted after DNA capture or PCR amplification (amplicon-seq) techniques as discussed above. Similarly as for variant discovery, application of these techniques to validation has certain drawbacks, the major of which is the differential efficiency of capture and amplification for alleles with and without mosaic variants, leading to a biased estimate of AF for indels and SVs.

5 Finding the Balance

When designing an experiment to detect mosaic variants and choosing a strategy and an experimental technique one has to consider characteristics of the variants to be found (i.e., type and frequency in studied cells), expected location of variants, practical limitations of dealing with the sample (i.e., whether isolation or culturing of single cell/nuclei is possible), and budget restrictions.

Two cases will outline the range of possible designs. In one case—the minimalistic scenario—one would need to ascertain the presence of one or a few already known mosaic variants in a given set of samples. In such a case, designing and conducting ddPCR reactions for each variant on bulk DNA will be the ideal solution. Once optimized, ddPCR reactions are relatively cost effective to run per sample while providing superb sensitivity. In another case—the global scenario—one would need to search for variants of all types that could be present anywhere in the entire genome and with an unknown frequency in the cell population of interest. Here, single-cell analysis with WGS is the ideal setup. Variants of all types could be found and at any frequency. The drawback is that WGS as well as single-cell culturing or single-cell/nucleus isolation and DNA amplification are costly procedures leading to potential confounds. Hence, extensive technical and biological validations are necessary to gain confidence in the results. Other scenarios would fall in between the outlined two and would provide ascertainment of more variants or more variant types than in the minimalistic scenario but at a drastically more cost-effective way as compared to the comprehensive scenario.

At the current cost of sequencing a general rule for good study design would be to rely on single-cell analyses when hunting for low- and very-low-frequency variants. Analysis of bulk tissue with enrichment for targeted DNA (i.e., certain genomic regions or elements) would make it possible to search for intermediate AF variants, while analysis of bulk tissue without DNA enrichment enables the discovery of relatively high AF variants.

References

1. Poduri A, Evrony GD, Cai X, Elhosary PC, Beroukhim R, Lehtinen MK, Hills LB, Heinzen EL, Hill A, Hill RS, Barry BJ, Bourgeois BF, Riviello JJ, Barkovich AJ, Black PM, Ligon KL, Walsh CA (2012) Somatic activation of AKT3 causes hemispheric developmental brain malformations. Neuron 74(1):41–48. doi:10.1016/j.neuron.2012.03.010

2. Matevossian A, Akbarian S (2008) Neuronal nuclei isolation from human postmortem brain tissue. J Vis Exp 20. doi:10.3791/914

3. Agerholm JS, Menzi F, McEvoy FJ, Jagannathan V, Drogemuller C (2016) Lethal chondrodysplasia in a family of Holstein cattle is associated with a de novo splice site variant of COL2A1. BMC Vet Res 12:100. doi:10.1186/s12917-016-0739-z

4. Bar DZ, Arlt MF, Brazier JF, Norris WE, Campbell SE, Chines P, Larrieu D, Jackson SP, Collins FS, Glover TW, Gordon LB (2017) A novel somatic mutation achieves partial rescue in a child with Hutchinson-Gilford progeria syndrome. J Med Genet 54(3):212–216. doi:10.1136/jmedgenet-2016-104295

5. Priest JR, Gawad C, Kahlig KM, Yu JK, O'Hara T, Boyle PM, Rajamani S, Clark MJ, Garcia ST, Ceresnak S, Harris J, Boyle S, Dewey FE, Malloy-Walton L, Dunn K, Grove M, Perez MV, Neff NF, Chen R, Maeda K, Dubin A, Belardinelli L, West J, Antolik C, Macaya D, Quertermous T, Trayanova NA, Quake SR, Ashley EA (2016) Early somatic mosaicism is a rare cause of long-QT syndrome. Proc Natl Acad Sci U S A 113(41):11555–11560. doi:10.1073/pnas.1607187113

6. Spier I, Drichel D, Kerick M, Kirfel J, Horpaopan S, Laner A, Holzapfel S, Peters S, Adam R, Zhao B, Becker T, Lifton RP, Perner S, Hoffmann P, Kristiansen G, Timmermann B, Nothen MM, Holinski-Feder E, Schweiger

MR, Aretz S (2016) Low-level APC mutational mosaicism is the underlying cause in a substantial fraction of unexplained colorectal adenomatous polyposis cases. J Med Genet 53(3):172–179.doi:10.1136/jmedgenet-2015-103468

7. Cibulskis K, Lawrence MS, Carter SL, Sivachenko A, Jaffe D, Sougnez C, Gabriel S, Meyerson M, Lander ES, Getz G (2013) Sensitive detection of somatic point mutations in impure and heterogeneous cancer samples. Nat Biotechnol 31(3):213–219. doi:10.1038/nbt.2514

8. Newman AM, Bratman SV, To J, Wynne JF, Eclov NC, Modlin LA, Liu CL, Neal JW, Wakelee HA, Merritt RE, Shrager JB, Loo BW Jr, Alizadeh AA, Diehn M (2014) An ultrasensitive method for quantitating circulating tumor DNA with broad patient coverage. Nat Med 20(5):548–554. doi:10.1038/nm.3519

9. Jacobs KB, Yeager M, Zhou W, Wacholder S, Wang Z, Rodriguez-Santiago B, Hutchinson A, Deng X, Liu C, Horner MJ, Cullen M, Epstein CG, Burdett L, Dean MC, Chatterjee N, Sampson J, Chung CC, Kovaks J, Gapstur SM, Stevens VL, Teras LT, Gaudet MM, Albanes D, Weinstein SJ, Virtamo J, Taylor PR, Freedman ND, Abnet CC, Goldstein AM, Hu N, Yu K, Yuan JM, Liao L, Ding T, Qiao YL, Gao YT, Koh WP, Xiang YB, Tang ZZ, Fan JH, Aldrich MC, Amos C, Blot WJ, Bock CH, Gillanders EM, Harris CC, Haiman CA, Henderson BE, Kolonel LN, Le Marchand L, McNeill LH, Rybicki BA, Schwartz AG, Signorello LB, Spitz MR, Wiencke JK, Wrensch M, Wu X, Zanetti KA, Ziegler RG, Figueroa JD, Garcia-Closas M, Malats N, Marenne G, Prokunina-Olsson L, Baris D, Schwenn M, Johnson A, Landi MT, Goldin L, Consonni D, Bertazzi PA, Rotunno M, Rajaraman P, Andersson U, Beane Freeman LE, Berg CD, Buring JE, Butler MA, Carreon T, Feychting M, Ahlbom A, Gaziano JM, Giles GG, Hallmans G, Hankinson SE, Hartge P, Henriksson R, Inskip PD, Johansen C, Landgren A, McKean-Cowdin R, Michaud DS, Melin BS, Peters U, Ruder AM, Sesso HD, Severi G, Shu XO, Visvanathan K, White E, Wolk A, Zeleniuch-Jacquotte A, Zheng W, Silverman DT, Kogevinas M, Gonzalez JR, Villa O, Li D, Duell EJ, Risch HA, Olson SH, Kooperberg C, Wolpin BM, Jiao L, Hassan M, Wheeler W, Arslan AA, Bueno-de-Mesquita HB, Fuchs CS, Gallinger S, Gross MD, Holly EA, Klein AP, LaCroix A, Mandelson MT, Petersen G, Boutron-Ruault MC, Bracci PM, Canzian F, Chang K, Cotterchio M, Giovannucci EL, Goggins M, Hoffman Bolton JA, Jenab M, Khaw KT, Krogh V, Kurtz RC,

McWilliams RR, Mendelsohn JB, Rabe KG, Riboli E, Tjonneland A, Tobias GS, Trichopoulos D, Elena JW, Yu H, Amundadottir L, Stolzenberg-Solomon RZ, Kraft P, Schumacher F, Stram D, Savage SA, Mirabello L, Andrulis IL, Wunder JS, Patino Garcia A, Sierrasesumaga L, Barkauskas DA, Gorlick RG, Purdue M, Chow WH, Moore LE, Schwartz KL, Davis FG, Hsing AW, Berndt SI, Black A, Wentzensen N, Brinton LA, Lissowska J, Peplonska B, McGlynn KA, Cook MB, Graubard BI, Kratz CP, Greene MH, Erickson RL, Hunter DJ, Thomas G, Hoover RN, Real FX, Fraumeni JF Jr, Caporaso NE, Tucker M, Rothman N, Perez-Jurado LA, Chanock SJ (2012) Detectable clonal mosaicism and its relationship to aging and cancer. Nat Genet 44(6):651–658. doi:10.1038/ng.2270

10. Laurie CC, Laurie CA, Rice K, Doheny KF, Zelnick LR, McHugh CP, Ling H, Hetrick KN, Pugh EW, Amos C, Wei Q, Wang LE, Lee JE, Barnes KC, Hansel NN, Mathias R, Daley D, Beaty TH, Scott AF, Ruczinski I, Scharpf RB, Bierut LJ, Hartz SM, Landi MT, Freedman ND, Goldin LR, Ginsburg D, Li J, Desch KC, Strom SS, Blot WJ, Signorello LB, Ingles SA, Chanock SJ, Berndt SI, Le Marchand L, Henderson BE, Monroe KR, Heit JA, de Andrade M, Armasu SM, Regnier C, Lowe WL, Hayes MG, Marazita ML, Feingold E, Murray JC, Melbye M, Feenstra B, Kang JH, Wiggs JL, Jarvik GP, McDavid AN, Seshan VE, Mirel DB, Crenshaw A, Sharopova N, Wise A, Shen J, Crosslin DR, Levine DM, Zheng X, Udren JI, Bennett S, Nelson SC, Gogarten SM, Conomos MP, Heagerty P, Manolio T, Pasquale LR, Haiman CA, Caporaso N, Weir BS (2012) Detectable clonal mosaicism from birth to old age and its relationship to cancer. Nat Genet 44(6):642–650. doi:10.1038/ng.2271

11. O'Huallachain M, Karczewski KJ, Weissman SM, Urban AE, Snyder MP (2012) Extensive genetic variation in somatic human tissues. Proc Natl Acad Sci U S A 109(44):18018–18023. doi:10.1073/pnas.1213736109

12. Forsberg LA, Rasi C, Razzaghian HR, Pakalapati G, Waite L, Thilbeault KS, Ronowicz A, Wineinger NE, Tiwari HK, Boomsma D, Westerman MP, Harris JR, Lyle R, Essand M, Eriksson F, Assimes TL, Iribarren C, Strachan E, O'Hanlon TP, Rider LG, Miller FW, Giedraitis V, Lannfelt L, Ingelsson M, Piotrowski A, Pedersen NL, Absher D, Dumanski JP (2012) Age-related somatic structural changes in the nuclear genome of human blood cells. Am J Hum Genet 90(2):217–228. doi:10.1016/j.ajhg.2011.12.009

13. Behjati S, Huch M, van Boxtel R, Karthaus W, Wedge DC, Tamuri AU, Martincorena I, Petljak M, Alexandrov LB, Gundem G, Tarpey PS, Roerink S, Blokker J, Maddison M, Mudie L, Robinson B, Nik-Zainal S, Campbell P, Goldman N, van de Wetering M, Cuppen E, Clevers H, Stratton MR (2014) Genome sequencing of normal cells reveals developmental lineages and mutational processes. Nature 513(7518):422–425. doi:10.1038/nature13448

14. Blokzijl F, de Ligt J, Jager M, Sasselli V, Roerink S, Sasaki N, Huch M, Boymans S, Kuijk E, Prins P, Nijman IJ, Martincorena I, Mokry M, Wiegerinck CL, Middendorp S, Sato T, Schwank G, Nieuwenhuis EE, Verstegen MM, van der Laan LJ, de Jonge J, IJ JN, Vries RG, van de Wetering M, Stratton MR, Clevers H, Cuppen E, van Boxtel R (2016) Tissue-specific mutation accumulation in human adult stem cells during life. Nature 538(7624):260–264. doi:10.1038/nature19768

15. Abyzov A, Mariani J, Palejev D, Zhang Y, Haney MS, Tomasini L, Ferrandino AF, Rosenberg Belmaker LA, Szekely A, Wilson M, Kocabas A, Calixto NE, Grigorenko EL, Huttner A, Chawarska K, Weissman S, Urban AE, Gerstein M, Vaccarino FM (2012) Somatic copy number mosaicism in human skin revealed by induced pluripotent stem cells. Nature 492(7429):438–442. doi:10.1038/nature11629

16. Lodato MA, Woodworth MB, Lee S, Evrony GD, Mehta BK, Karger A, Lee S, Chittenden TW, D'Gama AM, Cai X, Luquette LJ, Lee E, Park PJ, Walsh CA (2015) Somatic mutation in single human neurons tracks developmental and transcriptional history. Science 350(6256): 94–98. doi:10.1126/science.aab1785

17. Abyzov A, Tomasini L, Zhou B, Vasmatzis N, Coppola G, Amenduni M, Pattni R, Wilson M, Gerstein M, Weissman S, Urban AE, Vaccarino FM (2017) One thousand somatic SNVs per skin fibroblast cell set baseline of mosaic mutational load with patterns that suggest proliferative origin. Genome Res 27(4):512–523. doi:10.1101/gr.215517.116

18. Langer-Safer PR, Levine M, Ward DC (1982) Immunological method for mapping genes on Drosophila polytene chromosomes. Proc Natl Acad Sci U S A 79(14):4381–4385

19. Telenius H, Carter NP, Bebb CE, Nordenskjold M, Ponder BA, Tunnacliffe A (1992) Degenerate oligonucleotide-primed PCR: general amplification of target DNA by a single degenerate primer. Genomics 13(3):718–725

20. Gawad C, Koh W, Quake SR (2016) Single-cell genome sequencing: current state of the science. Nat Rev Genet 17(3):175–188. doi:10.1038/nrg.2015.16

21. McConnell MJ, Lindberg MR, Brennand KJ, Piper JC, Voet T, Cowing-Zitron C, Shumilina S, Lasken RS, Vermeesch JR, Hall IM, Gage FH (2013) Mosaic copy number variation in human neurons. Science 342(6158):632–637. doi:10.1126/science.1243472

22. Cai X, Evrony GD, Lehmann HS, Elhosary PC, Mehta BK, Poduri A, Walsh CA (2014) Single-cell, genome-wide sequencing identifies clonal somatic copy-number variation in the human brain. Cell Rep 8(5):1280–1289. doi:10.1016/j.celrep.2014.07.043

23. Dean FB, Hosono S, Fang L, Wu X, Faruqi AF, Bray-Ward P, Sun Z, Zong Q, Du Y, Du J, Driscoll M, Song W, Kingsmore SF, Egholm M, Lasken RS (2002) Comprehensive human genome amplification using multiple displacement amplification. Proc Natl Acad Sci U S A 99(8):5261–5266. doi:10.1073/pnas.082089499

24. Evrony GD, Lee E, Mehta BK, Benjamini Y, Johnson RM, Cai X, Yang L, Haseley P, Lehmann HS, Park PJ, Walsh CA (2015) Cell lineage analysis in human brain using endogenous retroelements. Neuron 85(1):49–59. doi:10.1016/j.neuron.2014.12.028

25. Zong C, Lu S, Chapman AR, Xie XS (2012) Genome-wide detection of single-nucleotide and copy-number variations of a single human cell. Science 338(6114):1622–1626. doi:10.1126/science.1229164

26. Langmore JP (2002) Rubicon Genomics, Inc. Pharmacogenomics 3(4):557–560. doi:10.1517/14622416.3.4.557

27. Gole J, Gore A, Richards A, Chiu YJ, Fung HL, Bushman D, Chiang HI, Chun J, Lo YH, Zhang K (2013) Massively parallel polymerase cloning and genome sequencing of single cells using nanoliter microwells. Nat Biotechnol 31(12):1126–1132. doi:10.1038/nbt.2720

28. Picher AJ, Budeus B, Wafzig O, Kruger C, Garcia-Gomez S, Martinez-Jimenez MI, Diaz-Talavera A, Weber D, Blanco L, Schneider A (2016) TruePrime is a novel method for whole-genome amplification from single cells based on TthPrimPol. Nat Commun 7:13296. doi:10.1038/ncomms13296

29. Saini N, Roberts SA, Klimczak LJ, Chan K, Grimm SA, Dai S, Fargo DC, Boyer JC, Kaufmann WK, Taylor JA, Lee E, Cortes-Ciriano I, Park PJ, Schurman SH, Malc EP, Mieczkowski PA, Gordenin DA (2016) The impact of environmental and endogenous damage on somatic mutation load in human skin fibroblasts. PLoS Genet 12(10):e1006385. doi:10.1371/journal.pgen.1006385

30. Hazen JL, Faust GG, Rodriguez AR, Ferguson WC, Shumilina S, Clark RA, Boland MJ, Martin G, Chubukov P, Tsunemoto RK, Torkamani A, Kupriyanov S, Hall IM, Baldwin KK (2016) The complete genome sequences, unique mutational spectra, and developmental potency of adult neurons revealed by cloning. Neuron 89(6):1223–1236. doi:10.1016/j.neuron.2016.02.004

31. Leung ML, Wang Y, Waters J, Navin NE (2015) SNES: single nucleus exome sequencing. Genome Biol 16:55. doi:10.1186/s13059-015-0616-2

32. McConnell MJ, Moran JV, Abyzov A, Akbarian S, Bae T, Cortes-Ciriano I, Erwin JA, Fasching L, Flasch DA, Freed D, Ganz J, Jaffe AE, Kwan KY, Kwon M, Lodato MA, Mills RE, Paquola ACM, Rodin RE, Rosenbluh C, Sestan N, Sherman MA, Shin JH, Song S, Straub RE, Thorpe J, Weinberger DR, Urban AE, Zhou B, Gage FH, Lehner T, Senthil G, Walsh CA, Chess A, Courchesne E, Gleeson JG, Kidd JM, Park PJ, Pevsner J, Vaccarino FM, Brain Somatic Mosaicism Network (2017) Intersection of diverse neuronal genomes and neuropsychiatric disease: the Brain Somatic Mosaicism Network. Science 356(6336). doi:10.1126/science.aal1641

33. Hill FS, Marchetti F, Liechty M, Bishop J, Hozier J, Wyrobek AJ (2003) A new FISH assay to simultaneously detect structural and numerical chromosomal abnormalities in mouse sperm. Mol Reprod Dev 66(2):172–180. doi:10.1002/mrd.10299

34. Kallioniemi A, Kallioniemi OP, Sudar D, Rutovitz D, Gray JW, Waldman F, Pinkel D (1992) Comparative genomic hybridization for molecular cytogenetic analysis of solid tumors. Science 258(5083):818–821

35. Haraksingh RR, Abyzov A, Gerstein M, Urban AE, Snyder M (2011) Genome-wide mapping of copy number variation in humans: comparative analysis of high resolution array platforms. PLoS One 6(11):e27859. doi:10.1371/journal.pone.0027859

36. Conrad DF, Pinto D, Redon R, Feuk L, Gokcumen O, Zhang Y, Aerts J, Andrews TD, Barnes C, Campbell P, Fitzgerald T, Hu M, Ihm CH, Kristiansson K, Macarthur DG, Macdonald JR, Onyiah I, Pang AW, Robson S, Stirrups K, Valsesia A, Walter K, Wei J, Tyler-Smith C, Carter NP, Lee C, Scherer SW, Hurles ME (2010) Origins and functional impact of copy number variation in the human genome. Nature 464(7289):704–712. doi:10.1038/nature08516

37. Le Caignec C, Spits C, Sermon K, De Rycke M, Thienpont B, Debrock S, Staessen C, Moreau Y, Fryns JP, Van Steirteghem A, Liebaers I, Vermeesch JR (2006) Single-cell chromosomal imbalances detection by array CGH. Nucleic Acids Res 34(9):e68. doi:10.1093/nar/gkl336

38. Fiegler H, Geigl JB, Langer S, Rigler D, Porter K, Unger K, Carter NP, Speicher MR (2007) High resolution array-CGH analysis of single cells. Nucleic Acids Res 35(3):e15. doi:10.1093/nar/gkl1030

39. Wang DG, Fan JB, Siao CJ, Berno A, Young P, Sapolsky R, Ghandour G, Perkins N, Winchester E, Spencer J, Kruglyak L, Stein L, Hsie L, Topaloglou T, Hubbell E, Robinson E, Mittmann M, Morris MS, Shen N, Kilburn D, Rioux J, Nusbaum C, Rozen S, Hudson TJ, Lipshutz R, Chee M, Lander ES (1998) Large-scale identification, mapping, and genotyping of single-nucleotide polymorphisms in the human genome. Science 280(5366):1077–1082

40. McCarroll SA, Kuruvilla FG, Korn JM, Cawley S, Nemesh J, Wysoker A, Shapero MH, de Bakker PI, Maller JB, Kirby A, Elliott AL, Parkin M, Hubbell E, Webster T, Mei R, Veitch J, Collins PJ, Handsaker R, Lincoln S, Nizzari M, Blume J, Jones KW, Rava R, Daly MJ, Gabriel SB, Altshuler D (2008) Integrated detection and population-genetic analysis of SNPs and copy number variation. Nat Genet 40(10):1166–1174. doi:10.1038/ng.238

41. Markello TC, Carlson-Donohoe H, Sincan M, Adams D, Bodine DM, Farrar JE, Vlachos A, Lipton JM, Auerbach AD, Ostrander EA, Chandrasekharappa SC, Boerkoel CF, Gahl WA (2012) Sensitive quantification of mosaicism using high density SNP arrays and the cumulative distribution function. Mol Genet Metab 105(4):665–671. doi:10.1016/j.ymgme.2011.12.015

42. Wang J, Fan HC, Behr B, Quake SR (2012) Genome-wide single-cell analysis of recombination activity and de novo mutation rates in human sperm. Cell 150(2):402–412. doi:10.1016/j.cell.2012.06.030

43. Zafar H, Wang Y, Nakhleh L, Navin N, Chen K (2016) Monovar: single-nucleotide variant detection in single cells. Nat Methods 13(6):505–507. doi:10.1038/nmeth.3835

44. Korbel JO, Urban AE, Grubert F, Du J, Royce TE, Starr P, Zhong G, Emanuel BS, Weissman SM, Snyder M, Gerstein MB (2007) Systematic prediction and validation of breakpoints associated with copy-number variants in the human genome. Proc Natl Acad Sci U S A 104(24):10110–10115. doi:10.1073/pnas.0703834104

45. Korbel JO, Abyzov A, Mu XJ, Carriero N, Cayting P, Zhang Z, Snyder M, Gerstein MB

(2009) PEMer: a computational framework with simulation-based error models for inferring genomic structural variants from massive paired-end sequencing data. Genome Biol 10(2):R23. doi:10.1186/gb-2009-10-2-r23

46. Abyzov A, Urban AE, Snyder M, Gerstein M (2011) CNVnator: an approach to discover, genotype and characterize typical and atypical CNVs from family and population genome sequencing. Genome Res 21(6):974–984. doi:10.1101/gr.114876.110

47. Falconer E, Hills M, Naumann U, Poon SS, Chavez EA, Sanders AD, Zhao Y, Hirst M, Lansdorp PM (2012) DNA template strand sequencing of single-cells maps genomic rearrangements at high resolution. Nat Methods 9(11):1107–1112. doi:10.1038/nmeth.2206

48. Jamuar SS, Lam AT, Kircher M, D'Gama AM, Wang J, Barry BJ, Zhang X, Hill RS, Partlow JN, Rozzo A, Servattalab S, Mehta BK, Topcu M, Amrom D, Andermann E, Dan B, Parrini E, Guerrini R, Scheffer IE, Berkovic SF, Leventer RJ, Shen Y, Wu BL, Barkovich AJ, Sahin M, Chang BS, Bamshad M, Nickerson DA, Shendure J, Poduri A, Yu TW, Walsh CA (2014) Somatic mutations in cerebral cortical malformations. N Engl J Med 371(8):733–743. doi:10.1056/NEJMoa1314432

49. Freed D, Pevsner J (2016) The contribution of mosaic variants to autism spectrum disorder. PLoS Genet 12(9):e1006245. doi:10.1371/journal.pgen.1006245

50. Upton KR, Gerhardt DJ, Jesuadian JS, Richardson SR, Sanchez-Luque FJ, Bodea GO, Ewing AD, Salvador-Palomeque C, van der Knaap MS, Brennan PM, Vanderver A, Faulkner GJ (2015) Ubiquitous L1 mosaicism

in hippocampal neurons. Cell 161(2):228–239. doi:10.1016/j.cell.2015.03.026

51. Baillie JK, Barnett MW, Upton KR, Gerhardt DJ, Richmond TA, De Sapio F, Brennan PM, Rizzu P, Smith S, Fell M, Talbot RT, Gustincich S, Freeman TC, Mattick JS, Hume DA, Heutink P, Carninci P, Jeddeloh JA, Faulkner GJ (2011) Somatic retrotransposition alters the genetic landscape of the human brain. Nature 479(7374):534–537. doi:10.1038/nature10531

52. Martincorena I, Roshan A, Gerstung M, Ellis P, Van Loo P, McLaren S, Wedge DC, Fullam A, Alexandrov LB, Tubio JM, Stebbings L, Menzies A, Widaa S, Stratton MR, Jones PH, Campbell PJ (2015) Tumor evolution. High burden and pervasive positive selection of somatic mutations in normal human skin. Science 348(6237):880–886. doi:10.1126/science.aaa6806

53. Badge RM, Alisch RS, Moran JV (2003) ATLAS: a system to selectively identify human-specific L1 insertions. Am J Hum Genet 72(4):823–838. doi:10.1086/373939

54. Evrony GD, Cai X, Lee E, Hills LB, Elhosary PC, Lehmann HS, Parker JJ, Atabay KD, Gilmore EC, Poduri A, Park PJ, Walsh CA (2012) Single-neuron sequencing analysis of L1 retrotransposition and somatic mutation in the human brain. Cell 151(3):483–496. doi:10.1016/j.cell.2012.09.035

55. Erwin JA, Paquola AC, Singer T, Gallina I, Novotny M, Quayle C, Bedrosian TA, Alves FI, Butcher CR, Herdy JR, Sarkar A, Lasken RS, Muotri AR, Gage FH (2016) L1-associated genomic regions are deleted in somatic cells of the healthy human brain. Nat Neurosci 19(12):1583–1591. doi:10.1038/nn.4388

Part II

Aneuploidy and Ploidy Variation

FISH-Based Assays for Detecting Genomic (Chromosomal) Mosaicism in Human Brain Cells

Yuri B. Yurov, Svetlana G. Vorsanova, Ilia V. Soloviev, Alexei M. Ratnikov, and Ivan Y. Iourov

Abstract

Genomic or chromosomal mosaicism in human brain cells is considered a source for neuronal diversity and a mechanism for neuropsychiatric diseases. However, there is still a lack of consensus concerning the extent and effects of mosaic chromosome abnormalities (i.e., aneuploidy) in the normal and diseased human brain. To solve this problem, a need for detailed description of single-cell techniques for chromosomal analysis of human brain cells appears to exist. In this chapter, FISH-based techniques for detecting genomic (chromosomal) mosaicism in the human brain are described.

Key words Human brain, Chromosomal mosaicism, Chromosome, Aneuploidy, Fluorescence in situ hybridization, Single cell, Molecular cytogenetics

1 Introduction

The human brain has long been found to show appreciable rates of chromosomal or genomic mosaicism [1, 2]. However, despite the achievements in molecular cytogenetics and single-cell biology, there are still a number of technical limitations hindering the evaluation of the intrinsic rates of somatic mosaicism in human postmitotic cells [3–5]. Alternatively, interphase molecular cytogenetics does provide a basis for uncovering chromosomal or genomic variations at molecular resolutions, at all stages of cell cycle and at single-cell level [6–9]. As a result, it has been repeatedly reported that the human brain appears to be highly affected by mosaic aneuploidy (gain or loss of chromosomes in a cell) [10–17]. Furthermore, it seems that these types of somatic genomic variations or chromosomal mosaicism are an integral component of the human brain development, neuronal diversity, and brain functioning [10–12, 14, 17–20]. However, taking into account the devastating effect of aneuploidy or similar types of chromosomal mosaicism (losses/gains of large chromosome

José María Frade and Fred H. Gage (eds.), *Genomic Mosaicism in Neurons and Other Cell Types*, Neuromethods, vol. 131, DOI 10.1007/978-1-4939-7280-7_2, © Springer Science+Business Media LLC 2017

parts, polyploidy—a gain of a haploid chromosome set) on cellular phenotype, it is considered that these somatic genome variations are likely to be candidate mechanisms for brain malfunctioning in neurological and psychiatric diseases [1–3, 15, 20]. To test it, advanced and specific molecular cytogenetic methodology for uncovering chromosomal mosaicism in the human brain is required. In a single-cell context, fluorescence in situ hybridization (FISH) appears to be a reliable basis for effective visualization approaches to detect chromosomal abnormalities in interphase.

Actually, schizophrenia [21–23], Alzheimer's disease [15, 24–29], Lewy body diseases [30], and ataxia-telangiectasia (including murine models and postmortem human brain samples) [15, 31–33] are considered to be associated with FISH-detectable somatic genomic variations in the affected brain. Additionally, autism is also hypothesized to result from brain-specific chromosomal mosaicism in an appreciable proportion of cases [34–37]. In the light of these studies, it was hypothesized that such genomic variations are likely to possess a kind of a general effect on behavior [38]. Furthermore, it appears that chromosomal mosaicism (mosaic aneuploidy) mediates aging of the brain [39–43] similarly to other human tissues [44–46]. Nonetheless, the incidence and consequences of somatic chromosome and gene mutations in the human brain remain a matter of debate, which can shape new paradigms in neuroscience and neurogenomic research. Still, there is a need for protocols of FISH-based assays with special reference to brain single-cell molecular cytogenetic analysis associated with specific technical and analytical (interpretational) problems [47–49]. Among these, a striking one is referred to the definition of non-pathogenic (background) levels of somatic mutations [1, 50–53]. Since chromosomal abnormalities are not unique type of somatic genome variations in the human brain detectable/confirmable by FISH [51, 52, 54–56], this problem becomes even more actual.

Here, we describe an interphase FISH protocol respecting technical and analytical (interpretational) problems arbitrarily designated as a molecular neurocytogenetic technology. The protocol describes basic procedure including brain sample preparation, DNA probe hybridization/detection, and microscopic visualization. Moreover, additional approaches to enhance the interpretation of FISH results and to increase the efficiency of interphase FISH (i.e., quantitative FISH (QFISH)) are mentioned.

2 Materials and Solutions

To perform FISH-based experiments standard molecular cytogenetic equipment, reagents, and solutions are required. For more details, please refer to [57–59].

1. Ethanol 100, 96, and 70%, water-diluted ethanol.

2. Glass tube and Teflon pestle (Cole-Parmer International; cat. Nos. A-04368-32 for Teflon pestle and A-04368-33 for glass tube).

3. EBBS (Earle's buffered saline solution).

4. PBS (phosphate-buffered saline; pH 7.3, containing 0.1% (w/v) of Nonidet P-40): Stored at room temperature. Prepare 10× stock water solution with 1.37 M NaCl, 27 mM KCl, 100 mM Na_2HPO_4, and 18 mM KH_2PO_4 (pH 7.4 is adjusted by 1 N HCl).

5. $PBS/MgCl_2$ solution: 1 volume of 2 M $MgCl_2$ in 38 volumes of 1× PBS.

6. $Formaldehyde/PBS/MgCl_2$ solution: Add 2.7 mL of formaldehyde (37%) in 100 mL of $PBS/MgCl_2$ solution (to produce 1% of formaldehyde in $PBS/MgCl_2$ solution).

7. Acetic acid glacial 60% (w/v).

8. SSC (20× saline-sodium citrate): 3 M sodium chloride, 0.3 M trisodium citrate stored at room temperature. The solution is prepared by dissolving one volume of 20× SSC in four volumes of water and adding Tween-20 to 0.5%.

9. Fixative solution of methanol: Glacial acetic acid (3:1, v:v) freshly prepared and stored at −20 °C. Attention: methanol is exceedingly toxic.

10. Pepsin solution as required (freshly prepared): Pepsin solution 10% (w/v) (stored at −20 °C) is diluted in prewarmed (37 °C) solution of 0.01 N HCl (chlorohydric acid).

11. NaSCN (sodium isothiocyanate) 1 M (w/v) for disruption of DNA–protein complexes. Attention: NaSCN is toxic.

12. Solution of Sudan black and ethanol–water (dissolve 0.7 g of Sudan black in 100 mL of 96% ethanol, then add and stir with 50 mL of water, stored at room temperature).

13. DAPI (Sigma).

14. Propidium iodide (Sigma).

15. Vectashield (antifade solution).

16. Rubber cement.

17. Xylene 100%.

18. RNase 0.5% (w/v).

19. Coplin jar (50 mL).

20. Microscope slides, 25 × 75 × 1 mm, plain.

21. 24 ×24 mm and 20 × 40 mm cover slips.

22. 15-mL sterile plastic (or glass) tube.

23. Epifluorescence Microscope Zeiss Axioskop (Carl Zeiss) with ×40 and ×100 Plan Fluo objectives and a kit with ×40 and ×100 Plan Fluo objectives and a triple filter (B/G/R filter).

24. Charge-coupled device (CCD) camera mounted on a fluorescence microscope equipped with a set of specific filters used for FISH (CoolCube 1; Meta Systems).

25. Image acquisition software (ImageJ; https://imagej.nih.gov/ij/).

3 Methods

Here, we describe basic FISH protocol that underlies almost all targeted/multicolor FISH-based assays for studying either specific chromosomal loci or whole homologous chromosomes in their integrity. The general outline of interphase FISH is depicted in Fig. 1.

3.1 Brain Tissue (Cell Suspension) Preparation for FISH Assays

The preparation of brain cells for a molecular cytogenetic analysis of chromosome complement and genome variations has special peculiarities quite different from the protocols used in interphase molecular cytogenetics. Here, we describe brain tissue processing to be used for FISH-based assays to study chromosomal mosaicism in interphase nuclei as proposed in [60].

3.1.1 Cell Suspension from Fresh-Frozen Tissue

1. Put the brain tissue in a dish and rinse it in 2 mL of Earle's buffered saline solution.

2. Take a part of the brain tissue of an approximate size of $3 \times 3 \times 3$ mm (3 mm^3) and place it into the homogenizer glass tube. Use the Teflon pestle to homogenize the piece of brain tissue by intensive rotating of pestle to produce the liquid-like material.

3. Append into glass tube, containing homogenized tissue, 2 mL of PBS and homogenize until the substance in the tube becomes a homogenous suspension.

4. Put the substance into a 15-mL plastic (or glass) tube and add 1 mL of 60% (w/v) glacial acetic acid. Leave the obtained mix for 3–5 min at room temperature.

5. Add 9 mL of fixative solution of methanol and centrifuge at 1000 g during 5 min.

6. Decant supernatant and add fixative mixture to 10 mL of total solution volume. Centrifuge at 3000 g for 8 min.

7. Repeat steps 5 and 6 three times.

8. The suspension can be stored for 6–12 months at −20 °C.

9. Place 100 μL of obtained suspension on microscope slide and leave to dry for 15–20 min.

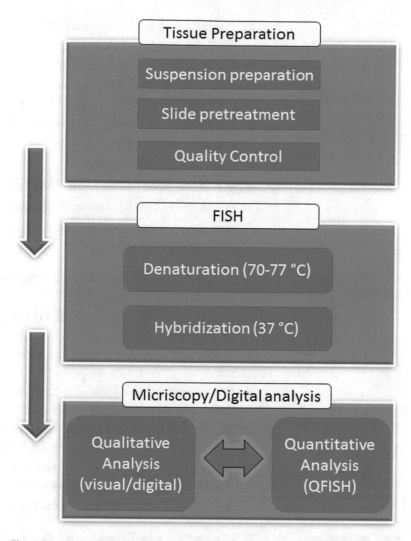

Fig. 1 A general schematic outline of a FISH-based assay (partially according to [60])

10. Put slides into pepsin solution (20–100 μL of pepsin) for 3–5 min.

11. Put slides into PBS for 5 min.

12. Dehydrate through a series of ethanol (100, 96, and 70%, 3 min each) and leave to dry (*see* **Note 1**).

3.1.2 Cell Suspension from Formalin-Fixed and Paraffin-Embedded Sections

1. Put fixed brain section slides (*see* **Note 2**) in 100% xylene at room temperature for 5 min.

2. Refresh xylene and repeat step 1.

3. Rehydrate sections in a series of ethanol (100, 96, and 70%, 2 min each) and wash them in SSC/detergent solution for 20 min mixed by inversion periodically at room temperature.

4. Put the slides into Coplin jar with 1 M NaSCN for 3–5 h (*see* **Note 3**).

5. Rinse slides in water for a few seconds (*see* **Note 4**).

6. Add 100 μL of RNase solution in 2× SSC to mount under coverslip and incubate at 37 °C for 15–30 min.

7. Put slides in pepsin solution (20–100 μL of pepsin) for 3–5 min.

8. Put slides in 2× PBS.

9. Put slides in PBS/MgCl₂.

10. Put slides in formaldehyde/PBS/MgCl₂.

11. Put slides in 1× PBS, subsequently each for 5 min.

12. Dehydrate through a series of ethanol (100, 96, and 70%, 3 min each) and leave to dry.

3.1.3 Quality Control (see Note 5)

1. Drop 5–12 μL of the suspension on the microscope slide and leave it to dry.

2. Check the distribution of nuclei through the light microscope using phase contrast.

3. If the distribution of nuclei is satisfactory, ignore next two steps.

4. If the distribution of nuclei is too low, centrifuge at 3500 g for 7 min and decrease the volume twice.

5. Mix by inversion and repeat steps 1–3. If the distribution of nuclei is satisfactory, ignore the next step.

6. If the distribution of nuclei is too dense to analyze, centrifuge at 3500 g for 7 min and decrease the volume twice.

7. Add 0.3–0.7 mL of fixative solution.

3.2 DNA Probes for FISH

To succeed in FISH-based analysis of chromosomal mosaicism in the human brain, a rigorous selection of DNA probes is mandatory. There are two types of probes that were found efficient for uncovering chromosomal mosaicism in neural cells: chromosome enumeration and microdissection-derived DNA probes [6–8, 13, 15, 19, 22, 28, 33, 62]. Chromosome enumeration (chromosome-specific) DNA probes (D1Z1, D2Z1, D3Z1, D4Z1, D6Z1, D7Z1, D8Z2, D9Z1, D10Z1, D11Z1, D12Z3, D13Z1/D21Z1, D14Z1/D22Z1, D15Z4, D16Z3, D17Z1, D18Z1, D20Z2, DXZ1, DYZ3; the number after "D" corresponds to chromosome designation number, i.e., D1Z1 is a probe for chromosome 1, D2Z1—for chromosome 2) are generally used for studying human chromosomes in interphase painting chromosome-specific pericentromeric DNAs [6, 8, 12, 57–59]. An example of multicolor interphase FISH using chromosome enumeration probes is given in Fig. 2. However, the use of DNA probes for specific chromosomal loci limits the potential of interphase FISH due to the lack of a view of the integral chromosome.

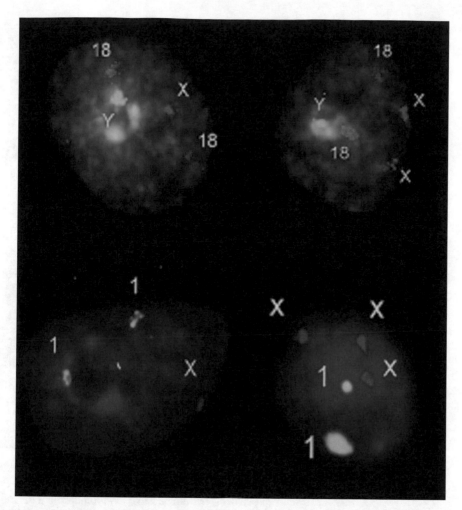

Fig. 2 Multiprobe interphase FISH using enumeration/centromeric probes for chromosomes 1, 18, X, and Y; **upper left nucleus**: a presumably normal nucleus with two chromosomes 18, one chromosome X, and one chromosome Y; **upper right nucleus**: an aneuploid nuclei with two chromosomes 18, *two chromosomes X,* and one chromosome Y; **lower left nucleus**: an aneuploid nuclei with two chromosomes 1 and *one chromosome X*; **lower right nucleus**: an aneuploid nuclei with two chromosomes 1 and *three chromosomes X*

Microdissection-based DNA probes designed for multicolor chromosomal banding (MCB) originally developed by Dr. Thomas Liehr and his colleagues [61] were also shown to be applicable for interphase molecular cytogenetics as shown in Fig. 3 [6, 7, 13, 15, 19, 22, 28, 33, 62]. These probes offer an opportunity for simultaneous visualization of all chromosome regions at once [13, 58, 61, 62]. FISH-based assays using MCB probes are known as interphase chromosome-specific multicolor chromosomal banding (ICS-MCB) [62].

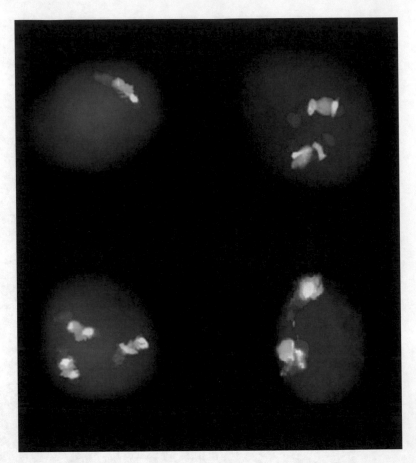

Fig. 3 ICS-MCB with DNA probes for chromosomes 9 and X showing monosomy of chromosome 9 in a nucleus of the developing human brain in the *upper left* nucleus, disomy (presumably normal nucleus) of chromosome 9 in a nucleus of the developing human brain in the *upper right* nucleus, trisomy of chromosome 9 in a nucleus of the developing human brain in the *lower left* nucleus, and disomy of chromosome X in the nucleus of the Alzheimer disease brain (from [16, 28], open-access articles distributed under the terms of the Creative Commons Attribution License (http://creativecommons.org/licenses/by/2.0)

3.3 Basic FISH Procedure

Basic FISH procedure is aimed at granting two key FISH processes (denaturation and hybridization). Here, a basic FISH procedure is described (for more details see books and articles dedicated to technical aspects of FISH [8, 57–59]).

1. Put 5 mL of the DNA probe on the pretreated slide and cover the suspension with an 18 × 18 mm coverslip.

2. Put the slide on a warming plate at 72–76 °C for 2–7 min.

3. Use rubber cement for sealing the sample.

4. Relocate it into a humid chamber at 37 °C overnight (*see* **Note 6**).

5. Take off the coverslip by putting distilled water or 2× SSC/0.2% Tween 20 on its edges (*see* **Note 7**).

6. Put the slide in 50% formamide solution in 2× SSC at 42–45 °C for 5–15 min.

7. Swap the washing solution for 2× SSC/Tween20 at 37–42 °C and leave it for 10–15 min in a 100 mL Coplin jar.

8. Append FITC solution on the slide and cover the whole slide with a coverslip.

9. Incubate the slides in a humid chamber at 37 °C for 40 min.

10. Put the slides in 2 × SSC/0.2% Tween for 20 min at room temperature.

11. Counterstain the slide by 24 mL of DAPI solution and cover the slide with a corresponding coverslip.

12. Proceed to microscopic analysis under fluorescence microscope.

3.4 QFISH

QFISH combines FISH and digital quantification of microscopic images for a variety of purposes. A detailed interphase QFISH protocol for evaluating signals for differing between false-positive signal appearance and chromosome loss [63, 64] is given below. Regardless of a variety of software for quantification microscopic images, we suggest ImageJ (*see* **Note 8**), inasmuch as it appears to represent one of the most familiar free software for these research purposes [65]. Figure 4 gives an example of QFISH on interphase nuclei derived from human brain cells.

1. Acquire FISH image by a CCD camera of a fluorescence microscope using a 100× (63×) objective and available software.

2. Capture the images using separate filters for each fluorochrome separately.

3. Save the images in 8-bit black and white images.

4. Load each FISH image into ImageJ software.

5. Selected FISH signal area by the "rectangular" selection tool.

6. Attribute the area to the First Lane using "Select First Lane," Analyze/Gels/Select First Lane, or simply press Ctrl + 1.

7. Obtain the signal appearance with reduced background by "Threshold" (Image/Adjust/Threshold...) or pressing Ctrl + Shift + T.

8. Obtain the plot of image containing the graph depicting intensity profiles (*see* **Note 9**) using "Plot profile" (Analyze/Plot Profile) or simply pressing Ctrl + K.

9. Remove the grid from the plot of image; Threshold is used (Image/Adjust/Threshold... or Ctrl + Shift + T).

10. Define graph borders corresponding to a signal (select area to be measured) by drawing a line (Draw -Edit/Draw or simply press Ctrl + D).

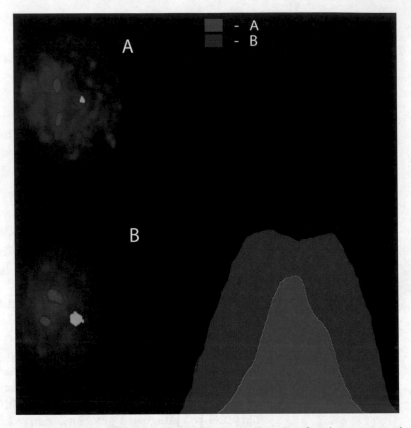

Fig. 4 QFISH with using enumeration/centromeric probes for chromosomes 1 (*red signals*/D1Z1) and X (*green signals*/DXZ1): nucleus A demonstrates a green signal with a relative intensity of 2120 pixels—true X chromosome monosomy; nucleus B demonstrates a green signal with a relative intensity of 4800 pixels— two overlapping chromosome X signals, but not a chromosome loss [partially according to Fig. 1 of [28], an open-access article distributed under the terms of the Creative Commons Attribution License)

11. Select the area resulting from drawing during the previous step by Wand (tracing) tool.

12. Measure the selection by Measure–Analyze/Measure or simply press Ctrl + M.

13. Output numerical values of the area (perimeter) in a separate window.

14. Compare numerical values of these different signals from the same image (Fig. 4).

4   Discussion

Technical solutions for studying chromosomal mosaicism in the interphase nuclei of the human brain are available in the biomedical literature [1, 3–5, 25, 33, 42, 48]. However, in regard to the problem of irreproducibility of data on chromosomal variations in the unaffected and diseased human brain, it appears that generalized/unified protocols are required. Furthermore, there is a need for determining the intrinsic rates of chromosomal mutations in human brain cell populations in the light of a panel of studies demonstrating a number of processes occurring in the human brain, which are likely to result in brain-specific chromosomal mosaicism [66–71]. Since it has been shown that genomic and chromosomal pathology confined to the brain is a mechanism for a variety of psychiatric and neurological disorders [21–33], there is a strong need for continuing single-cell analysis of the human brain aiming at the definition of cellular genome variability. Despite relatively high efficiency of current molecular cytogenetic techniques for diagnosing chromosomal mosaicism in postmitotic human cell populations [72, 73], interpretation and definition of a sample as a mosaic remains a challenge hard to manage, especially in case of neurocytogenetic studies [1, 47, 70, 72]. Nevertheless, developments in interphase molecular cytogenetics provide a basis for solving these problems [8] demonstrating the relevance of molecular neurocytogenetic studies to neuroscience (i.e., suggesting of molecular therapies for neurodegenerative diseases on the basis of molecular neurocytogenetic studies [48, 74] and uncovering molecular and cellular mechanisms for brain diseases [1, 8, 48, 70, 75]). Thus, chromosomal studies of the human brain (molecular neurocytogenetics) using FISH-based assays are to be recognized as an integral part of surveying genome variability that mediates interindividual/intercellular neuronal diversity in health and disease.

5 Notes

1. Cell suspensions are to be stored at −20 °C (for at least 1 year).

2. Brain tissue size is supposed to range from 7 to 20 μM.

3. NaSCN is highly toxic (wearing of gloves is indispensable).

4. Do not allow drying of slides.

5. Quality control procedure can be applied to cell suspensions only.

6. Larger chromosomal DNA sequences require longer periods of incubation (up to three nights relying on the experiment type).

7. This stage is mandatory if rubber cement was used to seal the sample.

8. https://imagej.nih.gov/ij/.

9. It is possible to measure relative DNA content in chromosomal loci painted by FISH, inasmuch as signal intensities are proportional to the DNA content within these chromosomal loci [59, 63, 64].

Acknowledgments

Professor YB Yurov is supported by a grant from the Russian Science Foundation (project #14-35-00060) at Moscow State University of Psychology and Education. Professors SG Vorsanova and IY Iourov are supported by a grant from the Russian Science Foundation (project #14-15-00411) at Mental Health Research Center. The study of the Alzheimer disease brain is supported by the ERA.Net RUS Plus program.

References

1. Iourov IY, Vorsanova SG, Yurov YB (2006) Chromosomal variation in mammalian neuronal cells: known facts and attractive hypotheses. Int Rev Cytol 249:143–191

2. Kingsbury MA, Yung YC, Peterson SE, Westra JW, Chun J (2006) Aneuploidy in the normal and diseased brain. Cell Mol Life Sci 63:2626–2641

3. Iourov IY, Vorsanova SG, Yurov YB (2012) Single cell genomics of the brain: focus on neuronal diversity and neuropsychiatric diseases. Curr Genomics 13(6):477–488

4. Bakker B, van den Bos H, Lansdorp PM, Foijer F (2015) How to count chromosomes in a cell: an overview of current and novel technologies. BioEssays 37(5):570–577

5. Harbom LJ, Chronister WD, McConnell MJ (2016) Single neuron transcriptome analysis can reveal more than cell type classification: does it matter if every neuron is unique? BioEssays 38(2):157–161

6. Iourov IY, Vorsanova SG, Yurov YB (2008) Recent patents on molecular cytogenetics. Recent Pat DNA Gene Seq 2(1):6–15

7. Vorsanova SG, Yurov YB, Iourov IY (2010) Human interphase chromosomes: a review of available molecular cytogenetic technologies. Mol Cytogenet 3:1

8. Yurov YB, Vorsanova SG, Iourov IY (eds) (2013) Human interphase chromosomes: biomedical aspects. Springer, New York

9. Riegel M (2014) Human molecular cytogenetics: from cells to nucleotides. Genet Mol Biol 37(1 Suppl):194–209

10. Rehen SK, Yung YC, McCreight MP, Kaushal D, Yang AH, Almeida BS, Kingsbury MA, Cabral KM, McConnell MJ, Anliker B, Fontanoz M, Chun J (2005) Constitutional aneuploidy in the normal human brain. J Neurosci 25(9):2176–2180

11. Kingsbury MA, Friedman B, McConnell MJ, Rehen SK, Yang AH, Kaushal D, Chun J (2005) Aneuploid neurons are functionally active and integrated into brain circuitry. Proc Natl Acad Sci U S A 102(17):6143–6147

12. Yurov YB, Iourov IY, Monakhov VV, Soloviev IV, Vostrikov VM, Vorsanova SG (2005) The variation of aneuploidy frequency in the developing and adult human brain revealed by an interphase FISH study. J Histochem Cytochem 53(3):385–390

13. Iourov IY, Liehr T, Vorsanova SG, Kolotii AD, Yurov YB (2006) Visualization of interphase chromosomes in postmitotic cells of the human brain by multicolour banding (MCB). Chromosom Res 14(3):223–229

14. Westra JW, Peterson SE, Yung YC, Mutoh T, Barral S, Chun J (2008) Aneuploid mosaicism in the developing and adult cerebellar cortex. J Comp Neurol 507(6):1944–1951

15. Iourov IY, Vorsanova SG, Liehr T, Yurov YB (2009) Aneuploidy in the normal, Alzheimer's disease and ataxia-telangiectasia brain: differential expression and pathological meaning. Neurobiol Dis 34(2):212–220

16. Westra JW, Rivera RR, Bushman DM, Yung YC, Peterson SE, Barral S, Chun J (2010) Neuronal DNA content variation (DCV) with regional and individual differences in the human brain. J Comp Neurol 518(19):3981–4000

17. Devalle S, Sartore RC, Paulsen BS, Borges HL, Martins RA, Rehen SK (2012) Implications of aneuploidy for stem cell biology and brain therapeutics. Front Cell Neurosci 6:36

18. Rehen SK, McConnell MJ, Kaushal D, Kingsbury MA, Yang AH, Chun J (2001) Chromosomal variation in neurons of the developing and adult mammalian nervous system. Proc Natl Acad Sci U S A 98(23):13361–13366

19. Yurov YB, Iourov IY, Vorsanova SG, Liehr T, Kolotii AD, Kutsev SI, Pellestor F, Beresheva AK, Demidova IA, Kravets VS, Monakhov VV,

Soloviev IV (2007) Aneuploidy and confined chromosomal mosaicism in the developing human brain. PLoS One 2(6):e558

20. Yurov YB, Vorsanova SG, Iourov IY (2010) Ontogenetic variation of the human genome. Curr Genomics 11(6):420–425

21. Yurov YB, Vostrikov VM, Vorsanova SG, Monakhov VV, Iourov IY (2001) Multicolor fluorescent in situ hybridization on post-mortem brain in schizophrenia as an approach for identification of low-level chromosomal aneuploidy in neuropsychiatric diseases. Brain Dev 23(Suppl 1):S186–S190

22. Yurov YB, Iourov IY, Vorsanova SG, Demidova IA, Kravetz VS, Beresheva AK, Kolotii AD, Monakchov VV, Uranova NA, Vostrikov VM, Soloviev IV, Liehr T (2008) The schizophrenia brain exhibits low-level aneuploidy involving chromosome 1. Schizophr Res 98(1–3): 139–147

23. Sakai M, Watanabe Y, Someya T, Araki K, Shibuya M, Niizato K, Oshima K, Kunii Y, Yabe H, Matsumoto J, Wada A, Hino M, Hashimoto T, Hishimoto A, Kitamura N, Iritani S, Shirakawa O, Maeda K, Miyashita A, Niwa S, Takahashi H, Kakita A, Kuwano R, Nawa H (2015) Assessment of copy number variations in the brain genome of schizophrenia patients. Mol Cytogenet 8:46

24. Mosch B, Morawski M, Mittag A, Lenz D, Tarnok A, Arendt T (2007) Aneuploidy and DNA replication in the normal human brain and Alzheimer's disease. J Neurosci 27(26): 6859–6867

25. Iourov IY, Vorsanova SG, Yurov YB (2011) Genomic landscape of the Alzheimer's disease brain: chromosome instability—aneuploidy, but not tetraploidy—mediates neurodegeneration. Neurodegener Dis 8(1–2):35–37

26. Yurov YB, Vorsanova SG, Iourov IY (2011) The DNA replication stress hypothesis of Alzheimer's disease. ScientificWorldJournal 11:2602–2612

27. Arendt T, Brückner MK, Lösche A (2015) Regional mosaic genomic heterogeneity in the elderly and in Alzheimer's disease as a correlate of neuronal vulnerability. Acta Neuropathol 130(4):501–510

28. Yurov YB, Vorsanova SG, Liehr T, Kolotii AD, Iourov IY (2014) X chromosome aneuploidy in the Alzheimer's disease brain. Mol Cytogenet 7:20

29. Hou Y, Song H, Croteau DL, Akbari M, Bohr VA (2017) Genome instability in Alzheimer disease. Mech Ageing Dev 161(Pt A):83–94. doi:10.1016/jmad2016.04.005

30. Yang Y, Shepherd C, Halliday G (2015) Aneuploidy in Lewy body diseases. Neurobiol Aging 36(3):1253–1260

31. Allen DM, van Praag H, Ray J, Weaver Z, Winrow CJ, Carter TA, Braquet R, Harrington E, Ried T, Brown KD, Gage FH, Barlow C (2001) Ataxia telangiectasia mutated is essential during adult neurogenesis. Genes Dev 15(5):554–566

32. McConnell MJ, Kaushal D, Yang AH, Kingsbury MA, Rehen SK, Treuner K, Helton R, Annas EG, Chun J, Barlow C (2004) Failed clearance of aneuploid embryonic neural progenitor cells leads to excess aneuploidy in the Atm-deficient but not the Trp53-deficient adult cerebral cortex. J Neurosci 24(37):8090–8096

33. Iourov IY, Vorsanova SG, Liehr T, Kolotii AD, Yurov YB (2009) Increased chromosome instability dramatically disrupts neural genome integrity and mediates cerebellar degeneration in the ataxia-telangiectasia brain. Hum Mol Genet 18(14):2656–2669

34. Vorsanova SG, Yurov IY, Demidova IA, Voinova-Ulas VY, Kravets VS, Solov'ev IV, Gorbachevskaya NL, Yurov YB (2007) Variability in the heterochromatin regions of the chromosomes and chromosomal anomalies in children with autism: identification of genetic markers of autistic spectrum disorders. Neurosci Behav Physiol 37(6):553–558

35. Yurov YB, Vorsanova SG, Iourov IY, Demidova IA, Beresheva AK, Kravets VS, Monakhov VV, Kolotii AD, Voinova-Ulas VY, Gorbachevskaya NL (2007) Unexplained autism is frequently associated with low-level mosaic aneuploidy. J Med Genet 44(8):521–525

36. Iourov IY, Yurov YB, Vorsanova SG (2008) Mosaic X chromosome aneuploidy can help to explain the male-to-female ratio in autism. Med Hypotheses 70(2):456

37. Vorsanova SG, Voinova VY, Yurov IY, Kurinnaya OS, Demidova IA, Yurov YB (2010) Cytogenetic, molecular-cytogenetic, and clinical-genealogical studies of the mothers of children with autism: a search for familial genetic markers for autistic disorders. Neurosci Behav Physiol 40(7):745–756

38. Charney E (2012) Behavior genetics and post-genomics. Behav Brain Sci 35(5):331–358

39. Yurov YB, Vorsanova SG, Iourov IY (2009) GIN'n'CIN hypothesis of brain aging: deciphering the role of somatic genetic instabilities and neural aneuploidy during ontogeny. Mol Cytogenet 2:23

40. Faggioli F, Wang T, Vijg J, Montagna C (2012) Chromosome-specific accumulation of aneuploidy in the aging mouse brain. Hum Mol Genet 21(24):5246–5253

41. Fischer HG, Morawski M, Brückner MK, Mittag A, Tarnok A, Arendt T (2012) Changes in neuronal DNA content variation in the

human brain during aging. Aging Cell 11(4): 628–633

42. Chow HM, Herrup K (2015) Genomic integrity and the ageing brain. Nat Rev Neurosci 16(11):672–684

43. Andriani GA, Vijg J, Montagna C (2017) Mechanisms and consequences of aneuploidy and chromosome instability in the aging brain. Mech Ageing Dev 161(Pt A):19–36. doi:10.1016/jmad2016.03.007

44. Abruzzo MA, Mayer M, Jacobs PA (1985) Aging and aneuploidy: evidence for the preferential involvement of the inactive X chromosome. Cytogenet Cell Genet 39(4):275–278

45. Russell LM, Strike P, Browne CE, Jacobs PA (2007) X chromosome loss and ageing. Cytogenet Genome Res 116(3):181–185

46. Iourov IY, Vorsanova SG, Yurov YB (2008) Chromosomal mosaicism goes global. Mol Cytogenet 1:26

47. Bushman DM, Chun J (2013) The genomically mosaic brain: aneuploidy and more in neural diversity and disease. Semin Cell Dev Biol 24(4):357–369

48. Iourov IY, Vorsanova SG, Yurov YB (2013) Somatic cell genomics of brain disorders: a new opportunity to clarify genetic-environmental interactions. Cytogenet Genome Res 139(3): 181–188

49. Insel TR (2014) Brain somatic mutations: the dark matter of psychiatric genetics? Mol Psychiatry 19(2):156–158

50. Fickelscher I, Starke H, Schulze E, Ernst G, Kosyakova N, Mkrtchyan H, MacDermont K, Sebire N, Liehr T (2007) A further case with a small supernumerary marker chromosome (sSMC) derived from chromosome 1—evidence for high variability in mosaicism in different tissues of sSMC carriers. Prenat Diagn 27(8):783–785

51. Iourov IY, Vorsanova SG, Yurov YB (2008) Molecular cytogenetics and cytogenomics of brain diseases. Curr Genomics 9(7):452–465

52. Iourov IY, Vorsanova SG, Yurov YB (2010) Somatic genome variations in health and disease. Curr Genomics 11(6):387–396

53. Hultén MA, Jonasson J, Iwarsson E, Uppal P, Vorsanova SG, Yurov YB, Iourov IY (2013) Trisomy 21 mosaicism: we may all have a touch of down syndrome. Cytogenet Genome Res 139(3):189–192

54. Coufal NG, Garcia-Perez JL, Peng GE, Yeo GW, Mu Y, Lovci MT, Morell M, O'Shea KS, Moran JV, Gage FH (2009) L1 retrotransposition in human neural progenitor cells. Nature 460(7259):1127–1131

55. McConnell MJ, Lindberg MR, Brennand KJ, Piper JC, Voet T, Cowing-Zitron C, Shumilina S, Lasken RS, Vermeesch JR, Hall IM, Gage FH (2013) Mosaic copy number variation in human neurons. Science 342(6158):632–637

56. van den Bos H, Spierings DC, Taudt AS, Bakker B, Porubský D, Falconer E, Novoa C, Halsema N, Kazemier HG, Hoekstra-Wakker K, Guryev V, den Dunnen WF, Foijer F, Tatché MC, Boddeke HW, Lansdorp PM (2016) Single-cell whole genome sequencing reveals no evidence for common aneuploidy in normal and Alzheimer's disease neurons. Genome Biol 17:116

57. Yurov YB, Soloviev IV, Vorsanova SG, Marcais B, Roizes G, Lewis R (1996) High resolution multicolor fluorescence in situ hybridization using cyanine and fluorescein dyes: rapid chromosome identification by directly fluorescently labeled alphoid DNA probes. Hum Genet 97(3):390–398

58. Liehr T (ed) (2009/2016) Fluorescence *in situ* hybridization (FISH)—application guide, 1st & 2nd edn. Springer Protocols. Springer, Heidelberg

59. Iourov IY, Vorsanova SG, Yurov YB (2016) Detection of nuclear DNA by interphase fluorescence *in situ* hybridization. Encyclopedia Anal Chem: 1–12

60. Iourov IY, Vorsanova SG, Pellestor F, Yurov YB (2006) Brain tissue preparations for chromosomal PRINS labeling. Methods Mol Biol 334:123–132

61. Liehr T, Heller A, Starke H, Rubtsov N, Trifonov V, Mrasek K, Weise A, Kuechler A, Claussen U (2002) Microdissection based high resolution multicolor banding for all 24 human chromosomes. Int J Mol Med 9(4):335–339

62. Iourov IY, Liehr T, Vorsanova SG, Yurov YB (2007) Interphase chromosome-specific multicolor banding (ICS-MCB): a new tool for analysis of interphase chromosomes in their integrity. Biomol Eng 24(4):415–417

63. Iourov IY, Soloviev IV, Vorsanova SG, Monakhov VV, Yurov YB (2005) An approach for quantitative assessment of fluorescence in situ hybridization (FISH) signals for applied human molecular cytogenetics. J Histochem Cytochem 53(3):401–408

64. Iourov IY (2017) Quantitative fluorescence in situ hybridization (QFISH). Methods Mol Biol 1541:143–149

65. Schneider CA, Rasband WS, Eliceiri KW (2012) NIH image to image J: 25 years of image analysis. Nat Methods 9(7):671–675

66. Yang AH, Kaushal D, Rehen SK, Kriedt K, Kingsbury MA, McConnell MJ, Chun J (2003)

Chromosome segregation defects contribute to aneuploidy in normal neural progenitor cells. J Neurosci 23(32):10454–10462

67. Peterson SE, Yang AH, Bushman DM, Westra JW, Yung YC, Barral S, Mutoh T, Rehen SK, Chun J (2012) Aneuploid cells are differentially susceptible to caspase-mediated death during embryonic cerebral cortical development. J Neurosci 32(46):16213–16222

68. Arendt T (2012) Cell cycle activation and aneuploid neurons in Alzheimer's disease. Mol Neurobiol 46(1):125–135

69. Granic A, Potter H (2013) Mitotic spindle defects and chromosome mis-segregation induced by LDL/cholesterol-implications for Niemann-pick C1, Alzheimer's disease, and atherosclerosis. PLoS One 8(4):e60718

70. Bajic V, Spremo-Potparevic B, Zivkovic L, Isenovic ER, Arendt T (2015) Cohesion and the aneuploid phenotype in Alzheimer's disease: a tale of genome instability. Neurosci Biobehav Rev 55:365–374

71. Iourov IY, Vorsanova SG, Zelenova MA, Korostelev SA, Yurov YB (2015) Genomic copy number variation affecting genes involved in the cell cycle pathway: implications for somatic mosaicism. Int J Genomics 2015:757680

72. Vorsanova SG, Yurov YB, Soloviev IV, Iourov IY (2010) Molecular cytogenetic diagnosis and somatic genome variations. Curr Genomics 11(6):440–446

73. Martin CL, Warburton D (2015) Detection of chromosomal aberrations in clinical practice: from karyotype to genome sequence. Annu Rev Genomics Hum Genet 16:309–326

74. Yurov YB, Iourov IY, Vorsanova SG (2009) Neurodegeneration mediated by chromosome instability suggests changes in strategy for therapy development in ataxia-telangiectasia. Med Hypotheses 73(6):1075–1076

75. Arendt T, Brückner MK, Mosch B, Lösche A (2010) Selective cell death of hyperploid neurons in Alzheimer's disease. Am J Pathol 177(1):15–20

Chapter 3

Flow Cytometric and Sorting Analyses for Nuclear DNA Content, Nucleotide Sequencing, and Interphase FISH

Gwendolyn E. Kaeser and Jerold Chun

Abstract

The study of genomic mosaicism among human brain cells is challenging. The human brain contains hundreds of billions of cells that are intricately connected and difficult to separate as intact, single cells. Additional challenges are encountered when interrogating small, seemingly random changes within single-cell genomes. Flow cytometric analysis (**FCM**), and fluorescence-activated *nuclear* sorting (**FANS**), has expanded our assessment capabilities for global and specific genomic and transcriptomic changes in human brain cells. The general approach is being utilized in a variety of downstream applications by many laboratories. Here we provide detailed methods of nuclear DNA content assessment and sorting that reports population averages as well as single-cell nuclear DNA content from cells of the human brain. We highlight protocol modifications that allow the same nuclear preparation to be used for subpopulation-specific FANS (also see chapter "Single-Cell Whole Genome Amplification and Sequencing to Study Neuronal Mosaicism and Diversity") in downstream analyses such as fluorescent in situ hybridization (**FISH**) (see chapters "FISH-Based Assays for Detecting Genomic (Chromosomal) Mosaicism in Human Brain Cells," "FISH Analysis of Aging-Associated Aneuploidy in Neurons and Non-neuronal Brain Cells" and "Using Fluorescence In Situ Hybridization (FISH) Analysis to Measure Chromosome Instability and Mosaic Aneuploidy in Neurodegenerative Diseases"), and single-cell genomic and transcriptomic sequencing (see chapters "Flow Cytometric Quantification, Isolation, and Subsequent Epigenetic Analysis of Tetraploid Neurons," "Single Cell CNV Detection in Human Neuronal Nuclei," "Multiple Annealing and Looping-Based Amplification Cycles (MALBAC) for the Analysis of DNA Copy Number Variation," and "Single-Cell Whole Genome Amplification and Sequencing to Study Neuronal Mosaicism and Diversity"). Other downstream techniques include, but are not limited to, single-cell qPCR (see chapter "Competitive PCR for Copy Number Assessment by Restricting dNTPs") and estimation of line-1 copy number (see chapters "Analysis of LINE-1 Retrotransposition in Neural Progenitor Cells and Neurons," "Estimation of LINE-1 Copy Number in the Brain Tissue and Isolated Neuronal Nuclei," and "Analysis of Somatic LINE-1 Insertions in Neurons").

Key words DNA content variation, Flow sorting, Flow cytometry, Neuron, Nuclei, NeuN, Sequencing, Fish, Genomic mosaicism, Somatic, Aneuploidy, Aneusomy

José María Frade and Fred H. Gage (eds.), *Genomic Mosaicism in Neurons and Other Cell Types*, Neuromethods, vol. 131, DOI 10.1007/978-1-4939-7280-7_3, © Springer Science+Business Media LLC 2017

1 Introduction

The initial demonstration that single neurons from the same, normal brain can show somatic genomic variability [1] has been expanded in subsequent years to encompass a vast range of genomic changes, including aneuploidy [2–10], other smaller copy number variations (**CNVs**) [11–15], and single-nucleotide variants (**SNVs**) [16, 17]. This led to the understanding that the vertebrate brain is a genomic mosaic, wherein each neuron—perhaps extending to non-neuronal cells—may be genomically distinct within a single brain. The functional roles for somatic genomic mosaicism are not yet known, but have been linked to pathological changes in sporadic Alzheimer's disease [15], and may contribute to high levels of transcriptional diversity [2] in human neurons of the cerebral cortex [18]. Genomic mosaicism may therefore contribute to the complex and poorly understood cellular diversity in the brain.

Each neuron in the human brain is thought to communicate via thousands of synaptic connections, creating an intermixed network that is virtually impossible to separate into single, complete cells. This high degree of complexity and inability to isolate intact brain cells make the study of genomic mosaicism especially challenging. Earlier interrogation methods that identified DNA changes in cells of the brain were primarily limited to fluorescent in situ hybridization (**FISH**), where whole chromosomes or specific chromosomal loci were interrogated using chromosome paints or point probes. These methods are labor intensive, limited by the number of cells that can be assessed, and are not easily adaptable for studying subpopulations of brain cells. By adapting methods used for the detection of cell cycle progression [19], apoptosis [20], and speciation changes in plants and animals [21] (reviewed in [22, 23]), our laboratory established methods for isolating and determining variable DNA content of brain nuclei. Use of this method led to the discovery that human cortical cells—particularly neurons—possess high levels of DNA content variation (**DCV**) compared to other cell types such as lymphocytes and cells of the cerebellum [11], and more recent analyses identified statistically significant DNA content increases in neurons of Alzheimer's disease (AD) brains compared to controls [15].

The first use of DNA content assessment of brain cell genomic mosaicism was employed to confirm relative levels of hypoploidy in developing mouse neuroblasts [1]. This approach was later modified to allow the first use of the neuronal nuclear antigen (NeuN) [24] with fluorescence-activated cell sorting (**FACS**) to isolate and study neurons by FANS [5]. We have since used this approach for multiple applications including the quantification of nuclear DNA content [11, 15], quantification of aneuploidies in developing mouse cerebral cortices [1, 10], single-cell DNA sequencing [12], and most recently, single-cell RNA sequencing in both mouse [25] and human [18] brain. Other groups also used similar methods for human single-cell DNA sequencing [13, 14, 26, 27] (also see Chaps. 4, 6, 7, and 13) and human DNA content measurements [28, 29] (also see Chap. 5).

The use of isolated nuclei is an important development for four reasons: (1) we are able to isolate intact nuclei from *postmortem* human brain tissue that possesses high-quality RNA and DNA, (2) nuclei are easily accessible to DNA dyes and antibodies and do not require fixation or cell permeabilization, (3) nuclei are free of other cytoplasmic components that may cause nonspecific DNA dye binding, and similarly (4) nuclei are free of mitochondria that possess DNA and therefore may alter DNA content.

DNA content assessments described here are based upon stoichiometric staining by DNA dyes. Multiple dyes can be used, each possessing its own strengths and weaknesses. We will focus on three DNA intercalators; 4′,6′-diamidino-2-phenylindole (**DAPI**), **DRAQ5**™, and propidium iodide (**PI**). Both DAPI and DRAQ5 are minor groove intercalators that bind strongly to A-T-rich regions of DNA, while PI is an intercalating agent that binds with no sequence preference at a stoichiometry of 1 molecule per 4 bases. Both DAPI and DRAQ5 can be used to establish DNA content in the presence of RNA. In contrast, pretreatment with RNase A must be used when PI is the dye chosen for quantitative DNA assessment. Additionally, when comparing among multiple species to calculate DNA content, it is important to consider the GC content ratios that can differ among species before choosing DRAQ5 or DAPI. The unbiased binding of PI makes it an ideal tool to measure DNA content linearly, and therefore, it will predominantly be used throughout this protocol; however it is recommended that total DNA content be established using multiple dyes.

Most importantly, intercalators do not bind covalently, but are reversible. Therefore, the intensity of the stain depends on the number of available binding sites and the number of available labeling molecules [22]. For accurate assessment of DNA content, there must be an equilibrium, or saturation point, where there are excess dye molecules compared to the number of DNA-binding sites. Therefore, it is of vital importance that each researcher independently determines the saturation point for their nuclear preparation.

For total DNA content determinations, it is also important to use appropriate reference samples and other controls. Chicken and/or trout erythrocytes that are nucleated (CEN and/or TEN) have become common internal reference standards for calculating DNA content [30]. Unlike mammalian erythrocytes, chicken and trout erythrocytes contain a nucleus and are commercially available as a flow cytometry standard (e.g., BioSure, Grass Valley, CA). We found that human blood lymphocytes/nuclei can also serve as a reference control for human brain cell nuclei; lymphocytes are approximately of the same size, and show a consistently low coefficient of variation and consistent levels of total DNA content.

The nuclear isolation method presented here for DNA content assessment differs from FANS for downstream sequencing, and FANS for downstream FISH in several key ways: use of fixation, composition of staining solutions, and sorting parameters. We present a detailed method protocol in four sections: (1) nuclear extraction from human brain cerebral cortices, (2) use of an iodixanol gradient, (3) nuclear staining, and (4) flow cytometry/FACS and gating. We also provide the equations necessary to calculate total DNA content in a population. Each step will note modifications of the protocol for different downstream applications.

2   Materials

2.1   Materials

1. Human brain tissue.

2. Ice bucket.

3. Cryostat.

4. Optional: Cut-resistant work gloves.

5. Weigh boats.

6. Razor blade.

7. Angled forceps.

8. Optional: Optimal cutting temperature (OCT) compound.

9. 1.5 mL Eppendorf tubes.

10. 40–50 μm Cell strainer.

11. 5 mL Dounce homogenizer with Teflon pestle.

12. 15 mL Conical tubes.

13. Optional: 96- or 384-well PCR plates.

14. Chicken erythrocyte nuclei (CENs) (BioSure).

15. Appropriate safety gear, precautions, and IRB approvals, including use of BSL-2 facilities.

2.2   Solutions

1. NEB buffer [20 mM Tris (pH 8), 320 mM sucrose, 5 mM $CaCl_2$, 3 mM $MgAc_2$, 0.1 mM EDTA, 0.1% Triton X-100].

2. 1× PBSE [1× PBS, 2 mM EGTA].

3. 6× Tricine stock [120 mM Tricine–KOH (pH 7.8), 150 mM KCl, 30 mM $MgCl_2$].

4. Solution D [0.25 M sucrose, 1× Tricine].

5. OptiPrep™ or other iodixanol solution.

6. 50% Iodixanol [5 volumes Optiprep, 1 volume 6× Tricine].

7. 35% Iodixanol [1 volume 50% iodixanol, 2.3 volumes solution D].

8. 10% Iodixanol [1 volume 50% iodixanol, 4 volumes solution D].

9. 2% BSA (fatty acid free) in 1× PBSE.

10. DAPI (4′,6-diamidino-2-phenylindole).

11. NeuN antibody (recommended: rabbit anti-NeuN monoclonal, MABN140, Millipore at 1:1500).

12. Propidium iodide (PI) 5 mg/mL stock solution.

13. RNase A.

14. Diethylpyrocarbonate (DEP-C)-treated water, or commercially available nuclease-free water (NFW) (*see* **Note 1**).

15. Neutral buffered formalin (NBF).

16. 4% Paraformaldehyde.

2.3 Equipment

1. Refrigerated centrifuge with swinging buckets.

2. Transilluminator.

3. For DNA content analysis: BD Biosciences LSRII, or comparable.

4. For FACS: BD Biosciences FACS Aria II, III, Beckman Coulter Astrios, or comparable.

2.4 Software

1. FlowJo or comparable software.

2. Statistical software (Excel, GraphPad Prism, etc.).

3 Methods

3.1 Before You Begin

1. All solutions should be cold.

2. Samples should be kept on ice as much as possible.

3. All spins are at 4 °C for 5 min unless otherwise noted.

4. All researchers must be trained in the proper safe handling of human tissues and all IRB and related approvals must be in place, along with access to BSL-2 or above safety environment, and related facilities for disposal of waste and sharp items.

3.2 Nuclear Extraction from Fresh-Frozen Human Cortices

1. Human brain tissue stored at −80 °C should first be allowed to equilibrate within the cutting box for 5–15 min at −25 to −28 °C to prevent fracturing during cutting. A cryostat system works well for this. It is recommended that you use cut-resistant gloves and care, while working in the cryostat with human specimens, along with appropriate eye and clothing protection.

2. Place the brain tissue in a weigh boat, hold the tissue in place with angled forceps, and use a razor blade to cut the desired piece of tissue. Alternatively, tissue can be frozen to a cutting

block using optimal cutting temperature (OCT) compound and thickly (50–60 µm) sectioned.

3. Place the isolated frozen tissue into 1 mL of ice-cold NEB buffer in a 1.5 mL Eppendorf tube and incubate for 10 min on ice.

4. Cut the tip from a p1000 filtered tip and transfer the tissue and 1 mL of NEB into an ice-cold, 5 mL glass Dounce homogenizer with Teflon pestle (*see* **Note 2**).

5. Extract nuclei with 10–20 gentle up-and-down strokes of the pestle; be cautious not to introduce bubbling.

6. Pass the homogenate through a 40–50 µm cell strainer into a 15 mL conical. Wash the glass homogenizer with 4 mL of NEB buffer and pass this through the same cell strainer.

7. Pellet nuclei at approximately 600 × g at 4 °C for 5 min (*see* **Note 3**).

8. Discard the supernatant and resuspend the pellet in 5 mL of 1× PBSE.

9. Pellet nuclei by spinning at approximately 600 × g at 4 °C for 5 min.

3.3 Iodixanol Gradient (Note 4)

1. Prepare solutions for the iodixanol gradient. For each sample, use approximately 2 mL of solution D, 700 µL of 50% iodixanol, 500 µL of 35% iodixanol, and 2.2 mL of 10% iodixanol *see* **Note 5**.

2. Discard the supernatant, resuspend nuclei in 200 µL solution D, and then mix with 200 µL of 10% iodixanol. Allow mixture to equilibrate while you prepare the gradient tubes.

3. For at least one sample, make a control tube by combining 100 µL of resuspended nuclei in 100 µL of DAPI (1:2000) in solution D.

4. In a 12 × 75 mm tube (one per sample) prepare the gradient (Fig. 1) by first adding 400 µL of 35% iodixanol directly to the bottom of the tube. Mark the volume height on the side of the tube; this is where your nuclei will separate. Then, slowly and without mixing, layer 2 mL of 10% iodixanol, followed by 300–500 µL of resuspended sample, to the top. If you have multiple samples, complete the first two layers for all tubes before adding the resuspended mixture.

5. Centrifuge at 1700 × g for 15 min at 4 °C.

6. Visualize the DAPI control samples on transilluminator to verify nuclear separation and location.

7. Save the nuclei at the 35%|10% boundary into new 15 mL conical and add 5 mL of 1× PBSE to wash nuclei.

8. Pellet nuclei at 780 × g at 4 °C for 5 min.

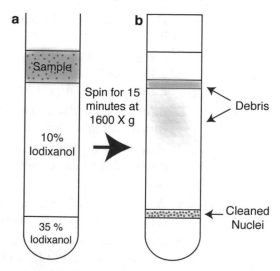

Fig. 1 Depiction of iodixanol gradient prior to separation (**a**) and after separation (**b**). (**a**) First, the sample containing both nuclei (colored in *blue* for depiction) and debris (*grey* clouds) is layered on top of a layer of 10% iodixanol and 35% iodixanol. The tubes are then centrifuged at 1700 × g at 4 °C for 15 min. (**b**) Nuclei are now located above the 35% iodixanol layer, while debris remained above or within the 10% iodixanol layer

9. Check to ensure that nuclei have pelleted, and then discard the supernatant.

10. *For downstream sequencing or DNA content assessment without fixation*: Resuspend in 1 mL of 2% BSA in PBSE and block for at least 20 min (can go up to 2 h) on ice with rotation.

11. *For downstream FISH*: Resuspend nuclei in 1 mL of neutral buffered formalin and fix for 5 min at room temperature with rotation. Next, add 9 mL of 1× PBSE and pellet nuclei at 780 × g at 4 °C for 5 min. Resuspend in 1 mL 2% BSA in 1× PBSE and block for at least 20 min (can go up to 2 h) on ice with rotation.

12. *For DNA content assessment with fixation*: Resuspend nuclei in 1 mL of 1× PBSE followed by 1 mL of 4% PFA added dropwise while gently vortexing. Fix for 40 min with rotation. Next, add 8 mL of 1× PBSE and pellet nuclei at 780 × g for 5 min. Resuspend in 1 mL 2% BSA in 1× PBSE and block for at least 20 min (can go up to 2 h) on ice with rotation (*see* **Note 6**).

3.4 Nuclear Staining   1. Prior to pelleting nuclei, separate unstained and single-color controls where necessary. Pellet nuclei at 780 × g at 4 °C for 5 min. Discard the supernatant and resuspend in 500 μL of primary antibody staining solution (*see* **Note 7**). If samples typically have a low yield, primary antibody staining solution can be added to blocking solution with final concentrations listed below.

For downstream RNA sequencing or FISH: Primary NeuN antibody; 20–60 min on ice with rotation.

For downstream DNA sequencing, FISH, or DNA content assessment with PI: RNase A, 50 µg/mL; primary NeuN antibody; 20–60 min on ice with rotation.

2. Add 9.5 mL of 2% BSA in 1× PBSE and rinse on ice with rotation for 10 min. Pellet nuclei at 780 × g at 4 °C for 5 min.

3. Resuspend in 500 µL of secondary antibody solution. The concentration of secondary antibody should be determined in-house.

4. Add 9.5 mL of 2% BSA in 1× PBSE and rinse on ice with rotation for 10 min. Pellet nuclei at 780 × g at 4 °C for 5 min.

5. Discard the supernatant and resuspend in appropriate volume of 2% BSA in 1× PBSE containing the appropriate DNA/RNA counterstain for flow cytometric analysis or sorting.

For DNA content analysis with PI: Add 1 mL of propidium iodide staining solution with 50 µg/mL of PI and the internal reference standard, CENs (*see* **Note 8**). Incubate for 1–3 h prior to FCM (*see* **Note 9**).

3.5 Flow Cytometry Gating Strategy

1. The gating strategy (Fig. 2) should be the same for nearly all applications, except that filter sets will change according to your selected secondary antibody and DNA counterstain.

2. For DNA content assessment: Record a minimum of 10,000 nuclear events after gating.

3. *Sorting for FISH*: Bulk sort nuclei into Eppendorf tubes containing 50 µL of 2% BSA in 1× PBSE. Following the sort procedure, drop ~20,000 nuclei on positively charged slides. Dry nuclei on a heating block at 55 °C for 5 min. Proceed with FISH protocol fixation and staining.

Sorting for DNA sequencing: For small population or single-cell sequencing which will require DNA amplification (*see* **Note 10**), sort samples into PCR tubes or 96-well plates containing a minimum of 2 µL of storage buffer (or more as recommended from amplification protocol) (*see* **Note 11**). Following sorting, centrifuge briefly and store nuclei at −80 °C for up to 4 months.

Sorting for RNA sequencing: For RNA sequencing of bulk samples, or downstream microfluidic assessment of single cells, nuclei are sorted into 1.5 mL Eppendorf containing 50 µL of 2% BSA in PBSE (*see* **Note 11**).

3.6 DNA Content Assessment

The DNA index of a sample is calculated by taking the mean fluorescence intensity of each population and taking a ratio compared to reference standards and control populations (Table 1). Controls may be another species (CENs or trout erythrocyte nuclei (TENs)),

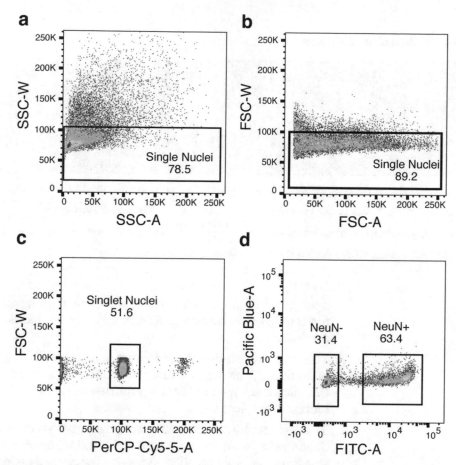

Fig. 2 Gating strategy for DNA content assessment and nuclear sorting. (**a**, **b**) Traditional doublet gating is used to eliminate debris and two nuclei stuck together. (**a**) Side scatter area (SSC-A) versus side scatter width (SSC-W) and (**b**) forward scatter area (FSC-A) and forward scatter width (FSC-W) are both used to eliminate most of the doublet nuclei. (**c**) The PI (PerCP-Cy5–5-A) positive singlet population is selected. The right part of this population is likely two nuclei stuck together. (**d**) NeuN+ and NeuN- populations (FIT-C) are selected. Pacific Blue-A is an unused filter with low levels of background, enabling accurate drawing of NeuN+ and NeuN- gates

other tissue types (lymphocytes (LYM) or cerebellum), or other cell types (NeuN negative nuclei). The mean fluorescence intensity of DNA in your sample can be calculated using FlowJo or other comparable software. DNA index can be represented as a fraction, or may be converted to the average pg of DNA per nucleus, total megabase change, or percent of change.

4   Notes

1. Solutions for downstream RNA sequencing should be made with DEP-C-treated water or commercially available NFW.

2. Our group has used both 1% NP40 [11, 15] and NEB buffer [18] to analyze and sort human tissue. No differences have

Table 1

Example calculation of the DNA index

	Mean PI internal reference CENs	Mean PI of sample	DNA index to CEN	DNA index to LYM	Average pg/ nucleus (pg)
Human cortex 1	164.8	458.1	2.780	1.014	6.794
Human cortex 2	169.7	489.6	2.885	1.052	7.048
Human LYM	168.6	462.3	2.742	1.000	6.700

In this example calculation, the mean CEN and mean sample fluorescence were determined following the gating strategy demonstrated in Fig. 2 using the FlowJo software. The DNA Index-to-CENs is determined by $\dfrac{\text{Mean sample}}{\text{Mean CEN}}$. The DNA index-to-lymphocytes (LYM) is determined by $\dfrac{DI\,CEN\,\text{Sample}}{DI\,CEN\,\text{LYM}}$. The approximation of average picograms per nucleus is determined using 6.7 pg/nucleus as a standard for lymphocytes

been detected for DNA content assessment; however, NEB tends to have less clumping of more dense brain tissues like the cerebellum.

3. This protocol can also be used for nuclear isolation from mouse brain tissue. However, we find that adult mouse nuclei need to be pelleted at a speed of $650 \times g$ for this first step, and all non-iodixanol gradient steps.

4. The iodixanol gradient will greatly improve your nuclei-to-debris ratio by removing excess lipid and white matter. This step is not required for all analyses and is frequently skipped for DNA content assessment. Extensive studies have shown that this does not impact the average total DNA content of a sample. Additionally, this step can generally be skipped for fresh fetal and adult mouse brain tissue.

5. Optiprep (Sigma) is a commercially available stock of 60% iodixanol.

6. Fixation of nuclei for DNA content assessment allows for nuclei to be collected and stored before analysis up to a few months. However, we have found that the preferred precipitating fixatives (e.g., alcohols) [22] often interfere with quality antibody staining. Cross-linking fixatives (e.g., formaldehyde) cross-link chromatin and compromise the stoichiometry of intercalating dyes [22], but preserve NeuN staining. It is therefore recommended to analyze total DNA content without fixation and post-fix for long-term storage.

7. This staining is tested to be optimal with $1–5 \times 10^6$ nuclei. For larger preparations, experimenters will need to optimize staining parameters.

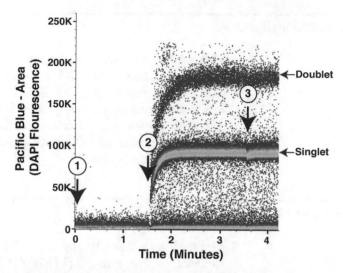

Fig. 3 Visualization of DAPI uptake in nuclear DNA. Flow cytometry scatter plot demonstrating DAPI uptake into human brain nuclei. (*1*) Nuclei were first recorded prior to incubation with DAPI. (*2*) The same sample was removed, and DAPI was added at a final concentration of 1 μg/mL, vortexed for 5 s, and placed back on the machine to record. (*3*) Next, the sample was removed, and DAPI was added to a final concentration of 5 μg/mL, vortexed for 5 s, and placed back on the machine to record events. This experiment demonstrates that DAPI uptake and saturation occur extremely quickly. However, PI (not shown) can require up to 1 h to reach equilibrium

8. Do not use internal reference standards when sorting. If necessary, standards can be run in separate tubes for gate setup and DNA content determination.

9. Other DNA content dyes may also be used. All dyes must be stained to saturation (equilibrium). The optimal concentration and staining time can be determined by keeping the number of cells with a sample constant, and fluctuating the concentration of PI (or other DNA counterstain) and the time of staining. A concentration of 50 μg/mL is sufficient for up to $1–5 \times 10^6$ human nuclei that must be stained at least for 1 h. Figure 3 demonstrates saturation of DAPI at 1 μg/mL within 5 min (it is suggested to stain for 20–30 min prior to analysis).

10. Downstream protocols (DNA amplification or library preparation) could either benefit from, or be impaired by, the presence of BSA or EGTA. It is the experimenter's responsibility to assess if 1× PBS, 1× PBSE, or 1% BSA will produce more consistent results with their specific protocol.

11. Take care that the PCR tubes and buffer are clean and sterile, and minimize the length of time the tubes are exposed to open air during the sort to minimize contamination artifacts.

5   Concluding Remarks

This protocol presents a detailed method for human brain cell nuclei extraction, DNA content assessment, and flow sorting for downstream sequencing or FISH. The use of flow cytometry and FACS in neuroscience has greatly increased our ability to study new aspects of cellular diversity and function, including the occurrence and significance of somatic genomic mosaicism in the developing and adult brain.

Acknowledgments

The authors are funded by the NIH Common Fund Single Cell Analysis Program (1U01MH098977), the NIAAA (4R01AA021402), and the Neuroplasticity of Aging Training Grant (5T32AG000216). We thank Danielle Jones, Dr. Ming Hsiang Lee, and Suzanne Rohrback for their contributions to this manuscript.

References

1. Rehen SK, McConnell MJ, Kaushal D, Kingsbury MA, Yang AH, Chun J (2001) Chromosomal variation in neurons of the developing and adult mammalian nervous system. Proc Natl Acad Sci U S A 98(23):13361–13366. doi:10.1073/pnas.231487398

2. Kaushal D, Contos JJ, Treuner K, Yang AH, Kingsbury MA, Rehen SK, McConnell MJ, Okabe M, Barlow C, Chun J (2003) Alteration of gene expression by chromosome loss in the postnatal mouse brain. J Neurosci 23(13):5599–5606

3. Yang AH, Kaushal D, Rehen SK, Kriedt K, Kingsbury MA, McConnell MJ, Chun J (2003) Chromosome segregation defects contribute to aneuploidy in normal neural progenitor cells. J Neurosci 23(32):10454–10462

4. Kingsbury MA, Friedman B, McConnell MJ, Rehen SK, Yang AH, Kaushal D, Chun J (2005) Aneuploid neurons are functionally active and integrated into brain circuitry. Proc Natl Acad Sci U S A 102(17):6143–6147. doi:10.1073/pnas.0408171102

5. Rehen SK, Yung YC, McCreight MP, Kaushal D, Yang AH, Almeida BS, Kingsbury MA, Cabral KM, McConnell MJ, Anliker B, Fontanoz M, Chun J (2005) Constitutional aneuploidy in the normal human brain. J Neurosci 25(9):2176–2180. doi:10.1523/JNEUROSCI.4560-04.2005

6. Yurov YB, Iourov IY, Monakhov VV, Soloviev IV, Vostrikov VM, Vorsanova SG (2005) The variation of aneuploidy frequency in the developing and adult human brain revealed by an interphase FISH study. J Histochem Cytochem 53(3):385–390. doi:10.1369/jhc.4A6430.2005

7. Yurov YB, Iourov IY, Vorsanova SG, Liehr T, Kolotii AD, Kutsev SI, Pellestor F, Beresheva AK, Demidova IA, Kravets VS, Monakhov VV, Soloviev IV (2007) Aneuploidy and confined chromosomal mosaicism in the developing human brain. PLoS One 2(6):e558. doi:10.1371/journal.pone.0000558

8. Peterson SE, Westra JW, Paczkowski CM, Chun J (2008) Chromosomal mosaicism in neural stem cells. Methods Mol Biol 438:197–204. doi:10.1007/978-1-59745-133-8_16

9. Westra JW, Peterson SE, Yung YC, Mutoh T, Barral S, Chun J (2008) Aneuploid mosaicism in the developing and adult cerebellar cortex. J Comp Neurol 507(6):1944–1951. doi:10.1002/cne.21648

10. Peterson SE, Yang AH, Bushman DM, Westra JW, Yung YC, Barral S, Mutoh T, Rehen SK, Chun J (2012) Aneuploid cells are differentially susceptible to caspase-mediated death during embryonic cerebral cortical development. J Neurosci 32(46):16213–16222. doi:10.1523/JNEUROSCI.3706-12.2012

11. Westra JW, Rivera RR, Bushman DM, Yung YC, Peterson SE, Barral S, Chun J (2010) Neuronal DNA content variation (DCV) with regional and individual differences in the human brain. J Comp Neurol 518(19):3981–4000. doi:10.1002/cne.22436

12. Gole J, Gore A, Richards A, Chiu YJ, Fung HL, Bushman D, Chiang HI, Chun J, Lo YH, Zhang K (2013) Massively parallel polymerase cloning and genome sequencing of single cells using nanoliter microwells. Nat Biotechnol 31(12):1126–1132. doi:10.1038/nbt.2720

13. McConnell MJ, Lindberg MR, Brennand KJ, Piper JC, Voet T, Cowing-Zitron C, Shumilina S, Lasken RS, Vermeesch JR, Hall IM, Gage FH (2013) Mosaic copy number variation in human neurons. Science 342(6158):632–637. doi:10.1126/science.1243472

14. Cai X, Evrony GD, Lehmann HS, Elhosary PC, Mehta BK, Poduri A, Walsh CA (2015) Single-cell, genome-wide sequencing identifies clonal somatic copy-number variation in the human brain. Cell Rep 10(4):645. doi:10.1016/j.celrep.2014.07.043

15. Bushman DM, Kaeser GE, Siddoway B, Westra JW, Rivera RR, Rehen SK, Yung YC, Chun J (2015) Genomic mosaicism with increased amyloid precursor protein (APP) gene copy number in single neurons from sporadic Alzheimer's disease brains. eLife 4. doi:10.7554/eLife.05116

16. Lodato MA, Woodworth MB, Lee S, Evrony GD, Mehta BK, Karger A, Lee S, Chittenden TW, D'Gama AM, Cai X, Luquette LJ, Lee E, Park PJ, Walsh CA (2015) Somatic mutation in single human neurons tracks developmental and transcriptional history. Science 350(6256):94–98. doi:10.1126/science.aab1785

17. Hazen JL, Faust GG, Rodriguez AR, Ferguson WC, Shumilina S, Clark RA, Boland MJ, Martin G, Chubukov P, Tsunemoto RK, Torkamani A, Kupriyanov S, Hall IM, Baldwin KK (2016) The complete genome sequences, unique mutational spectra, and developmental potency of adult neurons revealed by cloning. Neuron 89(6):1223–1236. doi:10.1016/j.neuron.2016.02.004

18. Lake BB, Ai R, Kaeser GE, Salathia NS, Yung YC, Liu R, Wildberg A, Gao D, Fung HL, Chen S, Vijayaraghavan R, Wong J, Chen A, Sheng X, Kaper F, Shen R, Ronaghi M, Fan JB, Wang W, Chun J, Zhang K (2016) Neuronal subtypes and diversity revealed by single-nucleus RNA sequencing of the human brain. Science 352(6293):1586–1590. doi:10.1126/science.aaf1204

19. Krishan A (1975) Rapid flow cytofluorometric analysis of mammalian cell cycle by propidium iodide staining. J Cell Biol 66(1):188–193

20. Riccardi C, Nicoletti I (2006) Analysis of apoptosis by propidium iodide staining and flow cytometry. Nat Protoc 1(3):1458–1461. doi:10.1038/nprot.2006.238

21. Dolezel J, Greilhuber J, Suda J (2007) Estimation of nuclear DNA content in plants using flow cytometry. Nat Protoc 2(9):2233–2244. doi:10.1038/nprot.2007.310

22. Darzynkiewicz Z, Huang X (2004) Analysis of cellular DNA content by flow cytometry. Curr Protoc Immunol Chapter 5:unit 5.7. doi:10.1002/0471142735.im0507s60

23. Darzynkiewicz Z (2011) Critical aspects in analysis of cellular DNA content. Curr Protoc Cytom Chapter 7:unit 7.2. doi:10.1002/0471142956.cy0702s56

24. Mullen RJ, Buck CR, Smith AM (1992) NeuN, a neuronal specific nuclear protein in vertebrates. Development 116(1):201–211

25. Fan J, Salathia N, Liu R, Kaeser GE, Yung YC, Herman JL, Kaper F, Fan JB, Zhang K, Chun J, Kharchenko PV (2016) Characterizing transcriptional heterogeneity through pathway and gene set overdispersion analysis. Nat Methods 13(3):241–244. doi:10.1038/nmeth.3734

26. Evrony GD, Cai X, Lee E, Hills LB, Elhosary PC, Lehmann HS, Parker JJ, Atabay KD, Gilmore EC, Poduri A, Park PJ, Walsh CA (2012) Single-neuron sequencing analysis of L1 retrotransposition and somatic mutation in the human brain. Cell 151(3):483–496. doi:10.1016/j.cell.2012.09.035

27. Evrony GD, Lee E, Mehta BK, Benjamini Y, Johnson RM, Cai X, Yang L, Haseley P, Lehmann HS, Park PJ, Walsh CA (2015) Cell lineage analysis in human brain using endogenous retroelements. Neuron 85(1):49–59. doi:10.1016/j.neuron.2014.12.028

28. Fischer HG, Morawski M, Bruckner MK, Mittag A, Tarnok A, Arendt T (2012) Changes in neuronal DNA content variation in the human brain during aging. Aging Cell 11(4):628–633. doi:10.1111/j.1474-9726.2012.00826.x

29. Arendt T, Bruckner MK, Losche A (2015) Regional mosaic genomic heterogeneity in the elderly and in Alzheimer's disease as a correlate of neuronal vulnerability. Acta Neuropathol 130(4):501–510. doi:10.1007/s00401-015-1465-5

30. Vindeløv LL, Christensen IJ, Nissen NI (1983) Standardization of high resolution flow cytometric DNA analysis by the simultaneous use of chicken and trout red blood cells as internal reference standards. Cytometry 3(5):328–331. doi:10.1002/cyto.990030504

Flow Cytometric Quantification, Isolation, and Subsequent Epigenetic Analysis of Tetraploid Neurons

Noelia López-Sánchez, Iris Patiño-Parrado, and José María Frade

Abstract

Different forms of genomic mosaicism can be detected in vertebrate neurons, including copy number variation, L1 transposition, and aneuploidy. In addition, some populations of vertebrate neurons can also show double the normal amount of DNA, a condition referred to as somatic tetraploidy. These neurons are generated during early stages of development, as they migrate to their adult locations in the adult nervous system, and constitute subpopulations of projection neurons. We have previously shown that neuronal tetraploidy can be characterized by flow cytometry using isolated cell nuclei from different mammalian and avian structures. In this chapter, we describe this procedure using a different model system: the rhombencephalic derivatives from adult zebrafish. In addition, tetraploid neuronal nuclei can be isolated by fluorescence-activated cell sorting and their genomic DNA used for further analyses, either used directly or after whole-genome amplification. Here we show as an example how to perform epigenetic analyses to characterize CpG methylation in differentially methylated regions controlling the *Rasgrf1*-imprinting domain in mice.

Key words Cell sorting, FACS, Tetraploid neuron, Cell nuclei, Zebrafish, Genomic imprinting, DNA methylation, Pyrosequencing

1 Introduction

Polyploidy can be defined as the presence in a cell of multiple copies of the haploid chromosome complement, which is usually referred to as N. This condition can affect all the cells in the body (i.e., "germinal polyploidy"), a strategy adopted by several organisms to increase their size. This type of polyploidy is widespread in plants [1–3], and it can often be observed in animals [4], including protozoa [5], insects [6], fishes [7], amphibians [8, 9], and reptiles [10].

With the occurrence of multicellularity and the acquisition of the soma, specific cell types or tissues can show changes in the amount of DNA proportional to the haploid complement, thus

Noelia López-Sánchez and Iris Patiño-Parrado are contributed equally to this work.

José María Frade and Fred H. Gage (eds.), *Genomic Mosaicism in Neurons and Other Cell Types*, Neuromethods, vol. 131, DOI 10.1007/978-1-4939-7280-7_4, © Springer Science+Business Media LLC 2017

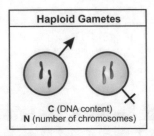

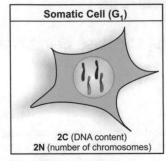

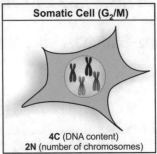

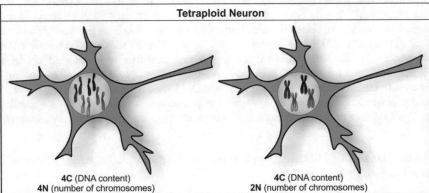

Fig. 1 *C* value and *N* value in germinal cells (gametes), somatic cells in G1 and G2/M phases of the cell cycle, and in tetraploid neurons. *C* value refers to the amount of DNA present in the gametes whereas *N* value denotes the chromosomal complement that characterizes the gametes. Since chromosomes may contain one or two chromatids the *C* values is duplicated during G2 while the *N* value is unaltered as the cell progresses through the cell cycle. Since neurons cannot divide, the number of chromosomes of those with 4C DNA content cannot be identified and they are referred to as somatic tetraploid neurons sensu *lato*

resulting in "somatic polyploidy." This term is used in a generic meaning since, in the absence of mitosis, the identification of the actual number of chromosomal sets is not feasible. In this case, somatic polyploid cells are usually referred to the amount of DNA that is present in their nuclei (measured in terms of the C value, i.e., the amount of DNA present in the haploid complement). It is important not to confound the C value with the N value: while the former increases during S phase and remains duplicated during G2, N is unaltered as the cell progresses through the cell cycle (Fig. 1).

Somatic polyploidy can affect different tissues in both vertebrates and invertebrates, where it may serve as a means to increase cell size and to promote gene expression and metabolic activity [11], which could facilitate better adaptation to metabolic and/or genotoxic stress and a lower sensibility to apoptotic stimuli [12–16]. In invertebrates, somatic polyploidy can be detected for instance in the glial cells making the blood–brain barrier [17] and in oocyte nurse cells [18] in Drosophila, and in Malpighian tubules and silk glands in the flour moth [19]. In mammals, a number of cell types have been shown to be polyploid, including megakaryocytes [20], trophoblast giant cells [21], hepatocytes [22], keratinocytes [23], vascular smooth cells [24], and cardiomyocytes [25].

As other tissues, the nervous system can also contain somatic polyploid cells. In invertebrates, several examples of somatic polyploid neurons have been described [26, 27]. The augment in the size of these neurons is likely required for increasing the speed at which action potentials move along their axons [28]. In some instances, the levels of ploidy may be really enormous as it is the case for nuclei of giant neurons from *Aplysia californica*, which has been shown to contain 200,000-fold the normal amount of haploid DNA (i.e., 200,000C) [29].

1.1 Somatic Tetraploidy in Vertebrate Neurons

For decades, it has been assumed that the vertebrate nervous system is homogenous from a genomic point of view, being constituted by diploid neurons (i.e., neurons showing 2C DNA content) [30]. This belief was challenged in the 1968 when Lapman [31] proposed that Purkinje cells contain tetraploid nuclei. This question has been readdressed, and current evidence indicates that a small proportion of neurons in the vertebrate nervous system contain tetraploid neurons (i.e., neurons with 4C DNA content) (Fig. 1). By using state-of-the-art techniques, Thomas Arendt's laboratory has shown that around 1% of neurons are tetraploid in the cerebral cortex of normal individuals [32]. In addition, we and others have demonstrated that tetraploid neurons exist in the normal vertebrate retina [33–35], as well as in the chick forebrain and mouse cerebral cortex [36]. In the chick retina, tetraploid neurons constitute subpopulations of horizontal cells [34, 35] and RGCs, the latter innervating lamina F at the *stratum-griseum-et-fibrosum-superficiale* of the tectal cortex [33]. These neurons are generated during the early stages of retinal development, soon after they acquire initial neuronal markers. In this regard, a subset of migrating RGCs expressing the transcription factor E2F1, and lacking Rb protein, undergoe S phase and remain in a permanent G2-like state, due to the inactivation of Cdk1 in these cells [37]. In these cells, $p27^{Kip1}$ expression prevents further round of replication, thus keeping these neurons in a tetraploid state [38]. Therefore, endoreduplication, but not alternative mechanisms such as aneuploidy or cell fusion [39, 40], seems to represent the major mechanism generating tetraploid RGCs in the vertebrate nervous system.

**1.2 Methods
to Characterize
Tetraploidy
in Vertebrate Neurons**

As indicated above, the initial attempts for DNA content quantification in vertebrate neurons took place in the late 1960s [31]. At this early period, DNA quantification was performed by using a histochemical method, the Feulgen stain [41]. This colorimetric procedure is based on the acid hydrolysis of DNA, which produces free aldehyde groups that are subsequently detected [42]. Then, the intensity of the Feulgen reaction, which is proportional to the DNA concentration, can be measured using microdensitometric procedures [43]. Although many laboratories were focused on the analysis of neuronal tetraploidy using this technology [44–46], no clear conclusion was reached at that time due to the limitations inherent to the Feulgen technique applied to cells with low levels of polyploidy [47].

More recently, a number of methods based on fluorescent labeling of DNA have been developed to identify and characterize neuronal tetraploidy. These methods allow the identification through fluorescent labeling of the cell types that become tetraploid in the vertebrate nervous system. In Chap. 5, Thomas Arendt and colleagues describe a procedure for DNA quantification in single cells in brain slices based on image cytometry. This method relies on laser scanning cytometry-based technology applied to tissue sections labeled with fluorescent DNA dyes such as 4′,6-diamidino-2-phenylindole (DAPI) or propidium iodide (PI) [48], which stoichiometrically bind to DNA. By using this method, these authors showed a positive correlation between the fluorescence intensity signal and both the number of dots detected with chromogenic in situ hybridization (CISH) using chromosome-specific probes and the amplification levels of *alu* repeats obtained by real-time PCR amplification in microdissected neurons [32].

A robust method for the analysis of neuronal tetraploidy is flow cytometry, a technique that allows the analysis of DNA content, protein expression, and other functional parameters in isolated cells or organelles in suspension [49–52]. An initial attempt for DNA content analysis in cerebral tissue using this technique was performed using fluorescent labeling of DNA through the acriflavine-Feulgen method [53]. This protocol uses formalin-fixed cell nuclei subjected to the acidic treatment inherent to the Feulgen stain, which includes acriflavine to provide fluorescence. This primitive protocol leads to a high frequency of nuclear doublets that cannot be removed from the analysis, thus impeding reliable quantification of 4C nuclei. Furthermore, the acriflavine-Feulgen method is too aggressive to allow parallel labeling to identify the nature of the tetraploid cell nuclei.

An important improvement for DNA analysis using flow cytometry derived from the use of fluorescent DNA dyes, such as PI [54]. The subsequent development of methods to remove doublets from the analysis, such as the standard pulse processing method [55] which we routinely use in our studies (see [36], for example), was an additional advance for the quantitative power of flow cytometry. Many studies have been performed since the development of this improved methodology. Embryonic neurons can be isolated through controlled trypsin digestion of developing neural tissues following a protocol similar to that used for cell culture. Cell suspensions are then subjected to a number of steps including ethanol fixation, cell immunolabeling with specific neuronal markers, RNA removal with RNase, and stoichiometric DNA labeling with PI, prior to flow cytometric analysis [56]. In the past, our laboratory has used this procedure for the analysis of neuronal tetraploidy in cell suspensions from different neural tissues obtained from the chick embryo [33, 57].

A further improvement to perform quantitative analyses by flow cytometry was the use of fresh cell nuclei isolated by tissue grinding [58]. This approach eliminates the need for fluorescent DNA dyes to cross a permeabilized plasma membrane and reduces nonspecific dye binding to cytoplasmic components, allowing very accurate DNA cell cycle analysis. Most protocols for cell nuclei analysis by flow cytometry use fixed cell nuclei, either by ethanol or formaldehyde [59, 60], although examples of flow cytometric analysis using fresh cell nuclei are also available in the literature [61]. It is important to point out that the analysis of DNA content with fresh cell nuclei avoids the disadvantages of formaldehyde, which triggers chromatin cross-linking that impairs stoichiometric analysis of DNA content [62], and of ethanol, which may lead to cell or organelle aggregation [63].

In our laboratory we have developed a flow cytometric method for DNA content quantification that uses fresh cell nuclei from neural tissues obtained with minimal manipulation [36, 64, 65]. This protocol, which is described in detail later in this chapter, results in narrow coefficients of variation that allows accurate DNA quantification in our samples. For additional information, see Chap. 2 in which this technology is used for nuclear DNA content analysis, nucleotide sequencing, and interphase FISH.

1.3 Methods for the Isolation of the Genome of Tetraploid Neurons

A simple way to isolate cell nuclei from hyperploid neurons relies on the use of laser capture microdissection from tissue sections, as shown by Mosch et al. [32]. Nevertheless, this method is time consuming, thus precluding its use for the analysis of a high number of cell nuclei. Therefore, other preparative procedures based on flow cytometry have been developed. Fluorescence-activated cell sorting (FACS) is a technical development of flow cytometry that enables sorting of a mixture of organelles or cells into two or more fractions using the scatter and fluorescence signals of each particle [66]. This methodology has been used by our laboratory to isolate neural precursors in different stages of the cell cycle [67] or live neurons out of cortical cells [68].

Immunolabeling of fresh cell nuclei with antibodies against nuclear markers specific for neuronal subtypes allows the isolation of rather pure neuronal cell populations through FACS. This approach yields sufficient amount of genomic DNA from cell populations enriched in tetraploid neurons, which can be used for genomic analyses either directly or after whole-genome amplification. Nascent RNA can also be obtained from these nuclear preparations, thus resulting in a convenient method for the analysis of gene expression in tetraploid neurons.

In this chapter, we describe the protocol routinely performed in our laboratory to reliably quantify the proportion of tetraploid neurons in specific neural tissues as well as a method for the purification of diploid and tetraploid nuclei to perform further analyses with their genomic DNA.

2   Materials

2.1 Mice and Zebrafish

Postnatal day 15 (P15) mice (C57BL/6J background) and 20-month-old zebrafish (WIK line) were used in this study to obtain cerebral hemicortices and rhombencephalic derivatives, respectively. These tissues were frozen and stored at −80 °C.

2.2 Oligonucleotides

Primers used in this chapter are those named as "inside forward" (TAGAGAGTTTATAAAGTTAG) and "inside reverse" (ACTAA AACAAAAACAACA) in the study by Li et al. [69]. For pyrosequencing, the inside forward primer was biotinylated at the 5′ end, and the pyrosequencing primer was TAATACAACAACAACAA TAACAATC. PCR reactions were performed as described in [69].

2.3 Reagents

- Triton X-100 (Sigma-Aldrich).
- Protease inhibitor cocktail tablets (cOmplete, Mini, EDTA-free) (Roche).
- Bovine serum albumin (BSA) (stock 30 mg/mL in PBS) (Sigma-Aldrich).
- Fetal calf serum (Life Technologies).
- Mouse Anti-NeuN antibody (clone A60) (EMD Millipore).
- Rabbit monoclonal Anti-NeuN antibody [clone A60] (Abcam).
- Alexa Fluor 488 Goat Anti-Mouse IgG (H + L) Antibody (Life Technologies).
- Alexa Fluor 488 Donkey Anti-Rabbit IgG (H + L) Antibody (Life Technologies).
- Propidium iodide (stock: 1 mg/mL) (Sigma-Aldrich).
- DAPI (stock 200 μg/mL) (Sigma-Aldrich).

- RNase A (10 mg/mL, inactivated by boiling as indicated by the manufacturer) (Sigma-Aldrich).
- PyroMark Gold Q26 reagents (Qiagen).
- DNeasy Blood & Tissue Kit (Qiagen).
- EpiTect Bisulfite Kit (Qiagen).
- DNA Clean & Concentrator-5 Kit (ZymoResearch).
- pGEM-T Easy (Promega).
- PureYield Plasmid Miniprep System (Promega).

2.4 Buffers

- Phosphate-buffered saline (PBS).
- Nuclear isolation buffer (NIB): DNase-free PBS containing 0.1% Triton X-100 and 1× complete Protease Inhibitor Cocktail (Roche).

2.5 Equipment

- 15 mL tube (Falcon).
- 1.5 mL Standard and DNA LoBind tubes (Eppendorf).
- 0.5 mL Protein LoBind tubes (Eppendorf).
- 7 mL Dounce homogenizer (WHEATON).
- Epifluorescence E80i Nikon microscope equipped with a DXM 1200 digital camera.
- Centrifuge 5415 R (Eppendorf).
- Autoclaved 40 μm nylon filters.
- FACSAria cytometer (BD Biosciences) equipped with a 488 nm Coherent Sapphire solid-state and 633 nm JDS Uniphase HeNe air-cooled laser [emission filters: BP 530/30 (FITC), BP 616/23 (PE-Texas Red) and BP 660/20 (APC) for Alexa 488, PI, and Alexa 647, respectively].
- PSQ96MA pyrosequencer (Biotage).

2.6 Software

- FACSDiva (BD Biosciences).
- Pyro Q CpG software (Biotage).
- QUMA software (http://quma.cdb.riken.jp).

2.7 Data Analysis and Statistics

Sanger sequencing results were analyzed for change in methylation pattern with the quantification tool for methylation analysis (QUMA) [70]. Data are shown as mean ± s.e.m. For pyrosequencing, statistics were calculated over all CpGs with passed quality. Statistical significance was determined using two-tailed Student's t test.

3   Methods

3.1   Flow Cytometric Characterization of Tetraploid Neurons in Vertebrates: General Considerations

The development of a flow cytometric procedure for the characterization of both cells and organelles based on size and/or fluorescent labeling has been a major scientific breakthrough in biological sciences. This technology was initially developed for the analysis of isolated cells that can easily be passed through a laser beam. This includes unicellular organisms [71], dissociated cell lines [72], or biological samples containing single cells including blood [73], sperm [74], gynecologic samples [75], or bladder washings [76]. Nevertheless, further developments of the technique made flow cytometry also useful for the analysis and isolation of cells or cell organelles derived from compact tissues, provided that they are previously dissociated through enzymatic and/or mechanic procedures. This was not an exception for the brain, and protocols for the analysis of DNA content were initially described already in the late 1970s, using formalin-fixed cell nuclei obtained by tissue grinding from solid brain tumors [77].

We routinely use flow cytometry for DNA quantification in isolated cell nuclei of neural origin and in living cells [33, 36, 57, 64, 65, 68, 78]. In this section, we describe the procedure developed by our laboratory for the quantification of tetraploid neurons in vertebrate neural tissues using unfixed cell nuclei immunolabeled with antibodies specific for nucleus-located neuronal markers [36]. The use of unfixed cell nuclei results in low coefficients of variation, thus facilitating the identification of low-abundance nuclear populations. By using this procedure, we have been able to detect a small proportion of tetraploid neurons in several neural structures from different species, including human (unpublished results), mouse [33, 36], chick [33, 57], and zebrafish (see below).

Our method is based on tissue grinding with a Dounce homogenizer followed by two centrifugation steps to first remove undissociated tissue and then to isolate fresh cell nuclei. These cell nuclei are subsequently immunolabeled with an anti-NeuN antibody to identify those that are derived from neurons. This method has the advantage of allowing the characterization of any neuronal type provided that it can be identified by a nuclear marker. We have used this method to show that the major population of tetraploid neurons in the mouse cerebral cortex express CTIP2 [36], a known marker of subcortical projection neurons located in layers V and VI [79]. This observation, together with the finding that retinal tetraploid neurons are mainly a subpopulation of retinal ganglion cells, indicates that neuronal tetraploidy is mainly associated with projection neurons. The use of antibodies against the early-response genes c-Fos and Erg-1, known to be expressed in active neurons, also allowed us to demonstrate that tetraploid neurons are functional [36].

Along this section we explain our protocol for the identification and quantification of tetraploidy in neurons using the rhombencephalic derivatives of the zebrafish brain as a model system. In addition, we describe an optimized protocol for the isolation by FACS of cell nuclei from diploid and tetraploid neurons obtained from P15 cerebral cortex. Then, we describe how to obtain genomic DNA from these neurons and, finally, we present an example of the genomic analysis that can be performed in these neurons. For this aim we have chosen the analysis of CpG methylation in genomic imprinting (i.e., monoallelic expression of genes depending on parental origin). Most imprinted genes are located in clusters distributed throughout the whole genome [80], and they are co-regulated by cis-acting imprinting control elements that contain chromosome-specific DMRs. Epigenetic methylation of the CpGs constituting these DMRs is established in the germline and then this methylation pattern is tightly maintained in all somatic tissues during adulthood [81, 82]. In this chapter we describe how to study the methylation pattern of the *Rasgrf1*-imprinted domain [69] in diploid versus tetraploid cortical neurons from P15 mice.

3.2 Analysis and Quantification of Neuronal Tetraploidy in Zebrafish Neural Tissues

Our analyses using freshly isolated cell nuclei have mainly been focused on the mouse cerebral cortex, but they can also be applied to other neural structures such as the hippocampus and the olfactory bulb in this species. In addition, we have succeeded in the characterization of tetraploid neurons in brain tissue from other species including zebrafish and chick as well as from postmortem human cerebral tissue using minor modifications of the basic protocol (see Fig. 2). In this section, we describe the protocol for the characterization of tetraploid neurons using the neural structures derived from the rhombencephalon in zebrafish as a model system. Whenever required, indications of the necessary modifications for other species are described in the Notes section.

3.2.1 Preparation of Nuclear Suspensions and Immunolabeling

There are a number of variables to consider when cell nuclei suspensions are being prepared, including the species, age, and neural tissue under analysis. In the case at hands, experiments are performed with the rhombencephalon-derived tissue from one 20-month-old zebrafish (one neural structure per sample), previously snap-frozen on dry ice and stored at −80 °C. The tissue is placed in 0.4 mL of ice-cold, nuclear isolation buffer (NIB) see Note 1), and cell nuclei are then released by mechanical disaggregation using a Dounce homogenizer (with sequential use of both loose and tight pestles). The characteristics of zebrafish brain allow a complete mechanical dissociation and the obtention of very clear nuclear suspensions, but removal of small amounts of undissociated tissue could be necessary for other tissues using a centrifugation step (see Note 2). In some instances, small debris present in the supernatants obtained after this centrifugation step can be removed through subsequent washes with NBT (see Note 3).

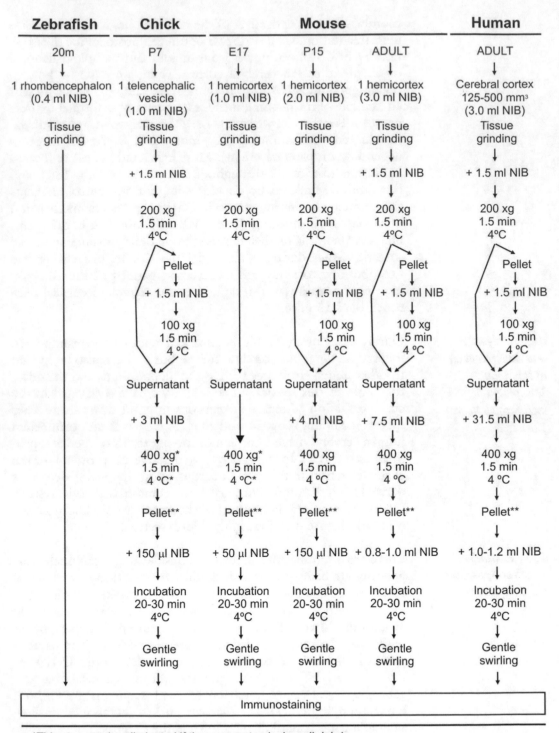

Fig. 2 Flow diagram describing the protocol for the isolation of fresh cell nuclei from the indicated tissues. These cell nuclei can be used for immunostaining with neuronal markers with nuclear location and flow cytometry analysis

Once the nuclear preparation is obtained, cell nuclei are immu-nostained with antibodies against specific nuclear markers (NeuN in this case). For this purpose, we recommend to use commercially available antibodies known to work for chromatin immunoprecipi-tation as they are likely to recognize the native structure of the antigen. An improvement of our methodology is that both primary and secondary antibodies are added together to the cellular prepa-rations (optimal dilutions have to be empirically deduced). This avoids unnecessary washing steps that may result in bias or general-ized loss of cell nuclei. This makes our protocol highly accurate and yielding representative results, something that is crucial for this type of analysis. As a first approximation, primary antibodies should be diluted around 0.5–1.0 times the recommended dilu-tion for immunohistochemistry. Secondary antibodies are usually employed at 1/500 dilution (*see* **Note 4**). For the example illus-trated in Fig. 3, an anti-NeuN rabbit monoclonal antibody was used at 1/600 dilution and an Alexa 488 anti-rabbit antibody was diluted 500 times. Immunostaining is performed in an Eppendorf tube containing the isolated unfixed nuclei described above plus 5% fetal calf serum (FCS) and 1.25 mg/mL bovine serum albumin (BSA) (final concentrations). In this way, labeling is specific as blocking proteins are co-added with the antibodies. It is important to prepare control samples in which the primary antibodies are excluded. When the availability of the samples is limited, this con-trol can be obtained by taking small aliquots from the experimental points, which are then joined together. Nuclear suspensions with antibodies are finally incubated O/N at 4 °C in the dark.

3.2.2 Flow Cytometric Analysis

NeuN-immunostained nuclei are filtered through a 40 μm nylon filter, and the volume adjusted to 0.5 mL (*see* **Note 5**) with PBS containing PI and DNase-free RNase I at a final concentration of 40 and 25 μg/mL, respectively, and then incubated for 30 min at room temperature (RT). The quality of the nuclei and specificity of immunostaining signal have to be checked by fluorescence micros-copy (Fig. 3a).

In our experiments, the analysis by flow cytometry is per-formed with a FACSAria cytometer. Nevertheless, any flow cytom-eter available in the market would be equally efficient for this kind of analysis. Immunostained samples are passed through the cytom-eter and data are collected. In our case, we use Alexa 488 to label NeuN and PI for DNA quantification. Therefore, the analysis is performed with the BP 530/30 and BP 616/23 emission filters (used for Alexa 488 and PI, respectively). Data are then analyzed with the FACSDiva package. To this aim, gating is adjusted to remove cellular debris, which is easily differentiated from nuclei due to its inability to incorporate PI. Aggregates are then removed after plotting DNA pulse area versus its corresponding pulse height following the procedure described in [55] (Fig. 3b). Then, the

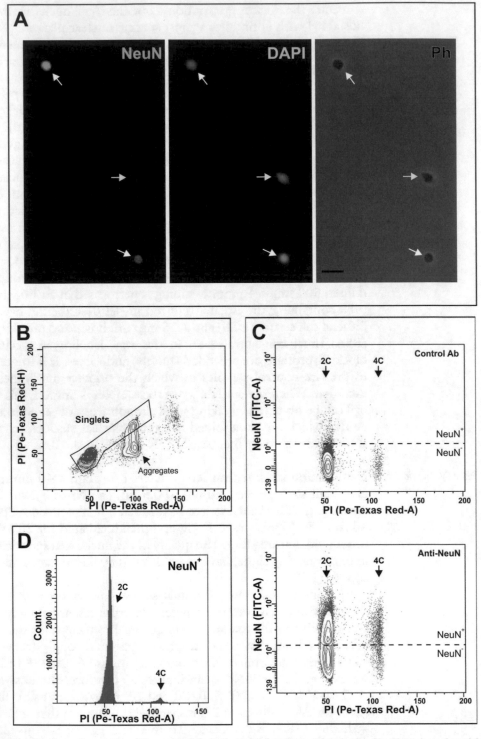

Fig. 3 Electronic gating method for zebrafish cell nuclei immunostained with NeuN and analyzed by flow cytometry. (**a**) NeuN-positive nuclei (*white arrows*) and NeuN-negative nucleus (*yellow arrow*) labeled with DAPI are shown. Ph: phase contrast. Bar: 10 μm. (**b**) Fresh cell nuclei were isolated from neural tissues derived from the rhombencephalon of 20 m zebrafish, stained with propidium iodide (PI), and subjected to flow

NeuN-positive nuclear population is selected and plotted against PI intensity to quantify the proportion of neuronal tetraploid and diploid nuclei (Fig. 3c). A histogram is finally created for the quantification of the tetraploid neuron population. Figure 3d illustrates a representative result obtained from this analysis, indicating that 4.21 ± 0.41% (mean ± s.e.m.) of neurons are tetraploid in the neural tissues derived from the rhombencephalon in the adult zebrafish ($n = 8$).

3.3 Isolation of Diploid and Tetraploid Neurons from Vertebrate Neural Tissues to Perform Epigenetic Studies

Besides the common use of flow cytometry for the characterization and quantification of tetraploid neurons in specific tissues, FACS technology can also be used for the isolation of genomic DNA from these neurons (i.e., 4C NeuN-positive cell nuclei). To this aim, we usually employ a nozzle with a 100 μm diameter, a size that does not perturb nuclear integrity. It is important to point out that cell sorting is a challenging approach when the abundance of the target population is low, as it is the case for tetraploid neurons. A single round of cell sorting is able to enrich the tetraploid nuclear population to around 20% as a maximum. Therefore, two rounds of cell sorting are required to enrich the tetraploid population to 80–90%. As a consequence, the final number of tetraploid nuclei that are obtained is quite low. Another reason for a reduced yield of tetraploid neuronal nuclei is the necessity of checking its proportion within the enriched sample. This analysis results in the loss of a significant number of isolated nuclei, which are passed through the flow cytometer following the protocol described above. As an average, we usually recover around 1000–9000 tetraploid nuclei out of four P15 mouse hemicortices, and about 100–500 cell nuclei are used for this validation step. The rest can be used for genomic DNA extraction. The isolation of basically pure diploid neurons is easier, as this population represents around 98–99% of total neurons and around half of all cell nuclei in the cerebral cortex. Therefore, a single round of FACS is sufficient for the isolation of highly enriched diploid neuron populations. We usually isolate genomic DNA from 200,000 diploid cell nuclei.

3.3.1 Isolation of Cell Nuclei from Diploid and Tetraploid Cortical Neurons

For the isolation of cortical neurons of diploid and tetraploid DNA content, one cortical hemisphere from P15 mice is homogenized using a Dounce homogenizer in 2 mL of ice-cold NIB (see Fig. 2). Undissociated tissue is then removed by centrifugation at $200 \times g$

Fig. 3 (continued) cytometric analysis. After gating on forward scattering area vs. side scattering area (not shown), DNA content was Fig. 3 (continued) assessed by plotting PI (Pe-Texas Red-H) vs. PI (Pe-Texas Red-A) levels. Diploid and tetraploid nuclei were subsequently gated from these plots (singlets), while the doublets of diploid nuclei were discarded (aggregates). (**c**) DNA content from the gated cell population containing both diploid (2C) and tetraploid (4C) events, plotted against Neu immunolabeling levels (FITC-A), is shown. A control analysis performed with nuclei labeled with secondary antibodies is also shown. (**d**) A histogram showing the count numbers of diploid (2C) and tetraploid (4C) nuclei is shown

for 1.5 min (4 °C) as described above. The supernatant is then diluted with 4 mL of NIB, while the pellet is resuspended with 1.5 mL NIB and centrifuged at $100 \times g$ for 1.5 min at 4 °C, and the resulting supernatant is added to the previous one (see Fig. 2). The mixed supernatant is then distributed in five Eppendorf tubes and centrifuged at $400 \times g$ to isolate cell nuclei and remove the cell debris from the sample. Each pellet is then incubated for 30 min with 30 µL NIB on ice, resuspended by gently swirling, joined into one single Eppendorf tube (150 µL in total), and directly used for immunolabeling. To this aim, the cell nuclei are incubated O/N in the dark at 4 °C with a mouse monoclonal antibody against NeuN diluted 1/1000 and a secondary Alexa 488 anti-mouse antibody used at a 1/500 dilution in NIB containing 5% fetal calf serum (FCS) and 1.25 mg/mL bovine serum albumin (BSA). In parallel, control samples are prepared lacking the primary antibody. After the incubation period, immunostained nuclei are filtered through a 40 µm nylon filter and, then, DNA staining is performed with 40 µg/mL PI, while RNA is removed with 25 µg/mL DNase-free RNase I, which is incubated for 30 min at RT. After analyzing the quality of the nuclei and the specificity of the immunostaining by fluorescence microscopy, cell nuclei are sorted through a FACSAria cytometer. Gating is set as described above to isolate diploid and tetraploid NeuN-positive nuclei on DNA LoBind and Protein LoBind Eppendorf tubes, respectively. Diploid neuronal cell nuclei are selected for further analysis while the tetraploid enriched fraction is subjected to a second round of cell sorting to isolate tetraploid neuronal cell nuclei. In this case, tetraploid events are collected in DNA LoBind tubes.

3.3.2 Genomic DNA Isolation

To obtain genomic DNA from diploid and tetraploid neurons, the isolated cell nucleus suspensions are centrifuged at $20{,}000 \times g$ for 5 min, and then most of the supernatant is removed (just 200 µL is kept). The genomic DNA is then isolated from this pellet by using the DNeasy Blood & Tissue Kit, a procedure that requires no phenol extraction and involves minimal handling. To this aim, 20 µL of the proteinase K solution and 200 µL of buffer AL (both included in the kit) are added to the remnant volume and then the mix is vortexed to assure cell nuclear lysis. This buffer allows selective binding of DNA to the DNeasy membrane located within a spin column also included in the kit. After two wash steps, DNA is eluted with 40 µL of the buffer AE included in the kit. This volume is optimal for DNA yields below 1 µg.

3.3.3 Bisulfite Treatment of Genomic DNA from Diploid and Tetraploid Neurons

The genomic DNA isolated in the previous step can be used for several types of analysis. In this chapter we used CpG methylation analysis in imprinting. The most common technique for the analysis of CpG methylation is the treatment of genomic DNA with bisulfite followed by sequencing [83]. Bisulfite converts cytosine residues to uracil, but leaves 5-methylcytosine residues unaffected. Therefore, only methylated cytosines are retained when genomic DNA has been treated with bisulfite.

For this analysis, 20 μL of the genomic DNA obtained in the previous step is bisulfite-converted using EpiTect Bisulfite Kit. To this aim, the DNA solution is combined with 85 μL of bisulfite mix solution and 35 μL of DNA protect buffer, and then the protocol is performed as indicated in the specifications of the kit.

3.3.4 Analysis of CpG Methylation Levels in Imprinting Control Regions of Diploid and Tetraploid Neurons through Sanger Sequencing

The Sanger sequencing procedure is the classical method for the analysis of CpG methylation [83]. For this analysis, a representative DNA sequence within the DMR of interest is PCR amplified from the bisulfite-converted DNA using specific oligonucleotides designed for the bisulfite-converted sequence (i.e., cytosines substituted by thymines). As a general rule, whenever possible the presence of GpGs should be avoided from the oligonucleotide sequences. In case this cannot be applied, degenerate oligonucleotides recognizing both cytidines and thymines should be used. In addition, when the CpG methylation levels are studied by pyrosequencing (see below) one of the oligonucleotides should be biotinylated in its 5′ end. If possible, primers should be designed flanking a single-nucleotide polymorphism (SNP), which is used to determine the parental origin of the specific strand when analyzed by Sanger sequencing. In this case, the analyzed tissue should be derived from mice obtained by crossing of two genetic backgrounds carrying the mentioned SNP. Finally, the analysis of the sequencing data is performed with the QUMA software [70], using its standard settings.

To study the CpG methylation levels of the DMR controlling the imprinted *Rasgrf1* region [69] (Fig. 4a) in diploid and tetraploid neurons, two primers flanking this DMR (see Fig. 4b) are used. These primers are specific for the bisulfite-converted sequence, and they encompass 29 CpG sites within the *Rasgrf1*-imprinted domain. PCR amplification is then performed using the PCR conditions described in [69], and the amplification product is cleaned with the DNA Clean & Concentrator-5 kit. After PCR product subcloning into the pGEM-Teasy vector, a significant number of positive clones are subjected to minipreparation of plasmids with the PureYield Plasmid Miniprep System, and these plasmids are subsequently sequenced through the Sanger method. Figure 5 illustrates the result obtained from this analysis, which indicates that at P15 both diploid and tetraploid neurons seem to show more than the expected 50% of methylated CpGs.

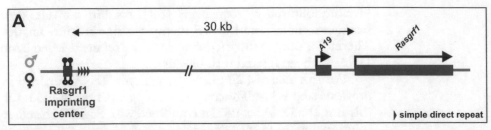

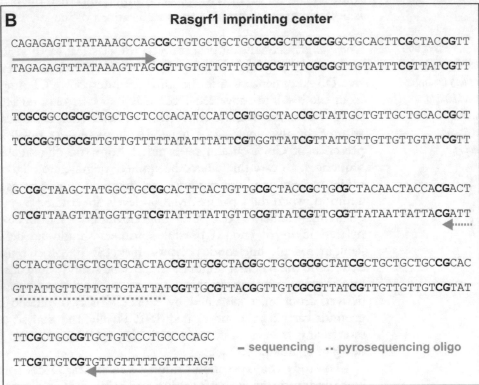

Fig. 4 Scheme showing the genomic structure of murine *Rasgrf1*-imprinted domain and the sequence analyzed within the *Rasgrf1*-imprinting center. (**a**) Genes constituting the *Rasgrf1*-imprinting domain are shown. These genes show paternal allele-specific expression. The *Rasgrf1*-imprinting center is 30 kb upstream of these genes. Methylated CpG sites from this imprinting center are represented by *solid circles*, while *open circles* represent non-methylated CpGs from the *Rasgrf1*-controlling DMR. (**b**) Genomic DNA sequence corresponding to the region analyzed in this chapter. *Solid arrows*: primers used for amplification (for pyrosequencing, the forward primer was 5′-biotinylated). *Dashed arrow*: primer used for pyrosequencing

3.3.5 Analysis of CpG Methylation Levels in Imprinting Control Regions of Diploid and Tetraploid Neurons through Pyrosequencing

Sanger sequencing depends on random incorporation of selective dideoxynucleotides lacking a 3′-OH group required for the DNA extension reaction, thus leading to products of different lengths that can be assigned to a particular sequence [84]. In contrast to this method, which needs a relatively limited number of clones that are then sequenced, pyrosequencing has a superior quantitative capacity based on the detection of the levels of pyrophosphate that is released when a deoxyribonucleotide triphosphate is added to

A

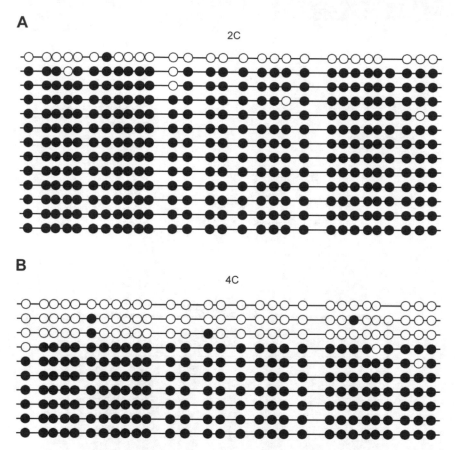

B

Fig. 5 Sanger sequencing methylation analysis of the *Rasgrf1*-imprinting control region shown in Fig. 4. Results were analyzed for change in methylation pattern with the QUMA software, which generates a panel of *solid* and *open circles* indicating methylated and unmethylated CpGs, respectively

the end of a nascent strand of DNA [85]. Although this technique can be used to reveal unknown DNA sequences, it is not as efficient as other available sequencing procedures. Nevertheless, pyrosequencing has proved to be highly efficient and sensitive for the analysis of single-nucleotide mutations in short DNA sequences (usually less than 80 nucleotides in length) [86]. Deoxyribonucleotide triphosphates are added to the reaction following the expected sequence. This results in the release of pyrophosphate (PPi), whose concentration can be monitored as a peak of light emission. This allows the quantification of the mutations that are present in the DNA sequence (see scheme in Fig. 6). This procedure is therefore quite convenient for the quantification of the proportion of cytidine conversion to thymine in bisulfite-treated DNA. The proportion of CpG methylation present within the DMRs of an imprinting DNA domain is therefore easily quantified by this procedure when DNA fragments initially enriched in CpGs are generated by PCR amplification from bisulfite-converted genomic DNA.

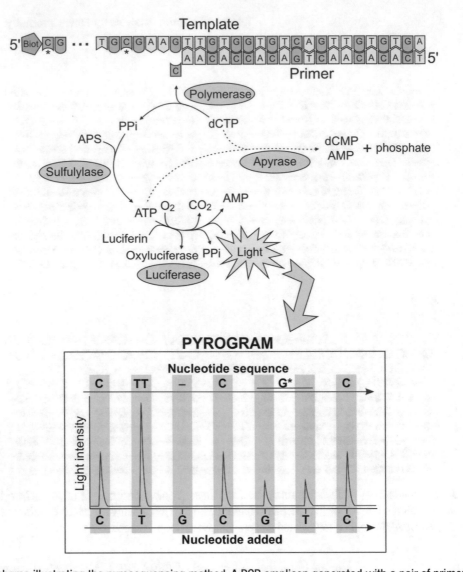

Fig. 6 Scheme illustrating the pyrosequencing method. A PCR amplicon generated with a pair of primers, one of them biotinylated at its 5′ end, is denatured, and then the biotinylated strand is immobilized through its biotinylated 5′ end. Asterisks indicate CpGs that may be methylated: if C is not methylated, C should be substituted by T (in the complementary strand, G should be substituted by A). This template is hybridized to a sequencing primer and then incubated with DNA polymerase, ATP sulfurylase, and luciferase together with the substrates adenosine 5′ phosphosulfate (APS) and luciferin. The addition of the deoxynucleoside triphosphate complementary to the one present in the template sequence (dCTP in this example) initiates the pyrosequencing reaction. dATPαS is used instead of dATP to avoid noise, since the former is not a substrate for a luciferase. The incorporation by DNA polymerase of the correct, complementary dNTP releases pyrophosphate (PPi). Then, ATP sulfurylase converts PPi to ATP in the presence of APS, and this ATP acts as a substrate for luciferase, which converts luciferin to oxyluciferin and generates light in proportion to the amount of ATP. This light is detected by a camera and shown as a pyrogram. Finally, the presence of apyrase results in the degradation of the unincorporated nucleotides and ATP, and the reaction can restart with another nucleotide. As an internal control of the pyrosequencing method, a reaction step can be performed with a noncomplementary nucleotide (dGTP in this example), thus resulting in lack of light signal. When the template contains more than one identical nucleotide (the pair TT in the example), the peak of light is proportional to the number of identical nucleotides present in the sequence. The percentage of methylated CpGs is calculated after successive addition of dGTP and dTTP (in the example 50% of the CpGs are methylated since the peak of light is similar for both dGTP and dTTP)

To confirm the observation that the *Rasgrf1* DMR contains a high proportion of methylated CpGs, bisulfite-treated DNA is PCR amplified using the same primers as above, the forward primer being biotinylated at the 5′ end. The PCR product is then bound to streptavidin-coated Sepharose beads (GE-Healthcare) through the biotin tag and denatured to generate single-stranded DNA to allow annealing of an internal sequencing primer (16 pmol of each per reaction) (see Materials). Pyrosequencing is then performed with a PSQ™96MA instrument using 25 μL of amplified DNA product and PyroMark Gold Q96 reagents. Pyrosequencing data analysis is finally done using the Pyro Q-CpG software. This technique can only resolve an average of 80 nucleotides upstream from the sequencing primer. Therefore, only five CpGs could be quantified with this technique (Fig. 7). This analysis indicated that $78.31 \pm 3.88\%$ ($n = 4$) of CpGs were methylated in diploid neurons whereas this proportion was $75.07 \pm 4.77\%$ ($n = 3$) in tetraploid neurons (Fig. 7). This result, which indicates that tetraploid neurons show similar proportion of methylated CpGs in the DMR of the *Rasgrf1*-imprinting domain, is consistent with the results obtained through Sanger sequencing illustrated in Fig. 5. For some unknown reason, this domain seems to be overmethylated in P15 cortical neurons.

4   Notes

1. This volume should be increased to 1 mL for one telencephalic hemisphere from P7 chick or E17 mouse embryos, to 2 mL for one hemicortex of P15 mice, and to 3 mL for either one hemicortex of adult mouse or one cube of 5–8 mm edge of human brain (see Fig. 2).

2. The standard procedure for the removal of the undissociated tissue (mainly fibers and blood vessels) is by centrifugation ($200 \times g$ for 1.5 min at 4 °C in 1.5 mL minifuge tubes) discarding pellet. If tissue from either adult mouse or human brain is processed, 1.5 mL of cold NIB should be added prior to the centrifugation step (see Fig. 2). Additionally, when quantification or high recovery is needed, the pellet of this centrifugation step can be resuspended with 1.5 mL of cold NIB, and then centrifuged at $100 \times g$ for 1.5 min at 4 °C to remove any dissociated cell nuclei that may remain among the undissociated tissue. The supernatant is then added to the previous one (see Fig. 2 for details).

3. For postnatal and adult mouse tissues the supernatant is further diluted with NIB (see Fig. 2 for details). In all cases (except for zebrafish rhombencephalon-derived neural structures) tissues are centrifuged at $400 \times g$ for 4 min at 4 °C.

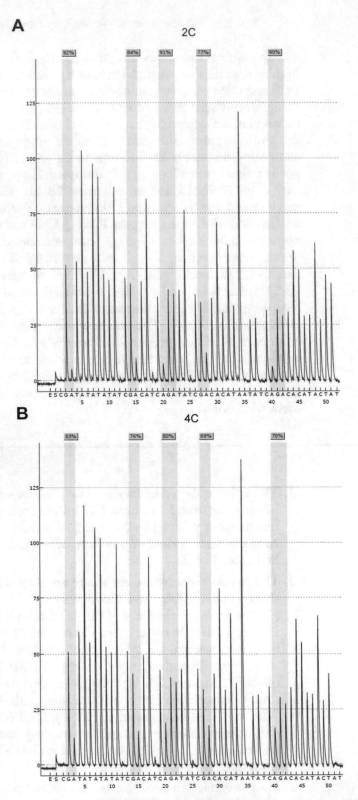

Fig. 7 Representative pyrograms from diploid (**a**) and tetraploid (**b**) neuronal cell nuclei, corresponding to the *Rasgrf1*-imprinting control region shown in Fig. 4, are shown. *Blue boxes*: Percentage of methylation in any particular CpG analyzed. This value is calculated as the ratio between the G peak (complementary to the bisulfite-protected ^{Met}C) and the G + A peaks (complementary, respectively, to the bisulfite-protected ^{Met}C plus the bisulfite-unprotected C)

Supernatants containing cellular debris are then discarded and the pellet is then incubated at 4 °C in cold NIB for at least 1 h, before mechanical resuspension by gentle swirl of the vial. The volume for this resuspension step differs from tissue to tissue (see Fig. 2 for details). A quality control that is required at this stage is the analysis under the microscope of a small aliquot of the isolated nuclei stained with 100 ng/mL DAPI. This analysis informs about the quality and purity of the nuclear preparation. It is convenient to use only cell nuclear preparations in which cell debris are kept to a minimum.

4. Fluorophores are chosen depending on the flow cytometer. It is imperative that both absorption and emission spectra of the fluorophores are compatible with the laser and filters of cytometer and that their emission spectra do not overlap.

5. This volume can be adjusted to 1–3 mL in other tissues to avoid high-concentrated cell nucleus suspensions.

Acknowledgments

We thank Gonzalo García de Polavieja and Aixa Morales for the zebrafish specimens used in this study. This research was supported by the Ministerio de Economía y Competitividad (grant numbers SAF2015-68488-R) and a R&D contract with Tetraneuron S.L.

References

1. Masterson J (1994) Stomatal size in fossil plants: evidence for polyploidy in majority of angiosperms. Science 264:421–424

2. Sugimoto-Shirasu K, Roberts K (2003) "big it up": endoreduplication and cell-size control in plants. Curr Opin Plant Biol 6:544–553

3. Madlung A, Wendel JF (2013) Genetic and epigenetic aspects of polyploid evolution in plants. Cytogenet Genome Res 140:270–285

4. Otto SP, Whitton J (2000) Polyploid incidence and evolution. Annu Rev Genet 34:401–437

5. Li S, Yin L, Cole ES, Udani RA, Karrer KM (2006) Progeny of germ line knockouts of ASI2, a gene encoding a putative signal transduction receptor in Tetrahymena Thermophila, fail to make the transition from sexual reproduction to vegetative growth. Dev Biol 295:633–646

6. Kolics B, Ács Z, Chobanov DP, Orci KM, Qiang LS, Kovács B, Kondorosy E, Decsi K, Taller J, Specziár A, Orbán L, Müller T (2012) Re-visiting phylogenetic and taxonomic relationships in the genus Saga (Insecta: Orthoptera). PLoS One 7:e42229

7. Leggatt RA, Iwama GK (2003) Occurrence of polyploidy in the fishes. Rev Fish Biol Fish 13:237–246

8. Vernon JA, Butsch J (1957) Effect of tetraploidy on learning and retention in the salamander. Science 125:1033–1034

9. Szaro BG, Tompkins R (1987) Effect of tetraploidy on dendritic branching in neurons and glial cells of the frog, Xenopus laevis. J Comp Neurol 258:304–316

10. Trifonov VA, Paoletti A, Caputo Barucchi V, Kalinina T, O'Brien PC, Ferguson-Smith MA, Giovannotti M (2015) Comparative chromosome painting and NOR distribution suggest a complex hybrid origin of triploid Lepidodactylus lugubris (Gekkonidae). PLoS One 10:e0132380

11. Frawley LE, Orr-Weaver TL (2015) Polyploidy. Curr Biol 25:R353–R358

12. Biesterfeld S, Gerres K, Fischer-Wein G, Böcking A (1994) Polyploidy in non-neoplastic tissues. J Clin Pathol 47:38–42

13. Edgar BA, Orr-Weaver TL (2001) Endoreplication cell cycles: more for less. Cell 105:297–306

14. Ullah Z, Lee CY, Depamphilis ML (2009) Cip/kip cyclin-dependent protein kinase inhibitors and the road to polyploidy. Cell Div 4:10

15. Ullah Z, Lee CY, Lilly MA, DePamphilis ML (2009) Developmentally programmed endoreduplication in animals. Cell Cycle 8:1501–1509

16. Pandit SK, Westendorp B, de Bruin A (2013) Physiological significance of polyploidization in mammalian cells. Trends Cell Biol 23:556–566

17. Unhavaithaya Y, Orr-Weaver TL (2012) Polyploidization of glia in neural development links tissue growth to blood-brain barrier integrity. Genes Dev 26:31–36

18. Lilly MA, Spradling AC (1996) The drosophila endocycle is controlled by cyclin E and lacks a checkpoint ensuring S-phase completion. Genes Dev 10:2514–2526

19. Buntrock L, Marec F, Krueger S, Traut W (2012) Organ growth without cell division: somatic polyploidy in a moth, Ephestia kuehniella. Genome 55:755–763

20. Vitrat N, Cohen-Solal K, Pique C, Le Couedic JP, Norol F, Larsen AK, Katz A, Vainchenker W, Debili N (1998) Endomitosis of human megakaryocytes are due to abortive mitosis. Blood 91:3711–3723

21. Ullah Z, Kohn MJ, Yagi R, Vassilev LT, DePamphilis ML (2008) Differentiation of trophoblast stem cells into giant cells is triggered by p57/Kip2 inhibition of CDK1 activity. Genes Dev 22:3024–3036

22. Vinogradov AE, Anatskaya OV, Kudryavtsev BN (2001) Relationship of hepatocyte ploidy levels with body size and growth rate in mammals. Genome 44:350–360

23. Zanet J, Freije A, Ruiz M, Coulon V, Sanz JR, Chiesa J, Gandarillas A (2010) A mitosis block links active cell cycle with human epidermal differentiation and results in endoreplication. PLoS One 5:e15701

24. Nagata Y, Jones MR, Nguyen HG, McCrann DJ, St Hilaire C, Schreiber BM, Hashimoto A, Inagaki M, Earnshaw WC, Todokoro K, Ravid K (2005) Vascular smooth muscle cell polyploidization involves changes in chromosome passenger proteins and an endomitotic cell cycle. Exp Cell Res 305:277–291

25. Kellerman S, Moore JA, Zierhut W, Zimmer HG, Campbell J, Gerdes AM (1992) Nuclear DNA content and nucleation patterns in rat cardiac myocytes from different models of cardiac hypertrophy. J Mol Cell Cardiol 24:497–505

26. Coggeshall RE, Yaksta BA, Swartz FJ (1970) A cytophotometric analysis of the DNA in the nucleus of the giant cell, R-2, in Aplysia. Chromosoma 32:205–212

27. Manfredi Romanini MG, Fraschini A, Bernocchi G (1973) DNA content and nuclear area in the neurons of the cerebral ganglion in Helix pomatia. Ann Histochim 18:49–58

28. Castelfranco AM, Hartline DK (2016) Evolution of rapid nerve conduction. Brain Res 1641(Pt a):11–33

29. Lasek RJ, Dower WJ (1971) Aplysia californica: analysis of nuclear DNA in individual nuclei of giant neurons. Science 172:278–280

30. Swift H (1953) Quantitative aspects of nuclear nucleoproteins. Int Rev Cytol 2:1–76

31. Lapham LW (1968) Tetraploid DNA content of Purkinje neurons of human cerebellar cortex. Science 159:310–312

32. Mosch B, Morawski M, Mittag A, Lenz D, Tarnok A, Arendt T (2007) Aneuploidy and DNA replication in the normal human brain and Alzheimer's disease. J Neurosci 27:6859–6867

33. Morillo SM, Escoll P, de la Hera A, Frade JM (2010) Somatic tetraploidy in specific chick retinal ganglion cells induced by nerve growth factor. Proc Natl Acad Sci U S A 107:109–114

34. Shirazi Fard S, Jarrin M, Boije H, Fillon V, All-Eriksson C, Hallböök F (2013) Heterogenic final cell cycle by chicken retinal Lim1 horizontal progenitor cells leads to heteroploid cells with a remaining replicated genome. PLoS One 8:e59133

35. Shirazi Fard S, All-Ericsson C, Hallböök F (2014) The heterogenic final cell cycle of chicken retinal Lim1 horizontal cells is not regulated by the DNA damage response pathway. Cell Cycle 13:408–417

36. López-Sánchez N, Frade JM (2013) Genetic evidence for p75[NTR]-dependent tetraploidy in cortical projection neurons from adult mice. J Neurosci 33:7488–7500

37. Ovejero-Benito MC, Frade JM (2013) Brain-derived neurotrophic factor-dependent cdk1 inhibition prevents G2/M progression in differentiating tetraploid neurons. PLoS One 8:e64890

38. Ovejero-Benito MC, Frade JM (2015) p27[Kip1] participates in the regulation of endoreplication in differentiating chick retinal ganglion cells. Cell Cycle 14:2311–2322

39. Kingsbury MA, Yung YC, Peterson SE, Westra JW, Chun J (2006) Aneuploidy in the normal and diseased brain. Cell Mol Life Sci 63:2626–2641

40. Ying QL, Nichols J, Evans EP, Smith AG (2002) Changing potency by spontaneous fusion. Nature 416:545–548

41. Feulgen R, Rossenbeck H (1924) Mikroskopisch-chemischer Nachweis einer Nukleinsäure von Typus der Thymonukleinsäure und die darauf beruhende selektive Färbung von Zellkernen in mikroskopischen Präparaten. Hoppe Seylers Z Physiol Chem 135:203–248

42. Chieco P, Derenzini M (1999) The Feulgen reaction 75 years on. Histochem Cell Biol 111:345–358

43. Woodard J, Gorovsky M, Swift H (1966) DNA content of a chromosome of Trillium erectum: effect of cold treatment. Science 151:215–216

44. Herman CJ, Lapham LW (1968) DNA content of neurons in the cat hippocampus. Science 160:537

45. Herman CJ, Lapham LW (1969) Neuronal polyploidy and nuclear volumes in the cat central nervous system. Brain Res 15:35–48

46. Museridze DP, Svanidze IK, Macharashvili DN (1975) Content of DNA and dry weight of the nuclei of neurons of the external geniculate body and retina of the eye in guinea pigs. Sov J Dev Biol 5:269–272

47. Swartz FJ, Bhatnagar KP (1981) Are CNS neurons polyploid? A critical analysis based upon cytophotometric study of the DNA content of cerebellar and olfactory bulbar neurons of the bat. Brain Res 208:267–281

48. Mosch B, Mittag A, Lenz D, Arendt T, Tárnok A (2006) Laser scanning cytometry in human brain slices. Cytometry A 69:135–138

49. Fulwyler MJ (1965) Electronic separation of biological cells by volume. Science 150:910–911

50. Kamentsky LA, Melamed MR, Derman H (1965) Spectrophotometer: new instrument for ultrarapid cell analysis. Science 150:630–631

51. Van Dilla MA, Trujillo TT, Mullaney PF, Coulter JR (1969) Cell microfluorometry: a method for rapid fluorescence measurement. Science 163:1213–1214

52. Hulett HR, Bonner WA, Barrett J, Herzenberg LA (1969) Cell sorting: automated separation of mammalian cells as a function of intracellular fluorescence. Science 166:747–749

53. Hoshino T, Nomura K, Wilson CB, Knebel KD, Gray JW (1978) The distribution of nuclear DNA from human brain-tumor cells. J Neurosurg 49:13–21

54. Bernocchi G, Barni S (1985) On the heterogeneity of Purkinje neurons in vertebrates. Cytochemical and morphological studies of chromatin during eel (Anguilla Anguilla L.) life cycle. J Hirnforsch 26:227–235

55. Nunez R (2001) DNA measurement and cell cycle analysis by flow cytometry. Curr Issues Mol Biol 3:67–70

56. Martinez-Morales PL, Quiroga AC, Barbas JA, Morales AV (2010) SOX5 controls cell cycle progression in neural progenitors by interfering with the WNT-beta-catenin pathway. EMBO Rep 11:466–472

57. López-Sánchez N, Ovejero-Benito MC, Borreguero L, Frade JM (2011) Control of neuronal ploidy during vertebrate development. Results Probl Cell Differ 53:547–563

58. Hasbold J, Hodgkin PD (2000) Flow cytometric cell division tracking using nuclei. Cytometry 40(3):230–237

59. Westra JW, Barral S, Chun J (2009) A reevaluation of tetraploidy in the Alzheimer's disease brain. Neurodegener Dis 6:221–229

60. Westra JW, Rivera RR, Bushman DM, Yung YC, Peterson SE, Barral S, Chun J (2010) Neuronal DNA content variation (DCV) with regional and individual differences in the human brain. J Comp Neurol 518:3981–4000

61. Nüsse M, Beisker W, Hoffmann C, Tarnok A (1990) Flow cytometric analysis of G1- and G2/M-phase subpopulations in mammalian cell nuclei using side scatter and DNA content measurements. Cytometry 11:813–821

62. Darzynkiewicz Z, Traganos F, Kapuscinski J, Staiano-Coico L, Melamed MR (1984) Accessibility of DNA in situ to various fluorochromes: relationship to chromatin changes during erythroid differentiation of friend leukemia cells. Cytometry 5:355–363

63. Darzynkiewicz Z (2011) Critical aspects in analysis of cellular DNA content. In: Robinson JP (ed) Current protocols in cytometry. Wiley, New York

64. López-Sánchez N, Frade JM (2013) Cell cycle analysis in the vertebrate brain using immunolabeled fresh cell nuclei. Bio-protocol 3:e973

65. López-Sánchez N, Frade JM (2015) Flow cytometric analysis of DNA synthesis and apoptosis in central nervous system using fresh cell nuclei. Methods Mol Biol 1254:33–42

66. Shapiro HM (2003) Practical flow cytometry, 4th edn. Wiley, New York

67. Murciano A, Zamora J, López-Sánchez J, Frade JM (2002) Interkinetic nuclear movement may provide spatial clues to the regulation of neurogenesis. Mol Cell Neurosci 21:285–300

68. Slaninová I, López-Sánchez N, Šebrlová K, Vymazal O, Frade JM, Táborská E (2016) Introduction of macarpine as a novel cell-permeant DNA dye for live cell imaging and flow cytometry sorting. Biol Cell 108:1–18

69. Li JY, Lees-Murdock DJ, GL X, Walsh CP (2004) Timing of establishment of paternal

methylation imprints in the mouse. Genomics 84:952–960

70. Kumaki Y, Oda M, Okano M (2008) QUMA: quantification tool for methylation analysis. Nucleic Acids Res 36(suppl 2):W170–W175

71. Hutter KJ, Eipel HE (1978) Flow cytometric determinations of cellular substances in algae, bacteria, moulds and yeasts. Antonie Van Leeuwenhoek 44:269–282

72. Gray JW, Carver JH, George YS, Mendelsohn ML (1977) Rapid cell cycle analysis by measurement of the radioactivity per cell in a narrow window in S phase (RCSi). Cell Tissue Kinet 10:97–109

73. Wilder ME, Cram LS (1977) Differential fluorochromasia of human lymphocytes as measured by flow cytometry. J Histochem Cytochem 25:888–891

74. Van Dilla MA, Gledhill BL, Lake S, Dean PN, Gray JW, Kachel V, Barlogie B, Göhde W (1977) Measurement of mammalian sperm deoxyribonucleic acid by flow cytometry. Problems and approaches. J Histochem Cytochem 25:763–773

75. Jensen RH (1977) Chromomycin A3 as a fluorescent probe for flow cytometry of human gynecologic samples. J Histochem Cytochem 25:573–579

76. Pedersen T, Larsen JK, Krarup T (1978) Characterization of bladder tumours by flow cytometry on bladder washings. Eur Urol 4:351–355

77. Frederiksen P, Reske-Nielsen E, Bichel P (1978) Flow cytometry in tumours of the brain. Acta Neuropathol 41:179–183

78. Morillo SM, Abanto EP, Román MJ, Frade JM (2012) Nerve growth factor-induced cell cycle reentry in newborn neurons is triggered by p38^{MAPK}-dependent E2F4 phosphorylation. Mol Cell Biol 32:2722–2737

79. Arlotta P, Molyneaux BJ, Chen J, Inoue J, Kominami R, Macklis JD (2005) Neuronal subtype-specific genes that control corticospinal motor neuron development in vivo. Neuron 45:207–221

80. Wan LB, Bartolomei MS (2008) Regulation of imprinting in clusters: noncoding RNAs versus insulators. Adv Genet 61:207–223

81. Leonhardt H, Page AW, Weier HU, Bestor TH (1992) A targeting sequence directs DNA methyltransferase to sites of DNA replication in mammalian nuclei. Cell 71:865–873

82. Constância M, Pickard B, Kelsey G, Reik W (1998) Imprinting mechanisms. Genome Res 8:881–900

83. Frommer M, McDonald LE, Millar DS, Collis CM, Watt F, Grigg GW, Molloy PL, Paul CL (1992) A genomic sequencing protocol that yields a positive display of 5-methylcytosine residues in individual DNA strands. Proc Natl Acad Sci U S A 89:1827–1831

84. Sanger F, Nicklen S, Coulson AR (1977) DNA sequencing with chain-terminating inhibitors. Proc Natl Acad Sci U S A 74:5463–5467

85. Ronaghi M, Karamohamed S, Pettersson B, Uhlén M, Nyrén P (1996) Real-time DNA sequencing using detection of pyrophosphate release. Anal Biochem 242:84–89

86. Langaee T, Ronaghi M (2005) Genetic variation analyses by pyrosequencing. Mutat Res 573:96–102

Chapter 5

A Cytomic Approach Towards Genomic Individuality of Neurons

Thomas Arendt, Birgit Belter, Martina K. Brückner, Uwe Ueberham, Markus Morawski, and Attila Tarnok

Abstract

Here, we describe an approach for the DNA quantification of single cells in brain slices based on image cytometry (IC) that allows mapping the distribution of neurons with DNA content variation (DCV) in the context of preserved tissue architecture. The method had been optimized for DNA quantification of identified neurons but could easily be adapted to other tissues. It had been validated against chromogenic in situ hybridization (CISH) with chromosome-specific probes and laser microdissection followed by quantitative PCR (qPCR) of alu repeats. It can be combined with immunocytochemical detection of specific marker proteins which allow for further specification of cellular identity in the context of defined brain pathology. The method can be applied in a high-throughput mode where it allows analyzing 500,000 neurons per brain in a reasonable time. The combination of cytometry with molecular biological characterization of single microscopically identified neurons as outlined here might be a promising approach to study molecular individuality of neurons in the context of its physiological or pathophysiological environment. It reflects the concept of cytomics and will forward our understanding of the molecular architecture and functionality of neuronal systems.

Key words Ageing, Alzheimer's disease, Aneuploidy, Cell death, Cellular individuality, Cytomics, DNA content variation, Genomic mosaic, Neurodegeneration, Polyploidy, Single-cell analysis

1 Introduction: DNA Content Variation in Neurons

All somatic mammalian cells with very few exceptions have conventionally been assumed to contain identical genomes corresponding to a diploid set of chromosomes [1]. More recent studies in the human brain, however, have indicated that structural variations in the human genome due to loss or gain of whole chromosomes or fragments thereof are a constant finding in both the healthy and diseased human brain [2–17].

A more than diploid level of neuronal DNA was first reported about 50 years ago in the cerebellum of rat [18] and human brain [19]. These initial observations, made on Purkinje cells [18]

José María Frade and Fred H. Gage (eds.), *Genomic Mosaicism in Neurons and Other Cell Types*, Neuromethods, vol. 131, DOI 10.1007/978-1-4939-7280-7_5, © Springer Science+Business Media LLC 2017

and neurons of the cerebellar dentate nucleus [19], prompted a number of subsequent studies, reporting on similar findings for a large variety of neurons of different mammalian species, including cerebellar Purkinje cells of human [20, 21], cat [22], mouse [23], rat [24–34], and chick [35]; hippocampal and cortical pyramidal cells of cat [22, 36], rat [37–39], and guinea pig [40]; nerve cells of the upper cervical ganglion in rabbit [41]; cat spinal motoneurons [22]; as well as neurons of the human cholinergic basal forebrain system (Arendt, unpublished observations) and bat olfactory bulb [42].

After a protracted controversy as to whether the DNA content of a very large part of neurons [30, 31] or even all neurons [32, 43, 44] exceeds the diploid level or not [45–50], consensus was reached eventually that the majority of neurons are diploid. Discrepant results were attributed to various technical limitations and cytophotometric artifacts of heterochromatin in interphase nuclei of postmitotic neurons [51–53] but also to large individual variations [54]. Still, a small but constant fraction of a few percent of neurons continuously escaped the diploid DNA amount, irrespectively of the analytical method or other confounding factors of tissue sampling and preparation. This argues in favor of the presence of a "low-frequency" somatic aneuploidy in neurons giving rise to a genomic mosaic in the brain [16, 17, 54–56].

Different experimental approaches have been developed to detect aneuploidy, chromosomal copy number variation (CNV), single-nucleotide variance (SNV), and DNA content variation (DCV). Thus, high-resolution genome-wide microarrays of small tissue samples [57] or next-generation sequencing (NGS) of even single cells allow a comprehensive genome-wide analysis [16]. In addition, more classical techniques for evaluating chromosome numbers in brain cells using multicolor fluorescence in situ hybridization (FISH) or chromogenic in situ hybridization (CISH) with site-specific and centromeric DNA probes [2–4, 58], or interphase high-resolution chromosome-specific multicolor banding (ICS-MCB) for the visualization of whole chromosomes [6–8, 10], have been constantly improved [59] and take advantage of in situ characterization of defined neurons.

Chromosome number variations (CNV) appear to be well tolerated in the normal brain, and might even be beneficial [60, 61]. Alternatively, DCV may contribute to pathogenetic cascades and thus promote neurodegenerative disease progression, particularly if it exceeds a certain level [8, 9, 12, 58, 62–67].

To address questions such as the functional significance of physiological CNV, how it can be distinguished from its pathological form, and what are the modes of its generation and its functional sequelae, more systematic studies both on its distribution throughout the brain and its potential changes over lifetime are required. Here, we describe an approach for the DNA quantification of single cells in brain slices based on image cytometry (IC) that allow to map the distribution of neurons with DNA content variation (DCV) [11] in the context of preserved tissue architecture. The method have been optimized for DNA quantification of identified neurons but could easily be adapted to other tissues. It have been validated against chromogenic in situ hybridization (CISH) with chromosome-specific probes and laser microdissection followed by quantitative PCR (qPCR) of alu repeats [5, 58, 68, 69]. It can be combined with immunocytochemical detection of specific marker proteins which allow for further specification of cellular identity in the context of defined brain pathology. The method can be applied in a high-throughput mode where it allows to analyze 500,000 neurons per brain in a reasonable time.

2 Materials

2.1 Tissue Probes and Neuropathological Diagnosis

Brains from patients with Alzheimer's disease (AD) and from age-matched healthy controls died without any history of neurological or psychiatric illness were used. All study protocols, including case recruitment, autopsy, and data handling, were performed in accordance with the ethical standards as laid down in the 1964 Declaration of Helsinki and its later amendments as well as with the convention of the Council of Europe on Human Rights and Biomedicine, and were approved by the responsible local Ethics Committee. Informed consent had been obtained from legal representatives of the patients.

The diagnosis of AD was made on the basis of both clinical and neuropathological evidence according to the criteria of the International Working Group (IWG) for New Research Criteria for the diagnosis of AD [70, 71] in the revision of 2014 (IWG-2) [72], the NIA-AA diagnostic criteria in the revision of 2011 [73–76], and the NIA-AA guidelines for the neuropathological assessment of AD [77, 78]. Only cases with typical AD according to the IWG-2 criteria were included. All cases had undergone neuropsychological assessment during the final 6 months of their lives. Clinical Dementia Rating (CDR) scale scoring was based on neuropsychological testing (CERAD) [79], MMSE [80], and rating scales [81]. The CDR scale score was used to identify patients with mild dementia (CDR 1) [82]. All cases were neuropathologically assessed for NFT stage according to Braak and Braak [83] and Braak et al. [84], for Aβ/amyloid plaque score according to Thal et al. [85], and for neuritic plaque score according to CERAD [86]. NFTs and Aβ/amyloid plaques were detected by immunocytochemical labeling of phosphotau (anti-human PHF-tau monoclonal antibody AT8; Thermo Scientific) and Aβ (beta-amyloid monoclonal antibody, 6E10; BioLegend), respectively. Severity of AD pathology was staged following the consensus guidelines for the neuropathologic evaluation of AD according to Hyman et al. [77] and Montine et al. [78].

2.2 Equipment

- Freezing microtome (microtome Reichert-Jung with Frigomobil freezing unit, Leica Microsystems, Germany).

- Orbital or horizontal shaker (Carl Roth GmbH & Co. KG, Karlsruhe, Germany).

- Microwave oven.

- Laser scanning cytometer (CompuCyte Corporation, Cambridge, MA, USA); the argon laser (wavelength 488 nm) was used for the excitation of PI and the helium-neon laser (633 nm) for the excitation of Cy5. Emission signals were detected with long-pass filters (650EFLP) for PI and Cy5. Analyses were performed with the 20× objective and the following software settings: "threshold" set to 1200 and adjusted individually for each section, "minimum area" set to 8 μm^2, "add pixels to threshold" set to 10.

- Thermal cycler (OmniGene, Hybaid GmbH, Heidelberg, Germany).

- Humidified chamber.

- Fluorescence microscope (Axiovert 200 M), equipped with 10× objective (Plan-Neofluar 10×) and 63× oil immersion objective (Apochromat 63×) and with the appropriate software AxioVision 4.7 with the MosaiX module (Carl Zeiss AG, Jena, Germany).

- Laser microdissection system (PALM microBeam, PALM Microlaser Technologies, Bernried, Germany); "laser focus" was set to 25 and "laser energy" set to 90 and both were adjusted individually for each section.

- Real-time PCR system (Rotor-Gene 2000, Corbett Research, Sydney, Australia).

- BioPhotometer (Eppendorf Vertrieb Deutschland GmbH, Wesseling-Berzdorf, Germany).

3 Methods

A detailed process chart illustrating the experimental procedures and potential points of entry and exit as described in the methods section is provided in Fig. 1.

3.1 Tissue Preparation

Tissue blocks about 3 × 3 cm in size were cut from the region of interest (entorhinal cortex, Brodmann area 28; frontal cortex, Brodmann area 10; parietal cortex, Brodmann area 7; temporal cortex, Brodmann area 22; occipital cortex, Brodmann area 17). Blocks were fixed in 4% phosphate-buffered paraformaldehyde (4% PFA in PBS), pH 7.4, for 9 days, with the fixation buffer renewed every second day. For cryoprotection, tissue was soaked in 30%

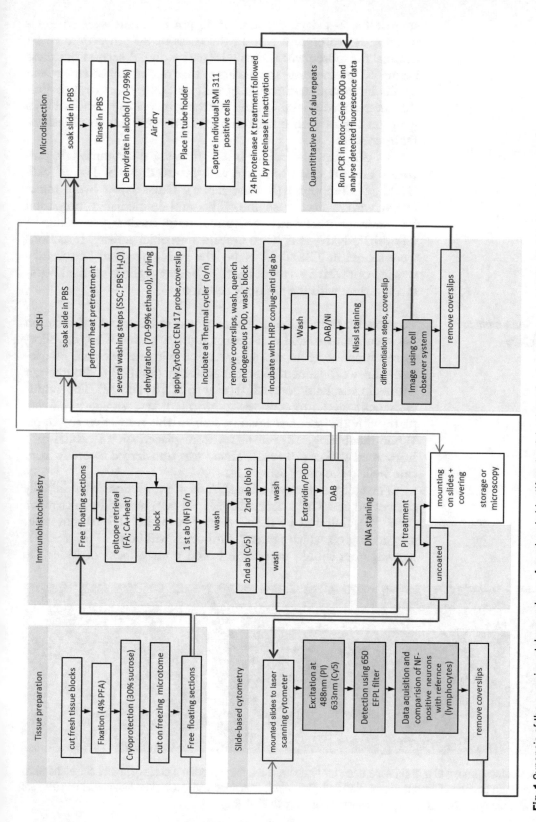

Fig. 1 Synopsis of the experimental procedures. Arrowheads identify potential points of entry, while *red arrowheads* pinpoint the major track and *blue arrowheads* show alternative tracks. Potential points of exit are marked (*white box*). The main Wet-Lab procedures are within the *grey boxes* as well as the applied methods are within the *yellow boxes*. The acquisition of images and software-based analysis is located within the *light green boxes*

sucrose for 2–4 days. Sections of 30 μm thickness were cut on a freezing microtome. A section thickness of 30 μm proved to be optimal for subsequent IC analysis (see below). A section thickness below 20 μm increases the likeliness that nuclei will be cut, which results in an underestimation of cellular DNA content. A section thickness above 30 μm, on the contrary, increases the risk of cells lying on top of each other resulting in overestimation of cellular DNA content.

To this end, the holder of the freezing microtome was pre-cooled and the tissue was fixed with 10–20 drops of Tissue-Tek® O.C.T.™ Compound (Sakura, Japan). Cut sections were transferred with a dry brush into TBS and immediately processed for immunofluorescence or stored until further processing in TBS/0.1% NaN$_3$ at 4 °C to prevent microbial activity. If sections were stored in TBS/0.1% NaN$_3$, the NaN$_3$ needs to be carefully washed out with TBS for at least 3 h or overnight as residual NaN$_3$ can interfere with the antibody reaction.

3.2 Immunohistochemistry

To allow for an identification of neurons, and a clear discrimination between neuronal and non-neuronal origin of the IC signal, slices were immunocytochemically processed with monoclonal mouse-anti-pan-neuronal neurofilament antibody (SMI-311, Sternberger Monoclonals, Baltimore, MD) prior to IC [5] (Fig. 2). The immunocytochemical protocol was performed in a standard 12-well plate with the sections floating free in the respective solutions. While incubating, 12-well plates were placed on an orbital or a horizontal shaker (10 rpm). Tissue was transferred carefully from one well to the other using a brush.

For epitope retrieval after PFA fixation, sections were initially pretreated in 10 mM citrate buffer (pH 6.0) heated by microwave (~85 °C), followed by three washing steps in TBS 10 min each. Depending on which primary antibodies will be used prior to IC, investigators should determine which pretreatment provides the

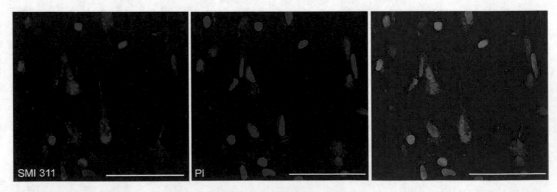

Fig. 2 Confocal image of a brain slice used for IC analysis (human entorhinal cortex). Immunohistochemical reaction with the neurofilament marker SMI 311 (Cy5; *blue*), combined with propidium iodide reaction (PI, *red*) for DNA staining. Scale bar: 200 μm. Modified after Mosch et al. [58]

strongest signal in their assay system. As an alternative, pretreatment can be performed using 50% formic acid in PBS or TBS pH 9.0 [87]. After blocking of unspecific binding sites with 0.3% milk powder, 0.1% gelatine, 1% bovine serum albumin, and 0.05% Tween 20 in 10 mM Tris-buffered saline (TBS, pH 7.4) for about 1 h, sections were incubated with the SMI-311 antibody (dilution 1:750) overnight at 4 °C in blocking solution. Sections were washed three times 10 min each in TBS-T between the single incubation steps. The SMI-311 antibody was labeled with a secondary cyanine 5 (Cy5™)-conjugated goat anti-mouse antibody (Dianova, Hamburg, Germany, 1:300). For DNA labeling, sections were subsequently treated with 50 µg/mL propidium iodide (PI) in TBS containing 100 µg/mL ribonuclease A (Sigma, St. Louis, MO), for 30 min at 37 °C in the dark (Fig. 2). Sections were mounted onto slides using DAKO fluorescent mounting medium, and covered and stored at 4 °C in the darkness until analysis. Uncoated slides were found to be best suited for the subsequent IC analysis, as they give rise to rather low background values for the PI staining [68].

Slices for microdissection and chromogenic in situ hybridization were processed with an avidin-biotin system. Bound SMI 311 antibody was labeled with secondary biotin-conjugated goat anti-mouse antibody (1:500; Dianova, Hamburg, Germany). Subsequently, sections were processed with ExtrAvidin® – Peroxidase conjugate (1:500; Sigma) and 0.04% 3,3'-diaminobenzidine (DAB)/0.015% H_2O_2 (Sigma) as chromogen. Finally, sections were mounted onto slides (MembraneSlide PEN 2.0 µm (Leica)), air-dried for microdissection, or further processed using CISH.

3.3 Image Cytometry

IC was performed using a laser scanning cytometer (LSC, CompuCyte, Cambridge, MA) with the software WinCyte version 3.4. The instrument settings were optimized by analysis of DNA content (low coefficient of variance values for the G0 peak) using peripheral blood lymphocytes for the present application as described [5, 68]. For excitation the 488 nm line of the argon laser (forward scatter, PI) and the 633 nm line of the helium-neon (HeNe) laser (Cy5) were used. The 650 EFLP (long-pass edge cut-on filter, 650 nm) filter was used for the detection of fluorescence light (Cy5 & PI). Triggering signal for cell detection was the PI signal. Threshold was typically set to 2000. Analysis was performed with 20× objective, numeric aperture 0.5. The minimum area to be necessary for a nucleus to be accepted was set to 8 µm in order to exclude debris and other smaller artifacts (e.g., cut nuclei) from the analysis. 10 pixels were added to threshold in order to include cytoplasm data and to avoid that only data of nuclei were gained which very likely may result in an underestimation of SMI-Cy5-positive cells.

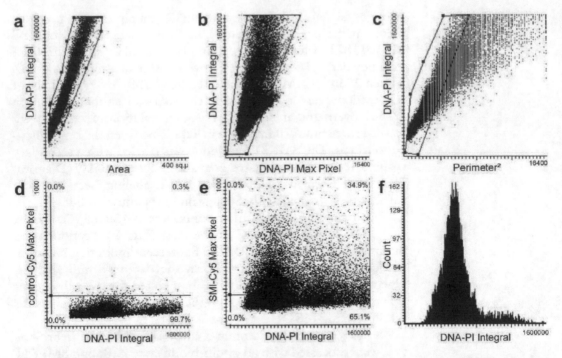

Fig. 3 Data acquisition by IC. To exclude doublets, artifacts, and cell debris from the analysis, a gating cascade was applied. Fluorescence events were gated by means of the area (**a**), the maximal pixel intensity (**b**), and the perimeter of single-fluorescence events (**c**). Using this gating strategy ensures that only single cells will be included in the analysis. To clearly discriminate the specific fluorescence signal from background fluorescence, control slices without primary antibodies were used to define a background fluorescence threshold (**d**). Applying this threshold to the fully stained slice allows the identification of neurons as Cy5-positive events (**e**). For each cell population, a PI histogram can be generated (**f**) and used to analyze the distribution of neurons according to their DNA content with lymphocytes serving as standards

The baseline for the threshold values was determined on sections labeled with PI and the secondary Cy5-labeled antibody alone in the absence of the primary antibody. PI generally produced a bright easily detectable nuclear signal. Here, the PMT (photomultiplier) signal was set so that all pixel values were below saturation. The Cy5 signal of the control was set by adapting the PMT voltage so that the background signal was above 0 value. From then on all instrument settings were not changed. Data acquisition was started and the scan data images which were produced for Cy5 and PI were checked for proper recognition of the software. If not all cells were recognized or too many cells were merged into one object, the threshold value was changed accordingly. Each fluorescent event was recorded with respect to size, perimeter, x–y position on the object slide, maximum (maximum pixel), and overall integral fluorescence intensity. To exclude doublets and other artifacts from the analysis, a gating cascade was created (Fig. 3a–c). Using the WinCyte software, a first window was created (Area vs. DNA-PI

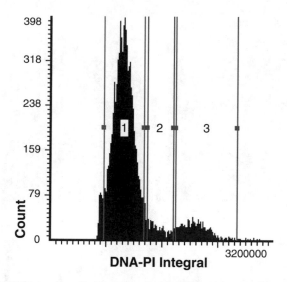

Fig. 4 DNA-PI histogram of cells, identified as neurons, gated from SMI-Cy5-positive events as displayed by WinCyte™. Cells in area 1 were considered as containing a DNA amount corresponding to 2n (diploid). Cells in area 3 were considered as containing a DNA amount corresponding to 4n (tetraploid). Cells in area 2 were considered as containing a DNA amount between 2n and 4n

Integral; Fig. 3a) to exclude artifacts, debris, and doublets. Singlets were gated into a second window (DNA-PI Max Pixel vs. DNA-PI Integral, Fig. 3b). From this window, cells were gated into a third window in order to exclude further doublets (Perimeter^2 vs. DNA-PI Integral, Fig. 3c). With the perimeter^2 feature it was possible to exclude further doublets and multiple events. To clearly discriminate the specific fluorescence signal from background fluorescence, control slices without primary antibodies were used to define a background fluorescence threshold (Fig. 3d, e). Only neurofilament-immunoreactive cells were considered for IC analysis. For each cell population, histograms of integral PI fluorescence values were generated (Figs. 3f and 4). These histograms were exported to the software ModfitLT (version 2.0; Verity Software House, Topsham, ME, USA) and used to analyze the distribution of neurons according to their DNA content with lymphocytes serving as standards.

The entire cortical depth of the respective cortical area was scanned with 80,000–120,000 cells analyzed for each specimen. Analyses were restricted to neurons with an amount of DNA corresponding to a diploid level or above. Neurons with a reduced amount of DNA were not considered for quantification, as a loss of DNA due to partial sectioning of nuclei is difficult to control for. A DNA content exceeding the mean value for diploid cells by two standard deviations was considered DNA content variation (DCV).

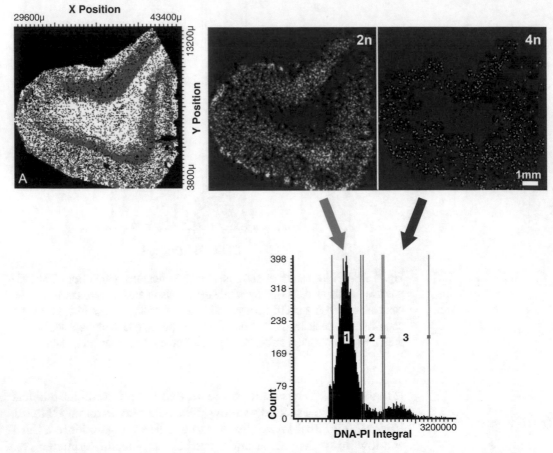

Fig. 5 Recording of the x-y position of each specific fluorescence event above background threshold allows to image the distribution of neurons (SMI-Cy5; color coded in *blue*) together with their PI value corresponding to the DNA amount (human parahippocampal gyrus, Alzheimer's disease). Distribution of neurons with a PI signal corresponding to a diploid (*middle, area 1, blue arrow*) or tetraploid (*right, area 3, red arrow*) amount of DNA

The numerical neuronal density was determined as a reference value using the tool of WinCyte to record the x–y position of each fluorescence event (Fig. 5). A square region with a side length of 1 mm was created and the number of neurons within this region randomly placed ten times through the entire cortical depth was determined.

3.4 Chromogenic In Situ Hybridization

The method used to quantify DNA using IC was validated by chromogenic in situ hybridization (CISH) on a subset of microscopically identified neurons from randomly selected cases of AD [12, 14, 58]. The specimens were initially processed for immunofluorescence and analyzed by IC. Subsequent sections were processed for chromogenic in situ hybridization (CISH) with a ZytoDotCEN 17 probe (ZytoVision, Bremerhaven, Germany), which targets alpha-satellite sequences of the centromere of chromosome 17. The digoxigenin-labeled probe was immunohistochemically visualized using peroxidase-conjugated Fab fragments of an anti-digoxigenin antibody from sheep (Roche Diagnostics, Mannheim, Germany) and nickel ammonium sulfate/DAB/0.015% H_2O_2 as chromogen. Fixed human lymphocytes dropped on object slides and HeLa cells cultured under standard conditions and grown on coverslips were used as controls.

Subsequent to IC, coverslips were carefully removed by soaking the object slide in a petri dish with pre-warmed PBS. Sections were incubated with "Heat Pretreatment Solution" (Zytovision) for 15 min at 45 °C and subsequently rinsed two times in 2× SSC for 30 s at room temperature. After additional washing steps with 2× SSC, PBS, and aqua dest., sections were dehydrated in alcohol (70%, 85%, 96%, 99%) and quickly air-dried. 15 µL of the probe (ZytoDot CEN 17 Probe) was mixed with 5 µL hybrisol, pipetted in the center of an appropriate coverslip (e.g., round ⌀ 22 mm) and everted onto the section. Coverslips were sealed with Fixogum (Marabu, Tamm, Germany) and left at 37 °C for 5 min. Slides were subsequently transferred into a thermal cycler (OmniGene) where they were incubated for 10 min at 94 °C, followed by an overnight incubation at 37 °C in a humidified formamide chamber. The next day, coverslips were carefully removed, and slides were washed in 50% formamide/2 × SSC (two times each 15 min at 45 °C), followed by 0.5 × SSC (30 min, at 40 °C) and PBS (10 min at room temperature). Endogenous peroxidase was quenched by incubation with 1% H_2O_2 in PBS for 30 min, sections were washed three times in PBS (each 10 min), and unspecific binding sites were blocked with 2% sheep serum in PBS. Subsequently, sections were incubated with the HRP-conjugated anti-digoxigenin antibody (Abcam, ab6212) diluted 1:200 for 1 h at room temperature, washed in 50 mM Tris, pH 8.0, and incubated with 10 mM nickel ammonium sulfate $(NH_4)_2Ni(SO_4)_2$/0.04% DAB/0.015% H_2O_2 to visualize the secondary antibody reaction under microscopic control. Incubation longer than 3 min should be avoided as nickel could precipitate and thus feign false-positive CISH signals. Slides were rinsed in 50 mM Tris, pH 8.0, followed by aqua dest.

To visualize the cytoarchitectonic context of both the PI/Cy5 and CISH signals, slides were counterstained according to Nissl. Sections were incubated in Cresyl (echt) violet acetate solution for 7 min with moderate shaking (20 rpm) and staining was differentiated by incubation in alcohol (70%, 80%, 96%). To avoid immoderate decoloration incubate only for a few seconds in 96% alcohol. Slides were rehydrated, washed in aqua dest., coverslipped with Aqua Poly Mount (Polyscience), and stored at 4 °C in the dark until analysis.

Regions of interest on tissue sections were imaged with the cell observer system (Axiovert 200 M, equipped with a motorized scanning table, with the AxioVision Software 4.7 containing the MosaiX module, Zeiss, Jena, Germany). Hybridization spots were

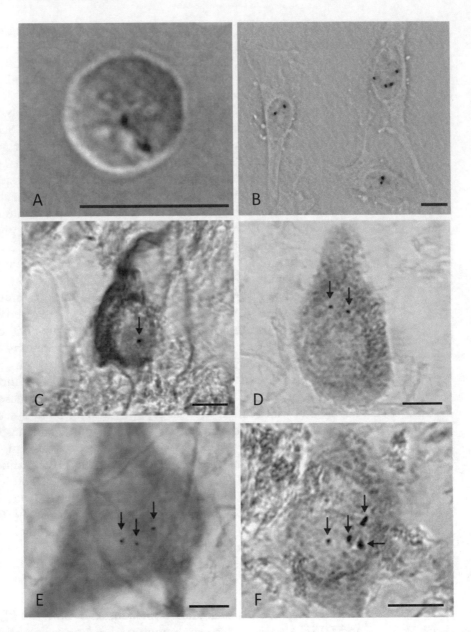

Fig. 6 Hybridization signals in lymphocytes, HeLa cells, and neurons of the entorhinal cortex in AD detected by CISH using a chromosome-specific probe for chromosome 17 which targets satellite sequences of the centromere (ZytoDotCEN probes, ZytoVision, Germany). (**A**) Two spots are present in 99% of all lymphocytes. (**B**) Proliferating HeLa cells show either two or four spots, indicating a diploid and tetraploid set of chromosome 17. (**C–F**) Pyramidal neurons in the brain of a patient with AD showing one, two, three, or four spots (*arrows*). Scale bars, 10 μm. Modified after Mosch et al. [58] and Arendt et al. [12]

recorded with a 63× oil immersion objective (Fig. 6). Cells were labeled in a digitized image by a color code according to their number of spots (Fig. 7). This image was used as "template image" for the subsequent laser microdissection. In total, 400–500 neuronal nuclei were analyzed.

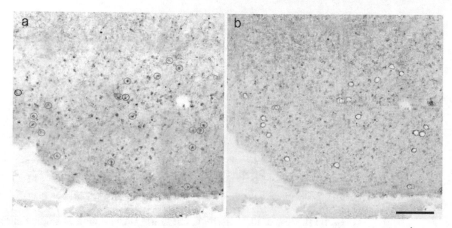

Fig. 7 Laser capture microdissection of microscopically identified neurons. (**a**) Color coding of neurons after CISH, according to the number of hybridization spots, serving as "template image" for subsequent sampling by laser capture microdissection. (**b**) Image after laser microdissection of marked single neurons. Scale bar: 100 μm

3.5 Laser Capture Microdissection of Identified Neurons

Single neurons, identified by immunoreactivity for neurofilaments (SMI 311-Cy5) and labeled by CISH, were cut from brain slices with a laser microdissector (PALM MicroBeam; P.A.L.M. Microlaser Technologies, Bernried, Germany) and subsequently subjected to DNA quantification by qPCR of alu repeats. Individual microscopically identified neurons were sampled according to the number of hybridization spots, obtained by CISH and recorded on the "template image."

Coverslips of brain slides were carefully removed by soaking the object slide in a petri dish with pre-warmed PBS. Slides were rinsed in fresh PBS, dehydrated in alcohol (70%, 85%, 96%, 99%), and air-dried. 200 μL PCR tubes were prepared for cell sampling by pipetting 10 μL PCR reaction buffer for laser microdissection in the center of the cap. Tubes were placed in the holder of the PALM system and neurons were captured according to their number of CISH spots recorded on the "template image" (Fig. 7). For each case, at least 20 SMI 311-immunoreactive cells were captured. Each single neuron was sampled into an individual PCR tube. Tubes were incubated at 37 °C for 24 h and proteinase K was inactivated by incubation at 90 °C for 10 min. Samples were further processed by quantitative PCR or stored at −20 °C until further analysis.

3.6 Quantitative PCR of Alu Repeats

DNA content of individual neurons was quantified through real-time PCR amplification of alu repeats [88], a class of short interspersed elements in the eukaryotic genome, which reach a copy number of one million in primates [89, 90]. Alu repeats were chosen because of their high copy number and low level of polymorphism compared with other short interspersed elements in the

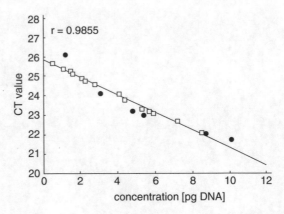

Fig. 8 Principle of the determination of the DNA content of single neurons using quantitative real-time PCR. *Black dots* show the DNA concentration and the corresponding CT value of the standard probes. The *black line* represents a linear regression through the standard probes with $r = 0.9855$. Using the CT value of the samples (*white squares*) the DNA concentration can be calculated

eukaryotic genome [91]. The residual risk of an artificial influence by different copy numbers or single-nucleotide polymorphisms in several individuals was avoided by the intraindividual comparison of two different brain areas of each patient. Alu oligonucleotide primers alu-for 5′-GTGGCTCACGCCTGTAATCCC-3′ and alu-rev 5′-ATCTCGGCTCACTGCAACCTC-3′, localized in conserved regions of the alu repeats, were designed using the Primer3 program (http://frodo.wi.mit.edu/cgi-bin/primer3/primer3_www.cgi).

Real-time PCR quantification was accomplished in a Rotor-Gene 2000 (Corbett Research, Sydney, Australia). A master mix was prepared for the appropriate number of samples containing 1× PCR reaction buffer, 0.5 μM each primer, 100 μM dNTP mix, SYBR Green diluted 1:40,000, and 0.5 U DyNAzyme II DNA Polymerase and 10 μL of it was added to each sample. Samples containing genomic DNA from lymphocytes were used as internal standards. PCR was run with the following settings: initial denaturation 95 °C, 10 min; 45 cycles with each 95 °C, 30-s denaturation; 72 °C, 45-s primer annealing and extension, 86 °C, 5-s detection of the fluorescence signal. A final extension step with 72 °C, 7 min, and products melted from 50 to 90 °C with 2 °C/min can be added. Melt curves and the CT values were analyzed by the Rotor-Gene 2000 software Rotorgene, version 4.6, and statistics were performed using PlotIT 3.2 (SPE Software, Ville de St-Georges, Quebec, Canada). The amount of DNA in pg was calculated for each sample by means of the CT values of human lymphocytes treated identically to human brain tissue and used as standard probes (Fig. 8). A DNA amount of 2.07 ± 0.6 pg (mean ± SD) and 4.06 ± 0.5 pg was obtained for one single and two lymphocytes, respectively.

3.7   Intermethod Reliability of DNA Quantification by IC and Quantitative PCR

To confirm our experimental approach we cross-validated the three methods of single-cell DNA quantification, i.e., IC, CISH, and qPCR, through subsequent application one by one to a subset of microscopically identified neurons and obtained a remarkably high intermethod reliability (Fig. 9) [12, 14, 58]. The specimen was initially processed for immunofluorescence and analyzed by IC with a laser scanning cytometer. Subsequently, the slice was subjected to CISH, combined with conventional Nissl staining. Thereafter, identified cells were captured with laser microdissection and subjected to quantitative PCR. For each cell, the PI integral obtained by LSC was plotted against the number of spots and the amount of DNA calculated from the PCR (Fig. 9). The combination of the three methods in a row can be easily adapted to other issues. When the PI integral (IC) is plotted against the number of hybridization spots (CISH), a highly significant correlation will be obtained (Fig. 9a). Similarly, the PI integral shows a highly significant correlation with the DNA content determined by PCR amplification of alu repeats (Fig. 9b). The DNA amount determined by PCR amplification of alu repeats also correlates highly significantly with the number of hybridization spots (Fig. 9c).

Based on these correlations, the DNA amount (mean ± SD) of a single diploid neuron identified by two hybridization spots was determined as 2.52 ± 0.87 pg, while the DNA content of a neuron with four spots amounts to 5.94 ± 1.28 pg. For comparison, the mean single-cell DNA content (±SD) of lymphocytes treated identically to brain tissue and used for control experiments was determined at 2.07 pg ± 0.6 pg. Comparable values for single-neuron DNA content were obtained by qPCR of alu repeats.

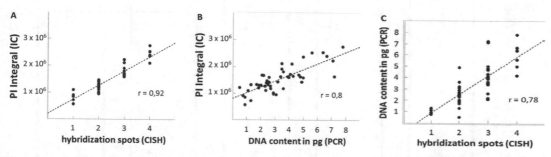

Fig. 9 Intermethod reliability of three independent methods for single-cell DNA quantification. A set of 48 microscopically identified neurons of the entorhinal cortex in a patient with early AD was evaluated through subsequent application of IC, CISH, and PCR amplification of alu repeats. Tissue sections were first processed for IC, followed by hybridization with a chromosome 17-specific probe. Subsequently, identified neurons were captured through laser microdissection and subjected to PCR amplification of alu repeats. Regression analyses reveal correlation coefficients according to Bravais-Pearson of (**A**) $r = 0.92$ for the IC data versus hybridization results (CISH), (**B**) of $r = 0.80$ for LSC versus PCR amplification, and (**C**) $r = 0.78$ for PCR amplification versus hybridization. All correlation coefficients are significant at $p < 0.001$. Modified after Mosch et al. [58]

As in our examples the DNA content of single neurons determined by IC is far better correlated to CISH (Fig. 9a) than is the qPCR measurement (Fig. 9b, c), we conclude that IC is a very reliable technique for studies on the single-cell DNA content in tissue sections.

4   Application of the Cytomic Analysis of Single-Cell DNA Content to the Healthy and Diseased Human Brain

4.1   The Normal Human Brain Is a Genomic Mosaic

The DNA content of single neurons was quantified by IC in frontal, temporal, parietal, entorhinal, and occipital cortices of mentally healthy individuals aged 75.8 ± 8.2 years (Fig. 10). For each brain, the DNA content of about 500,000 neurons was analyzed [14].

In all cortical areas, DNA analysis by IC identified the vast majority of neurons as containing an amount of DNA corresponding to a diploid set of chromosomes. Still, on average about 10–12% of neurons showed a DNA content above the diploid level [14, 15, 58] (Fig. 10). No indications for an effect of gender were observed (mean values ± SD, men: 11.41 ± 2.95%; women: 11.59 ± 2.57%) [15]. After chromogenic in situ hybridization (CISH) with a chromosome 17 probe, two hybridization spots were obtained for the majority of neurons in the entorhinal cortex of control brain [58]. In addition, a constant number of about 6–7% of neurons contained three hybridization spots while a small number of less than 0.4% contained four hybridization spots or more (Figs. 6 and 11).

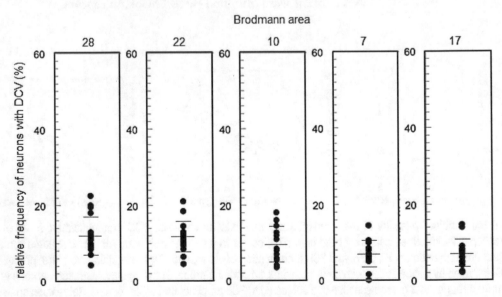

Fig. 10 Relative frequency of neurons with an amount of DNA exceeding the diploid level (DCV) in five different cortical areas in the normal human brain (n = 16 normal elderly, mean age: 75.8 ± 8.2 years; quantification of single-neuron DNA content by IC; mean values ± SD). Modified after Arendt et al. [14]

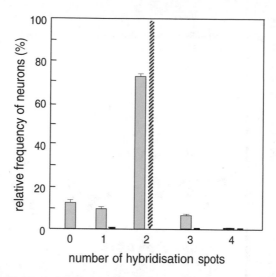

Fig. 11 Quantification of CISH signals in neurons (entorhinal cortex) of the normal human brain (hatched column: PFA-fixed lymphocytes treated identically to brain slices served as standard; mean values ± SEM). Modified after Mosch et al. [58]

4.2 Neuronal DNA Content Variation in the Normal Elderly Shows Systematic Regional Differences Throughout the Cerebral Cortex

The frequency of neurons with a DNA content above the diploid level showed regional differences and varied between 7% (primary visual cortex) and 12% (entorhinal cortex) (Fig. 10). When cortical regions were arranged according to their frequency of DCV in decreasing order, the following sequence was obtained: area 28 > area 22 > area 10 > area 7 > area 17 (Fig. 10). The values obtained for the two regions with the highest frequencies of DCV (area 28 and area 22) were significantly different ($p < 0.01$; Student's t test) from those obtained for the two regions with the lowest frequencies of DCV (area 7 and area 17).

4.3 Neuronal DNA Content Variation Decreases with Age

Individual mean values of DCV, averaged over five cortical areas, showed an age-related decline between 30 and 90 years of age (Fig. 12a) [15]. When individuals were grouped according to their age of either below or above 60 years, mean values differed by 21% (Fig. 12b). When age-related changes were analyzed separately for each region, there was a similar tendency for most regions (Fig. 13). This age-related decline was most pronounced in the occipital cortex while hardly present, however, in the entorhinal cortex.

4.4 The Frequency of Neurons with DNA Content Variation Is Elevated in Patients with Alzheimer's Disease Throughout the Cerebral Cortex

In AD, a significant increase in the number of neurons with DCV above the levels in brains of normal elderly can be observed throughout the entire cortex [5, 14] (Fig. 14). Compared to age-matched elderly, the frequency of neurons with DCV was roughly doubled in AD. Still, the frequency of DCV showed constant regional differences, ranging from about 10% in the primary visual cortex to above 30% in the entorhinal cortex. It showed intermediate values in the temporal, frontal, and parietal association cortices.

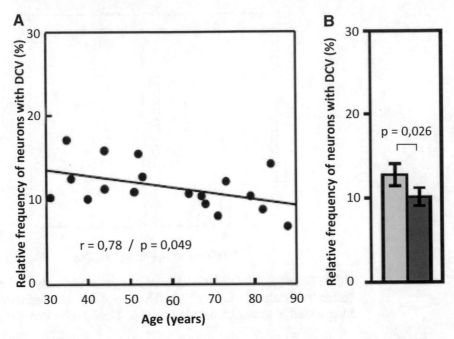

Fig. 12 Age-related changes in the relative frequency of cortical neurons with an amount of DNA exceeding the diploid level. (**A**) Changes in individual mean values, averaged over five cortical areas in persons between the fourth and ninth decades of life. (**B**) Age-related alterations are displayed as mean values for subjects below (*light grey*) and above (*dark grey*) 60 years of age (group size: nine subjects; ± SD). *r* Pearson's product–moment correlation coefficient, *p* level of significance (*t*-test). Modified after Fischer et al. [15]

When regions were arranged according to their relative increases in the frequency of DCV in decreasing order, the following sequence was obtained: area 28 > area 22 > area 10 > area 7 > area 17 (Fig. 14).

In agreement with data obtained by IC, the number of nuclei with three hybridization spots was about doubled in AD compared to controls (controls: ~7%; AD ~12%; Fig. 6). The relative number of these neurons with four spots or more was increased almost by factor four (control: 0.4%; AD: ~1.5%). This increase in the number of neurons with DCV in AD could further be validated by PCR amplification of alu repeats after laser capture microdissection of identified neurons [58].

The frequency distribution of single-cell DNA content obtained by this method is displayed in Fig. 14. While the distributions obtained from normal brains show a single maximum at 2.5–3.5 pg per cell which corresponds a 2n DNA content, AD patients show a second maximum at 6.5–7.5 pg per cell corresponding to a 4n DNA content (Fig. 15) [58].

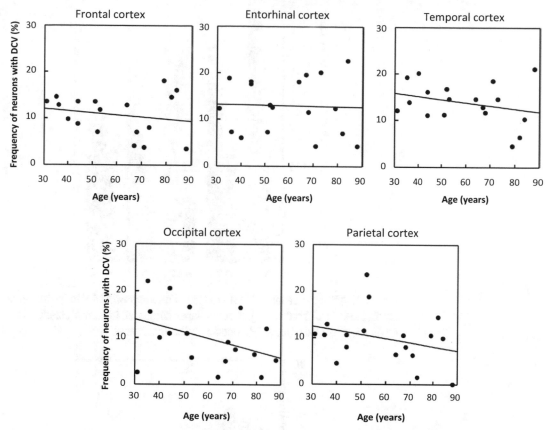

Fig. 13 Age-related changes in the relative frequency of cortical neurons with an amount of DNA exceeding the diploid level, analyzed separately for frontal (Brodmann area 10, $r = 0.19$; $p = 0.46$), temporal (Brodmann area 22, $r = 0.28$; $p = 0.26$), parietal (Brodmann area 7, $r = 0.29$; $p = 0.24$), entorhinal (Brodmann area 28, $r = 0.00$; $p = 0.97$), and occipital (Brodmann area 18, $r = 0.40$; $p = 0.10$) cortices. r, Pearson's product–moment correlation coefficient; p, level of significance (t-test). Modified after Fischer et al. [15]

4.5   Elevation of the Number of Neurons with DCV in AD Is an Early Event that Occurs at Preclinical Stages

The AD-associated increase in the number of neurons with DCV can be observed already at very early, i.e., preclinical stages of the disease. The number of neurons with DCV was found more than doubled in the entorhinal cortex of patients at preclinical AD stages, defined as CDR0, Braak stages I–II/B, rated "possible" on CERAD criteria and "low to intermediate" on NIA criteria (Fig. 16). While neurons with DCV were also significantly elevated in patients with mild AD, their number decreased significantly during progression of the disease, remaining only marginally different from controls at most severe stages of AD. This decrease in the number of neurons with DCV from early to more advanced stages

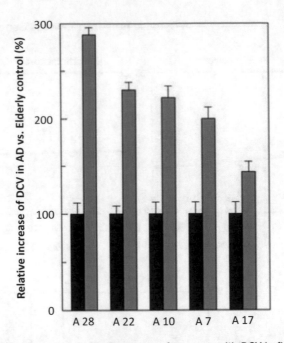

Fig. 14 Relative increases in the frequency of neurons with DCV in five different cortical areas in AD patients (grey columns) as compared to normal elderly (black columns, mean values ± SEM). Modified after Arendt et al. [14]

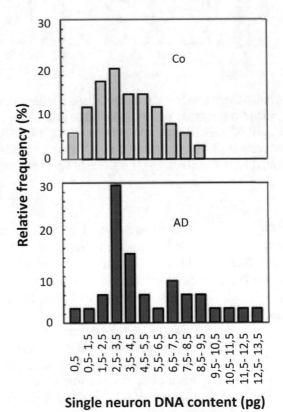

Single neuron DNA content (pg)

Fig. 15 Frequency distribution of single-neuron DNA content determined by PCR amplification of *alu* repeats. Note in AD the shift towards higher DNA content and the appearance of a second peak corresponding to a tetraploid DNA content. Modified after Mosch et al. [58] and Arendt et al. [69]

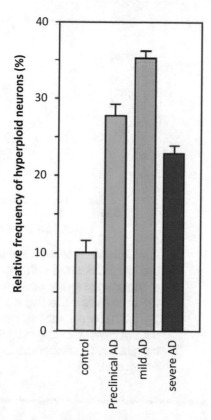

Fig. 16 Quantification of neurons with DCV in the entorhinal cortex by IC in healthy controls and cases with preclinical, mild, and severe AD (mean values ± SEM). Modified after Arendt et al. [12]

of AD can only be explained by a loss of these neurons during progression of AD as generation of DCV most likely is an irreversible process.

4.6 DNA Copy Number Variation Accounts for the Majority of Neuron Loss in AD

To analyze the fate of neurons with DCV in the entorhinal cortex during progression of the disease, we plotted the changes in the number of these neurons against the total loss of neurons (Fig. 17). The regression analysis for the number of neurons with a 2n content of DNA against the total neuronal number did not reveal a significant relationship, indicating that the numerical density of diploid neurons remained stable during the process of cell loss. The loss of neurons with a DNA content above 2n (hyperploid neurons), however, correlated significantly with the total loss of neurons. This indicates that 89.5% of the total loss of neurons in the entorhinal cortex might be explained by a loss of neurons with DVC. At the next step, we analyzed the transition between healthy controls and preclinical and mild AD in more detail. To visualize the relationship to cell death, the number of hyperploid neurons was plotted against total neuronal number (Fig. 18). At preclinical stages of AD, the elevation of neurons with DCV was not associated

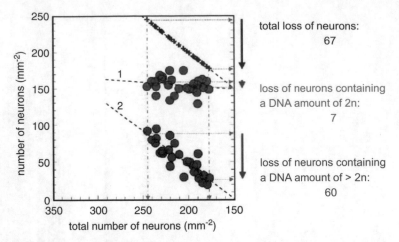

Fig. 17 Relation between changes in the number of neuron with a 2n DNA content and neurons with a DNA content above 2n and total neuronal loss. Total loss of neurons during progression of AD highly correlates with changes in the number of neurons containing more than 2n DNA (regression line 2: $r = 0.86$; $p < 0.001$) but not with neurons containing 2n DNA. (regression line 1: $r = 0.18$; n.s.). Neurons containing more than 2n DNA (*red*) account for a loss of 60 of 67 neurons, which corresponds to 89% of the total neuronal loss. Modified after Arendt et al. [12]

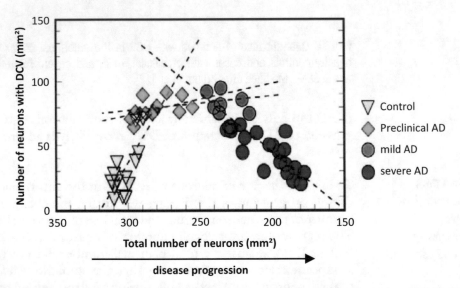

Fig. 18 Relation between changes in the number of neurons with DVC and total neuronal loss in healthy controls and preclinical, mild, and severe AD. Neurons with DCV are increased already in preclinical AD. They remain stable in number during transition from preclinical to mild AD and decrease with further progression of the disease. Modified after Arendt et al. [12]

with neuronal loss. First cell loss became detectable at the transition from preclinical to mild AD. Neurons with DCV remained stable in number at this point and subsequently decreased during disease progression from mild towards severe AD indicating a decreased viability of these neurons (Fig. 18).

References

1. Thomson RY, Frazer SC (1954) The deoxyribonucleic acid content of individual rat cell nuclei. Exp Cell Res 6:367–383

2. Rehen SK, McConnell MJ, Kaushal D, Kingsbury MA, Yang AH, Chun J (2001) Chromosomal variation in neurons of the developing and adult mammalian nervous system. Proc Natl Acad Sci U S A 98: 13361–13366

3. Rehen SK, Yung YC, McCreight MP, Kaushal D, Yang AH, Almeida BSV, Kingsbury MA, Cabral KMS, McConnell MJ, Anliker B, Fontanoz M, Chun J (2005) Constitutional aneuploidy in the normal human brain. J Neurosci 25:2176–2180

4. Kingsbury MA, Friedman B, McConnell MJ, Rehen SK, Yang AH, Kaushal D, Chun J (2005) Aneuploid neurons are functionally active and integrated into brain circuitry. Proc Natl Acad Sci U S A 102:6143–6147

5. Mosch B, Mittag A, Lenz D, Arendt T, Tárnok A (2006) Laser scanning cytometry in human brain slices. Cytometry Part A 69:135–138

6. Yurov YB, Iourov IY, Vorsanova SG, Liehr T, Kolotii AD, Kutsev SI, Pellestor F, Beresheva AK, Demidova IA, Kravets VS, Monakhov VV, Soloviev IV (2007) Aneuploidy and confined chromosomal mosaicism in the developing human brain. PLoS One 2:e558

7. Iourov IY, Vorsanova SG, Liehr T, Yurov YB (2009) Aneuploidy in the normal, Alzheimer's disease and ataxia-telangiectasia brain: differential expression and pathological meaning. Neurobiol Dis 34:212–220

8. Iourov IY, Vorsanova SG, Liehr T, Kolotii AD, Yurov YB (2009) Increased chromosome instability dramatically disrupts neural genome integrity and mediates cerebellar degeneration in the ataxia-telangiectasia brain. Human Mol Genet 18:2656–2669

9. Iourov I, Vorsanova S, Yurov Y (2011) Genomic landscape of the Alzheimer's disease brain: chromosome instability-aneuploidy, but not tetraploidy—mediates neurodegeneration. Neurodegener Dis 8:35–37; discussion 38–40.

10. Vorsanova SG, Yurov YB, Iourov IY (2010) Human interphase chromosomes: a review of available molecular cytogenetic technologies. Mol Cytogen 3:1

11. Westra JW, Rivera RR, Bushman DM, Yung YC, Peterson SE, Barral S, Chun J (2010) Neuronal DNA content variation (DCV) with regional and individual differences in the human brain. J Comp Neurol 518:3981–4000

12. Arendt T, Brückner MK, Mosch B, Lösche A (2010) Selective cell death of hyperploid neurons in Alzheimer's disease. Am J Pathol 177:15–20

13. Arendt T (2012) Cell cycle activation and aneuploid neurons in Alzheimer's disease. Mol Neurobiol 46:125–135

14. Arendt T, Brückner MK, Lösche A (2015) Regional mosaic genomic heterogeneity in the elderly and in Alzheimer's disease as a correlate of neuronal vulnerability. Acta Neuropathol 130(4):501–510

15. Fischer HG, Morawski M, Brückner MK, Mittag A, Tarnok A, Arendt T (2012) Changes in neuronal DNA content variation in the human brain during aging. Aging Cell 11:628–633

16. McConnell MJ, Lindberg MR, Brennand KJ, Piper JC, Voet T, Cowing-Zitron C et al (2013) Mosaic copy number variation in human neurons. Science 342(6158):632–637. doi:10.1126/science.1243472

17. Cai X, Evrony GD, Lehmann HS, Elhosary PC, Mehta BK, Poduri A, Walsh CA (2014) Single-cell, genome-wide sequencing identifies clonal somatic copy-number variation in the human brain. Cell Rep 8(5):1280–1289. doi:10.1016/j.celrep.2014.07.043

18. Brodsky VJ, Kusc AA (1962) Changes in the number of polyploid cells during postnatal development of rat tissue. Sci USSR 147: 713–716

19. Müller H (1962) Cytophotometrische DNS-Messungen an Ganglienzellkernen des Nucleus dentatus beim Menschen. Naturwiss 49:243

20. Lapham LW (1965) The tetraploid DNA content of normal human Purkinje cells and its development during the perinatal period: a quantitative cytochemical study. Excerpta Med (Amst) Cong Ser 100:445–449

21. Lapham LW (1968) Tetraploid DNA content of Purkinje neurons of human cerebellar cortex. Science 159:310–312

22. Herman CJ, Lapham LW (1969) Neuronal polyploidy and nuclear volumes in the cat central nervous system. Brain Res 15:35–48

23. Mares V, Lodin Z, Sácha J (1973) A cytochemical and autoradiographic study of nuclear DNA in mouse Purkinje cells. Brain Res 53:273–289

24. Sandritter W, Novokova V, Pilny J, Kiefer G (1967) Cytophotometrische Messungen des Nukleinsäure- und Proteingehaltes von Ganglienzellen der Ratte während der

postnatalen Entwicklung und im Alter. Zeitschr Zellforsch 80:145–152

25. Lentz RD, Lapham LW (1969) A quantitative cytochemical study of the DNA content of neurons of rat cerebellar cortex. J Neurochem 16:379–384

26. Lentz RD, Lapham LW (1970) Postnatal development of tetraploid DNA content in rat purkinje cells: a quantitative cytochemical study. J Neuropathol Exp Neurol 29:43–56

27. Brodsky VJ, Sokolova GA, Manakova TE (1971) Multiple increase of DNA in the Purkinje cells of cerebellum in ontogenesis of rats. Ontogeny (USSR) 2:33–36

28. Brodsky VY, Agroskin LS, Lebedev EA, Marshak TL, Papayan GV, Segal OL, Sokolova GA, Yarygin KN (1974) Stability and variations of amount of DNA in a population of cerebellum cells. Z Obshchei. Biol 35:917–925

29. Bernocchi G (1975) Contenuto in DNA e area nucleare dei neuroni durante l'istogenese cerebellare del ratto. Instituto Lombardo (Rend Sc) 109:143–161

30. Bernocchi G, Manfredi Romanini MG (1977) Chromatin Cytochemistry as a tool for the functional interpretation of Purkinje's cells in rat cerebellum. Riv Istochim Norm Patol 21:131–142

31. Bernocchi G, Redi CA, Scherini E (1979) Feulgen-DNA content of the Purkinje neuron: "diploid" or "tetraploid"? Basic Appl Histochem 23:65–70

32. Bregnard A, Knüsel A, Kuenzle CC (1975) Are all the neuronal nuclei polyploid? Histochemistry 43:59–61

33. Marshak TL, Petruchuk EM, Aref'eva AM, Shalunova NV, Brodskii VI (1976) O soderzhanii DNK V kletkakh Purkin'e mozzhechka krys, spontanno infitsirovannykh virusom Kilkhema. Biull Eksp Biol Med 82:1274–1276

34. Marshak TL, Maresh V, Brodskii V (1978) [Number of Purkinje cells with an increased DNA content in rat cerebellum]. Kolichestvo kletok Purkin'e s uvelichennym soderzhaniem DNK v mozzhechke krysy. Tsitolog 20: 651–656

35. Magkian I, Karalova EM (1975) [Basic factors in the polyploidization of cerebellar Purkinfe cells in chick embryogenesis. III. Kinetics of RNA and DNA in Purkinje cell nuclei]. Osenovnye faktory poliploidizatsii kletok Purkin'e mozzhechka v embriogeneze kur IIIKinetika RNK i DNK v iadrakh kletok Purkin'e. Tsitolog 17:653–659

36. Herman CJ, Lapham LW (1968) DNA content of neurons in the cat hippocampus. Science 160:537

37. Svanidze IK (1967) Cytophotometry of DNA amounts in nuclei of neurons and gliacytes of the cerebral cortex of rats at different ontogenetic stages. J Gen Biol (USSR) 28: 697–708

38. Nováková V, Sandritter W, Schlueter G (1970) DNA content of neurons in rat central nervous system. Exp Cell Res 60:454–456

39. Bregnard A, Kuenzle CC, Ruch F (1977) Cytophotometric and autoradiographic evidence for post-natal DNA synthesis in neurons of the rat cerebral cortex. Exp Cell Res 107: 151–157

40. Museridze DP, Svanidze IK (1972) Quantitative analysis of DNA in the neurons of the visual cortex of the brain at the early stages of postnatal ontogenesis of guinea pigs. Sov J Dev Biol 3:430–434

41. Kut AA, Iarygin VN (1965) Polipliodiia odnoiadernykh i dvuiadernykh neĭronov v verkhnem sheĭnom uzle krolika. [polyploidy of mononuclear and binuclear neurons in the upper cervical ganglia of the rabbit; article in Russian]. Tsitologiia 7(2):228–233

42. Swartz FJ, Bhatnagar KP (1981) Are CNS neurons polyploid? A critical analysis based upon cytophotometric study of the DNA content of cerebellar and olfactory bulbar neurons of the bat. Brain Res 208:267–281

43. Kuenzle CC, Bregnard A, Hübscher U, Ruch F (1978) Extra DNA in forebrain cortical neurons. Exp Cell Res 113:151–160

44. Bregnard A, Ruch F, Lutz H, Kuenzle CC (1979) Histones and DNA increase synchronously in neurons during early postnatal development of the rat forebrain cortex. Histochemistry 61:271–279

45. Morselt AF, Braakman DJ, James J (1972) Feulgen-DNA and fast-green histone estimations in individual cell nuclei of the cerebellum of young and old rats. Acta Histochem 43: 281–286

46. Cohen J, Mares V, Lodin Z (1973) DNA content of purified preparations of mouse Purkinje neurons isolated by a velocity sedimentation technique. J Neurochem 20:651–657

47. Mann DM, Yates PO (1973) Polyploidy in the human nervous system: part 1. The DNA content of neurones and glia of the cerebellum. J Neurol Sci 18:183–196

48. Mann DM, Yates PO, Barton CM (1978) The DNA content of Purkinje cells in mammals. J Comp Neurol 180:345–347

49. Fujita S (1974) DNA constancy in neurons of the human cerebellum and spinal cord as revealed by Feulgen cytophotometry and cytofluorometry. J Comp Neurol 155:195–202

50. Fukuda M, Bohm N, Fujita S (1978) Cytophotometry and its biological application. Prog Histochem Cytochem 11:1–119

51. Fujita S, Fukuda M, Kitamura T, Satoru Y (1972) Two-wave-length-scanning method in Feulgen Cytophotometry. Acta Histochem Cytochem 5:146–152

52. Duijndam WA, Smeulders AW, van Duijn P, Verweij AC (1980) Optical errors in scanning stage absorbance cytophotometry. I. Procedures for correcting apparent integrated absorbance values for distributional, glare, and diffraction errors. J Histochem Cytochem 28:388–394

53. Duijndam WA, van Duijn P, Riddersma SH (1980) Optical errors in scanning stage absorbance cytophotometry. II. Application of correction factors for residual distributional error, glare and diffraction error in practical cytophotometry. J Histochem Cytochem 28:395–400

54. Brodsky VJ, Marshak TL, Mares V, Lodin Z, Fülöp Z, Lebedev EA (1979) Constancy and variability in the content of DNA in cerebellar Purkinje cell nuclei. A cytophotometric study. Histochemistry 59:233–248

55. Mares V, van der Ploeg M (1980) Cytophotometric re-investigation of DNA content in Purkinje cells of the rat cerebellum. Histochemistry 69:161–167

56. Marshak TL, Mares V, Brodsky VY (1985) An attempt to influence DNA content in postmitotic Purkinje cells of the cerebellum. Acta Histochem 76:193–200

57. Costain G, Lionel AC, Ogura L, Marshall CR, Scherer SW, Silversides CK, Bassett AS (2016) Genome-wide rare copy number variations contribute to genetic risk for transposition of the great arteries. Int J Cardiol 204:115–121. doi:10.1016/j.ijcard.2015.11.127

58. Mosch B, Morawski M, Mittag A, Lenz D, Tarnok A, Arendt T (2007) Aneuploidy and DNA replication in the normal human brain and Alzheimer's disease. J Neurosci 27:6859–6867

59. Cui C, Shu W, Li P (2016) Fluorescence in situ hybridization: cell-based genetic diagnostic and research applications. Front Cell Dev Biol 4:89. doi:10.3389/fcell.2016.00089

60. Morillo SM, Escoll P, La HA d, Frade JM (2010) Somatic tetraploidy in specific chick retinal ganglion cells induced by nerve growth factor. Proc Natl Acad Sci U S A 107:109–114

61. Peterson SE, Yang AH, Bushman DM, Westra JW, Yung YC, Barral S, Mutoh T, Rehen SK, Chun J (2012) Aneuploid cells are differentially susceptible to caspase-mediated death during embryonic cerebral cortical development. J Neurosci 32:16213–16222

62. Potter H (1991) Review and hypothesis: Alzheimer disease and down syndrome--chromosome 21 nondisjunction may underlie both disorders. Am J Human Gen 48:1192–1200

63. Geller LN, Potter H (1999) Chromosome missegregation and trisomy 21 mosaicism in Alzheimer's disease. Neurobiol Dis 6:167–179

64. Iourov IY, Vorsanova SG, Yurov YB (2006) Chromosomal variation in mammalian neuronal cells: known facts and attractive hypotheses. Int Rev Cytol 249:143–191

65. Iourov IY, Vorsanova SG, Yurov YB (2008) Chromosomal mosaicism goes global. Mol Cytogen 1:26

66. Boeras DI, Granic A, Padmanabhan J, Crespo NC, Rojiani AM, Potter H (2008) Alzheimer's presenilin 1 causes chromosome missegregation and aneuploidy. Neurobiol Aging 29:319–328

67. Granic A, Padmanabhan J, Norden M, Potter H (2010) Alzheimer Abeta peptide induces chromosome mis-segregation and aneuploidy, including trisomy 21: requirement for tau and APP. Mol Biol Cell 21:511–520

68. Lenz D, Mosch B, Bocsi J, Arendt T, Tarnok A (2004) Assessment of DNA replication in central nervous system by laser scanning cytometry. In: Nicolau DV, editor. Imaging, manipulation, and analysis of biomolecules, cells, and tissues II. 27–28 January 2004, San Jose, California, USA, Bellingham, WA, USA. Proc SPIE 5322:146–156

69. Arendt T, Mosch B, Morawski M (2009) Neuronal aneuploidy in health and disease: a cytomic approach to understand the molecular individuality of neurons. Int J Mol Sci 10:1609–1627

70. Dubois B, Feldman HH, Jacova C, Cummings JL, Dekosky ST, Barberger-Gateau P et al (2010) Revising the definition of Alzheimer's disease: a new lexicon. Lancet Neurol 9:1118–1127

71. Dubois B, Feldman HH, Jacova C, Dekosky ST, Barberger-Gateau P, Cummings J et al (2007) Research criteria for the diagnosis of Alzheimer's disease: revising the NINCDS-ADRDA criteria. Lancet Neurol 6:734–746

72. Dubois B, Feldman HH, Jacova C, Hampel H, Molinuevo JL, Blennow K et al (2014) Advancing research diagnostic criteria for Alzheimer's disease: the IWG-2 criteria. Lancet Neurol 13:614–629

73. Albert MS, DeKosky ST, Dickson D, Dubois B, Feldman HH, Fox NC et al (2011) The

diagnosis of mild cognitive impairment due to Alzheimer: disease: recommendations from the National Institute on Aging—Alzheimer's association workgroups on diagnostic guidelines for Alzheimer's disease. Alzheimers Dement 7:270–279

74. Jack CR Jr, Albert MS, Knopman DS, McKhann GM, Sperling RA, Carrillo MC et al (2011) Introduction to the recommendations from the National Institute on Aging–Alzheimer's association workgroups on diagnostic guidelines for Alzheimer's disease. Alzheimers Dement 7:257–262

75. McKhann GM, Knopman DS, Chertkow H, Hyman BT, Jack CR Jr, Kawas CH et al (2011) The diagnosis of dementia due to Alzheimer's disease: recommendations from the National Institute on Aging–Alzheimer's association workgroups on diagnostic guidelines for Alzheimer's disease. Alzheimers Dement 7: 263–269

76. Sperling RA, Aisen PS, Beckett LA, Bennett DA, Craft S, Fagan AM et al (2011) Toward defining the preclinical stages of Alzheimer's disease: recommendations from the National Institute on Aging–Alzheimer's association workgroups on diagnostic guidelines for Alzheimer's disease. Alzheimers Dement 7:280–292

77. Hyman BT, Phelps CH, Beach TG, Bigio EH, Cairns NJ, Carrillo MC et al (2010) National Institute on Aging–Alzheimer's association guidelines for the neuropathologic assessment of Alzheimer's disease. Alzheimers Dement 8:1–13

78. Montine TJ, Phelps CH, Beach TG, Bigio EH, Cairns NJ, Dickson DW et al (2012) National Institute on Aging–Alzheimer's association guidelines for the neuropathologic assessment of Alzheimer's disease: a practical approach. Acta Neuropathol 123:1–11

79. Morris JC, Heymann A, Mohs RC et al (1989) The consortium to establish a registry for Alzheimer's disease (CERAD). Part I. Clinical and neuropsychological assessment of Alzheimer's disease. Neurology 39:1159–1165

80. Folstein MF, Folstein SE, McHugh PR (1975) Mini-mental state (a practical method for grading the state of patients for the clinician). J Psychiatr Res 12:189–198

81. Reisberg B, Ferris SH, de Leon MJ, Crook T (1982) The global deterioration scale for assessment of primary degenerative dementia. Am J Psychiatry 139(9):1136–1139

82. Hughes CP, Berg L, Danziger WL, Coben LA, Martin RL (1982) A new clinical scale for the staging of dementia. Br J Psychiatry 140: 566–572

83. Braak H, Braak E (1991) Neuropathological stageing of Alzheimer-related changes. Acta Neuropathol 82:239–259

84. Braak H, Alafuzoff I, Arzberger T, Kretzschmar H, Del Tredici K (2006) Staging of Alzheimer disease-associated neurofibrillary pathology using paraffin sections and immunocytochemistry. Acta Neuropathol 112:389–404

85. Thal DR, Rub U, Orantes M, Braak H (2002) Phases of a betadeposition in the human brain and its relevance for the development of AD. Neurology 58:1791–1800

86. Mirra SS, Heyman A, McKeel D, Sumi SM, Crain BJ, Brownlee LM et al (1991) The consortium to establish a registry for Alzheimer's disease (CERAD). Part II. Standardization of the neuropathologic assessment of Alzheimer's disease. Neurology 41:479–486

87. Evers P, Uylings HB (1997) An optimal antigen retrieval method suitable for different antibodies on human brain tissue stored for several years in formaldehyde fixative. J Neurosci Methods 72:197–207

88. Walker JA, Kilroy GE, Xing J, Shewale J, Sinha SK, Batzer MA (2003) Human DNA quantitation using Alu element-based polymerase chain reaction. Anal Biochem 315: 122–128

89. Houck CM, Rinehart FP, Schmid CW (1979) A ubiquitous family of repeated DNA sequences in the human genome. J Mol Biol 132:289–306

90. Batzer MA, Deininger PL (2002) Alu repeats and human genomic diversity. Nat Rev Genet 3:370–379

91. Roy-Engel AM, Carroll ML, Vogel E, Garber RK, Nguyen SV, Salem AH, Batzer MA, Deininger PL (2001) Alu insertion polymorphisms for the study of human genomic diversity. Genetics 159:279–290

Part III

DNA Copy Number Variation

Chapter 6

Single-Cell CNV Detection in Human Neuronal Nuclei

Margaret B. Wierman, Ian E. Burbulis, William D. Chronister, Stefan Bekiranov, and Michael J. McConnell

Abstract

Genomic mosaicism is prevalent throughout human somatic tissues and is much more common than previously thought. Here, we describe step-by-step methods to isolate neuronal nuclei from human brain and identify megabase-scale copy number variants (CNVs) in single nuclei. The approach detailed herein includes use of CellRaft technology for single-nucleus isolation, the PicoPLEX approach to whole-genome amplification and library preparation, and a pooled library purification protocol, termed Gel2Gel, which has been developed in our laboratory. These methods are focused toward neuroscience research, but are adaptable to many biomedical fields.

Key words Copy number variation, Whole-genome amplification, Single-cell genome sequencing, Neurons, And somatic mosaicism

1 Introduction

Somatic mosaicism refers to the presence of two or more genetically distinct cell lines constituting the somatic tissues of a monozygotic individual. The original observation that genotype can vary between cells of an individual animal was reported by Curt Stern in 1931 when he showed variable recombination events between tissues of the same fruit fly [1]. He also demonstrated that genetic recombination can take place during mitosis in addition to meiosis [1–4]. When these events occur, it results in clonal lineages of unique genotypes distributed throughout somatic (body) tissues, i.e., an organism that contains two or more genetically distinct cell lineages. The term "somatic mosaicism," used to describe organisms in which genomic content varies on a cell-to-cell basis, was first used by C. W. Cotterman in his seminal 1956 paper on antigenic variation [5], fairly remarkable given that these findings were postulated before the karyotype for humans was determined in 1956 [6].

Margaret B. Wierman, Ian E. Burbulis, and William D. Chronister contributed equally to this work.

José María Frade and Fred H. Gage (eds.), *Genomic Mosaicism in Neurons and Other Cell Types*, Neuromethods, vol. 131, DOI 10.1007/978-1-4939-7280-7_6, © Springer Science+Business Media LLC 2017

Somatic mosaicism is prevalent in human tissues [7–11]. In the brain, where long-lived neurons control behavior, the consequences of somatic mosaicism might be significant, making brain somatic mosaicism a growing area of research [8–10, 12–19]. Collectively, these studies indicate that every human neuron is likely to have a unique genome [20]. Each neuron's genome contains ~1000 somatic single-nucleotide variants and 1 de novo retrotransposon insertion [21, 22]. In addition, between 10 and 40% of neurons contain at least 1 megabase (Mb)-scale de novo copy number variant (CNV) affecting >10 genes [23–25]. Given these findings and their potential physiological impact, single-cell approaches, such as whole-genome amplification [26–29], are vital to continue the study of brain somatic mosaicism [30–32].

A variety of molecular biology techniques may be applied to individual cells or nuclei depending on the biological question. Where single-nucleus sequencing is concerned, whole-genome amplification (WGA) is applied to create quantities of DNA sufficient for preparing sequencing libraries. We have had success using both homemade and commercial whole-genome amplification recipes to detect CNVs. Here, we recommend the PicoPLEX system (a hybrid of the common DOP-PCR and MDA genome amplification techniques) for routine CNV detection because of the ease of use and consistent data quality. The PicoPLEX protocol amplifies genomic material and appends the i5 and i7 index adaptors to library fragments necessary for sequencing in a simple three-step process that takes less than 3 h. Overall, the biochemistry of this approach reduces the frequency of technical artifacts known to distort genomic data and reduce confidence of CNV detection.

We found that the stringency of size selection applied during library cleanup affected sequence quality and data usability. Size selection of DNA fragments and purification of libraries are critical steps in the workflow. Even the best made libraries may yield unreadable data if the cleanup steps are not done correctly. To this end, we present a polyacrylamide-based size selection method that we term "Gel2Gel," which has routinely produced more readable DNA fragments than the commonly used Ampure magnetic bead system recommended by Illumina. Sequence data are processed using one or more publicly and/or commercially available platforms for data mining and hypothesis testing. Here we present our lab-developed analytics to detect and visualize genome-wide CNVs in single-cell human data sets in relation to these other platforms.

These procedures yield data with excellent bioinformatic statistics and are suitable for many types of sequencing projects beyond the scope of single-cell analysis. The DNA libraries are sequenced on the Illumina platform in pools designed to produce at least 10^6 reads per cell. Meeting this threshold typically ensures sufficient genomic coverage to accurately identify copy number states at loci across the genome and avoid distortion of the copy number profile of the original genome that can be caused by smaller, and more highly variable, read counts. Following sequencing, we used a custom bioinformatics pipeline to carry out read alignment, read binning, and copy number segmentation, the details of which are described in this chapter. This analysis generates copy number estimates from read depth for several thousand genomic bins into which reads are mapped; however, the number of bins varies depending on bin size, which is determined by the bioinformatician. These are then used to identify CNV regions. As an alternative or a complementary approach, the online tool Ginkgo can also be employed to detect CNVs following the creation of BED files. Ginkgo's analysis is performed in a similar fashion to ours but with a few key differences that are explained in section "CNV Detection Using Ginkgo."

2   Materials

Innovation in our laboratory is focused on developing methods for achieving better, faster, and cheaper isolation, amplification, and analysis of single-neuronal genomes. As such, the materials required to complete these procedures through data analysis have been selected from an inventory of basic equipment and common supplies that even modest labs possess.

2.1   Nuclei Isolation and Immunostaining

Several buffer solutions are required to perform these methods that include nuclei isolation media (NIM): 25 mM KCl, 5 mM $MgCl_2$, 10 mM Tris-Cl (pH 8.8), 250 mM sucrose, and 1 mM dithiothreitol (DTT). Throughout the procedures, the use of proteinase inhibitors will be routine. We recommend using complete EDTA-free protease inhibitor cocktail by dissolving 1 tablet in 1 mL H_2O to yield a 50× stock solution (available in glass vials from Roche, Basel, Switzerland). During the isolation of nuclei we used OptiPrep Diluent for Nuclei (ODN): 150 mM KCl, 30 mM $MgCl_2$, 60 mM Tris-Cl (pH 8.8), and 250 mM sucrose. For distinguishing live from dead cells we used a trypan blue solution of 0.4% (Thermo Fisher Scientific, Waltham, MA). We stored isolated nuclei in nuclei storage buffer (NSB): 5 mM $MgCl_2$, 50 mM Tris-Cl (pH 8.8), 166 mM sucrose, and 1 mM dithiothreitol (DTT). We blocked nonspecific binding sites in samples from interfering substances by using blocking buffer (BB): 1× PBS, 1.0% bovine serum albumin (BSA), and 0.1% Tween 20. We used the mouse monoclonal anti-NeuN IgG (clone A60) antibody conjugated to Alexa Fluor 555 (MAB377A5, EMD Millipore) to detect neuronal nuclei. We used SYTO 13 green fluorescent nucleic acid stain (Thermo Fisher Scientific, Waltham, MA) at a final concentration of 500 nM to visualize DNA inside nuclei. This measurement ensures that the genomic material is intact within the isolated nuclei. We preferred using a Triton X-100 stock pre-diluted to a 10% solution for accurate measure. We used OptiPrep Density Gradient Medium (ODGM; 60% iodixanol solution; Sigma-Aldrich, St. Louis, MO) to cushion nuclei during centrifugation. We used the following equipment and supplies to perform these investigations: polypropylene Falcon tube (5 mL; round bottom), Polytron PT 1300 D Manual Disperser (handheld tissue homogenizer; Kinematica Inc., Bohemia, NY), Dounce tissue grinder set (2 mL, glass, 2 pestles; Sigma-Aldrich, St. Louis, MO), 1 mL syringe without needle, ultracentrifuge with compatible rotor, and 3–5 mL, transparent, round-bottom tubes (Our lab uses a Beckman L8-M ultracentrifuge with SW55 Ti rotor and 5 mL polyallomer, thin-wall tubes, but any ultracentrifuge with swinging buckets accommodating 3–5 mL round-bottom tubes will suffice).

2.2 Nuclei Capture, Whole-Genome Amplification, and Library Preparation

A standard thermocycler suitable for routine PCR is sufficient to perform these experiments. We generally use 48-well PCR plates or thin-wall PCR strip tubes with plastic covers or caps, respectively. For imaging fluorescently stained nuclei, we used the EVOS FL Cell Imaging System (Thermo Fisher Scientific, Waltham, MA). The 100 μm × 100 μm CellRaft arrays are available from Cell Microsystems Inc. (Research Triangle Park, NC). The CellRaft array release device is affixed to an inverted benchtop microscope with 4× or 10× objective according to the manufacturer's instructions (Cell Microsystems Inc.). The magnetic raft retrieval wand and release platform are available from Cell Microsystems Inc. as part of their standard platform kit. The PicoPLEX DNA sequencing kit is required for some of the library preparation steps described in this protocol: cell extraction buffer—green cap, extraction enzyme dilution buffer—violet cap, cell extraction enzyme—yellow cap, pre-amp buffer—red cap, pre-amp enzyme—white cap, amplification buffer—orange cap, amplification enzyme—blue cap, nuclease-free water—clear cap, and dual-index plate (Rubicon Genomics, Ann Arbor, MI). For procedures requiring vertical polyacrylamide electrophoresis we used the Mini-Protean Tetra Handcast system (10-well, 0.75 mm thickness) with the Mini-Protean Tetra Handcast Vertical Electrophoresis Cell (Bio-Rad Laboratories, Hercules, CA). Any well-functioning gel electrophoresis power supply may suffice for both vertical and horizontal gel electrophoresis. We composed acrylamide gels according to the following recipe: 1 × TAE buffer (40 mM Tris, 20 mM acetic acid, 1 mM EDTA), 7.5% acrylamide, 0.05% ammonium persulfate, and 0.002% TEMED. We used TrackIt Cyan/Orange 6× DNA Loading Buffer (Thermo Fisher) to gauge electrophoresis in real time. We routinely used the 100 bp DNA ladder from New England BioLabs during electrophoresis, but any standard DNA ladder may be used for this purpose. We recommend using 500 nM

SYBR gold nucleic acid gel stain (10,000× in DMSO; Thermo Fisher Scientific, Waltham, MA) to visualize DNA in polyacrylamide gels during the Gel2Gel procedure. This dye yields superior signal-to-noise compared to other DNA stains; however it is important not to over-stain the gel (>5 min) and to wash away the talcum powder used during the manufacture of disposable latex (or nitrile) gloves before handling the gel because this dusty residue will leave fingerprints on the gel surface that cannot be removed. A UV light (either a handheld UV light or a stationary UV transilluminator box will work) will be required for visualizing DNA in gels. We used the wide Mini-Sub Cell GT horizontal electrophoresis system, 15 cm × 7 cm tray with casting gates (15 wells; Bio-Rad Laboratories, Hercules, CA) for casting horizontal agarose gels. The agarose gel composition we used throughout this investigation was 1.0% ultra-pure low-melting-point agarose (Thermo Fisher Scientific, Waltham, MA), 1 × TAE buffer (see above), and 0.2 μg/mL ethidium bromide. For isolating DNA from low-melt agarose slices we used QIAquick PCR columns and the "gel extraction" method described in the standard instructions for these columns from Qiagen (Hilden, Germany). Stock solutions of general chemical reagents and common disposable lab equipment can be found at Sigma-Aldrich (St. Louis, MO) or Thermo Fisher Scientific (Waltham, MA).

2.3 Bioinformatic Pipeline

High-performance computing environment with at least 10 GB of memory is recommended. More memory may be required when working with larger files.

For full details on specific computer programs, please refer to Table 1.

Table 1

Bioinformatics software and programs

Software	Version used in pipeline	Most recent version
Bedtools	2.17.0	2.26.0
BWA	0.75a	0.7.15
FastQC	0.11.3	0.11.5
FASTX Toolkit	0.0.13	0.0.13
Java	1.6.0_43	8u121
Picard Tools	1.105	2.8.2
Python	2.6.6	2.7.13 / 3.6.0*
R	3.0.2	3.3.2
Samtools	0.1.18	1.3.2

3   Methods

Single-cell CNV detection in neurons is a multistep process with many technical challenges. First, many cell types comprise neural tissue, necessitating means of clearly distinguishing neuronal from non-neuronal nuclei. Second, limited availability of primary tissue necessitates methods that isolate nuclei from extremely small biopsies. Third, nuclear contents must be enzymatically amplified via whole-genome amplification (WGA) to generate enough material for constructing DNA libraries. Fourth, WGA must be even across genome space and efficient so as to limit generation of CNV artifacts that distort genomic contents. Fifth, compatibility with multiplex-based platforms is necessary for sequencing large numbers of samples simultaneously. Sixth, WGA products must be cleaned and selected to make fragment libraries conducive to maximizing the amount of usable data. Seventh, information must be extracted from the sequencing data to visualize genomic structure, interpret measurements in the context of the tissue, and yield useful conclusions about the biological system. The methods we detail herein (Fig. 1) provide a step-by-step analysis of human

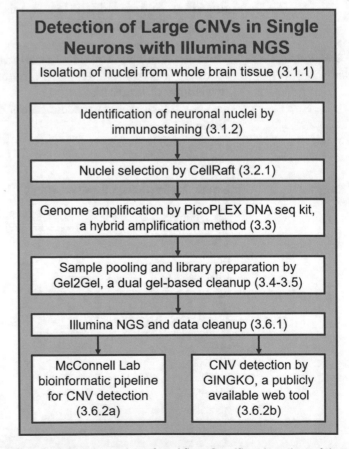

Fig. 1 Flowchart: broad overview of workflow. Specific subsections of the protocol are noted in parentheses

postmortem neurons, but are applicable to many lines of research. These methods are complementary to other molecular-based approaches to assess CNVs in single neurons as described in Chaps. 7 and 8.

3.1 Preparation and Staining of Nuclei from Postmortem Brain Tissue Overview

Long neural processes and abundant extracellular material preclude high-efficiency isolation of intact single neurons from frozen brain samples. Fortunately, our capture and amplification methods sequence individual nuclei as effectively as whole cells. This facilitates CNV analysis in single-neuronal genomes in that whole-cell isolation, per se, is not necessary to sequence the genome of individual neurons. Due to the limited availability of human neuronal material, we maximized nuclei yield per mass brain sample when developing our protocol, a stepwise mechanical disruption of brain tissue to release nuclei.

3.1.1 Disruption of Tissue and Isolation of Single Nuclei

To begin, transfer 50–100 mg of tissue to a 1.5 mL microcentrifuge tube containing 1 mL of nuclei isolation media (NIM). It is important that all reagents and samples be kept on ice throughout the protocol. Using a razor blade, cut off the end of a disposable 1000 μL pipette tip to create an opening large enough to allow the tissue to be taken up. Gently triturate the NIM solution to reduce the tissue to smaller pieces. If needed, the process can be repeated with another pipette tip cut to a slightly smaller size to further dissociate the tissue. Proceed once the tissue can be transferred with an uncut 1000 μL pipette tip to a 5 mL round-bottom, polypropylene Falcon tube. After transferring the entire volume of tissue and NIM, homogenize the sample with a Polytron tissue homogenizer. The sample solution should appear opaque and homogeneous by this point. Add 10 μL 10% Triton X-100 solution to the sample to attain a 0.1% final concentration and transfer to a Dounce homogenizer. It is ideal to use a Dounce with two pestle sizes, one small enough to disrupt tissue further and the other large enough to lyse individual cells. After using the Dounce homogenizer, transfer the sample to a new 1.5 mL microfuge tube and spin at $1000 \times g$ for 8 min at 4 °C. Remove the supernatant and resuspend pellet in 1 mL 6:5:1 NIM: OptiPrep Density Gradient Medium: OptiPrep Diluent for Nuclei. This will yield a final concentration of 25% iodixanol. To separate the nuclei from other cell components, layer the sample onto 1 mL 29% iodixanol solution (29:31 ODGM:ODN) in an ultracentrifuge tube. To avoid mixing the different layers during loading, use a 1 mL syringe without a needle to slowly load and suspend the sample on top. Centrifuge the tube at $10,300 \times g$ for 20 min at 4°. After the spin, remove the supernatant, being sure to remove the visible residual cell debris near the interface and leaving ~50 μL in the bottom of the ultracentrifuge tube. The pelleted nuclei may not be visible to the naked eye, but are easily disturbed; be careful not to disrupt the pellet.

Depending on the immediacy of use, the bulk of the nuclei can be resuspended in either blocking buffer for immunostaining (described below) or nuclei storage buffer (NSB). To confirm the presence of nuclei and check the purity of the prep by microscope, a vital dye such as trypan blue can be used. In NSB and stored at 4°, intact nuclei can be preserved for several weeks.

3.1.2 Immunostaining for the Identification of Neuronal Nuclei

To ensure that the isolated nuclei contain genomic material and distinguish neuronal from non-neuronal nuclei, the preparation is stained for DNA content and a neuron-specific nuclear splicing factor, NeuN [33–35]. Specifically, suspend the bulk nuclei in PBS containing 0.1% Tween and 1.0% BSA and incubate for 1 h at 4 °C with gentle agitation. Add syto-13 nucleic acid stain (50 nM) and AF555-conjugated anti-NeuN antibody (1:500, 20 µg/mL) and incubate with gentle agitation overnight at 4 °C. It should be noted that this combination of fluorophores was chosen for its compatibility with the GFP and RFP filter modules of the EVOS FL Cell Imaging System, allowing for independent determination of both DNA content and presence of NeuN by fluorescent visualization. We have also used other compatible fluorophore combinations and other filter modules for the same purpose, such as AF488-conjugated anti-NeuN antibody (1:500, 20 µg/mL) and DAPI (10 µg/mL). Nuclear markers for cell types in other lines of research may be used here according to research needs.

3.2 Single-Nuclei Isolation

The standard approach to obtain single cells is fluorescence-activated cell sorting (FACS); in the case of nuclei sorting this process is sometimes referred to as fluorescence-activated nuclei sorting (FANS). Regardless, the instrument and procedure are the same. There are additional descriptions of nuclei isolation methodologies in Chaps. 3, 4, and 13. Given the broad familiarity with FACS, we prefer to avoid confusion and use the more general term. FACS is a straightforward approach to obtain single cells or nuclei dispensed into a 96- or 384-well microtiter plate. This works well when the experimental design requires hundreds of samples. However, it also requires excess sample as 50,000 cells or nuclei must be expended to establish sorting parameters. Furthermore, FACS is not practical in assay development experiments requiring small numbers of samples or multiple, sequential experiments in the course of a day.

To reduce cost, increase flexibility, and maximize the number of nuclei captured from a single-nuclei prep, we adapted the CellRaft system from Cell Microsystems (Research Triangle Park, NC). Briefly, the CellRaft is a petri dish-sized annotated array made of thousands of magnetic microwells designed to disperse and capture single cells (Fig. 2a). These microwells can be released one at a time through a release device which can be affixed to most inverted lab

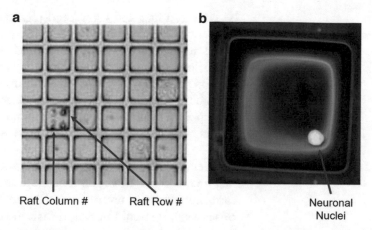

Fig. 2 The CellRaft is an effective method of single-nuclei isolation. (**a**) The CellRaft (100 μm × 100 μm array) is an annotated grid made up of individual magnetic, polystyrene bins. Consisting of 40,000 rafts total, these bins are easily visible by 10× objective of a benchtop microscope under bright field. (**b**) Once a nuclei suspension has been dispersed over the surface of the CellRaft, individual nuclei settle into bins and are identifiable by immunostaining. This photo shows a composite image (GFP and RFP fluorescence) of an individual well containing a single-neuronal nuclei that has been stained with SYTO 13 green nucleic acid stain and Alexa Fluor 555-conjugated anti-NeuN antibody

microscopes with a 4× or 10× objective. The raft with the nuclei of interest is then manually retrieved and removed by a small magnetic probe. It can then be released into a buffer/tube of choice (described in greater detail below). Although the sizes of the wells can vary depending on the array, the 100 μm × 100 μm array contains 40,000 individual wells and is ideal for separating and isolating single cells or nuclei. Unlike FACS, the CellRaft system can be used to select single samples from as few as 2000 total nuclei.

Additionally, the ability to visualize and selectively pick single nuclei (or cells) holds great advantage over methods like flow sorting or limiting dilution because there is an extra level of quality control in choosing which nuclei get sequenced. With microscopic observation, the experimenter may directly see if there are any defects in the subject or whether cell debris has been mistaken as a nucleus due to similar forward and side scatter. This approach also affords application of immunofluorescence to increase discriminating cell types. Lastly, this path does not obligate the researcher to handle 100s to 1000s of nuclei; just a few nuclei may be analyzed to determine which experimental treatments are most appropriate for the subject material. In this manner, the costs of troubleshooting methods are highly reduced. However it should be mentioned that the methods detailed here are not mutually exclusive. Nuclei that have been flow-sorted based on any number of immuno-epitopes, or separated/enriched based on other morphological features, may be collected first and then hand-picked as a second level of visual scrutiny to ensure analysis of the desired cellular subtype. An additional advantage of using the CellRaft system compared to traditional flow sorting, or limited dilution, is that each individual nucleus may be photographed and curated during the experimental procedure. These photographic inventories are extremely helpful during retrospective analysis of sequencing data in regard to unanticipated results obtained after sequence analysis is finished.

3.2.1　Nuclei Capture by CellRaft

To prepare the rafts for nuclei, a protective layer of glucose must first be removed. Cover and incubate the raft with 1 mL of ddH$_2$O for 5 min at room temperature. Remove the water and repeat the rinse twice. It should be noted that unlike adherent cells that spontaneously attach to the polystyrene rafts after application and remain attached upon release, nuclei must be "adhered" to the raft so as not to fall off during the transfer to individual tubes. We create a "sticky" surface on the rafts by building a layer of purified recombinant protein adhesive, originally discovered to "glue" barnacles to solid surfaces and now commercialized by Corning as Cell-Tak cell and tissue adhesive. Apply the Cell-Tak to the raft in 1 mL alkalized PBS (PBS, 2.5 mM NaOH, 15 μg/mL Cell-Tak) according to the basic absorption coating protocol detailed by Corning and incubate at room temperature for 45 min. Remove the Cell-Tak solution and rinse the raft three more times with ddH$_2$O. Nuclei should then be applied to the raft. Pipette anywhere between 2,000 and 40,000 nuclei in 2 mL of PBS onto the raft to achieve a workable distribution. Spreading less than 2,000 nuclei results in too sparse a distribution to efficiently locate and retrieve microwells with single nuclei (Fig. 2), and spreading over 40,000 results in multiple nuclei per well. Incubate for an hour at 4 °C to allow the nuclei to settle. Stored at 4 °C, the nuclei are stable on the raft for 1–2 weeks.

To transfer nuclei from the raft surface to a PCR tube, identify a well containing a neuronal nucleus using the appropriate fluorescent filters (Fig. 2b) and position the raft over the point of the release device. Under bright field, move the magnet-containing plastic retrieval wand over the microwell. Activate the needle on the release device to free the microwell from the raft. The raft will then attach to the tip of the retrieval wand. To release the well from the wand, place the tip of the wand holding the well into a PCR tube containing the appropriate buffer. (For use with the WGA method described below, the appropriate buffer will be the cell extraction buffer (5 μL) from the PicoPLEX DNA-seq kit.) Move the wand and tube over an opposing magnet to push the wand magnet away from the tip and release the well into the buffer. Leaving the tube over the magnet, remove the wand from the buffer. Confirm the presence of the well visually by microscope. Rinse the wand tip in sterile PBS before selecting the next well.

3.3   Whole-Genome Amplification Overview

Consisting of approximately 6.3 pg of DNA, the genomic content of a single human cell must be amplified before it can undergo NGS. Most methods of creating Illumina-compatible sequencing libraries require >1 ng of total DNA. Whole-genome amplification can be achieved by three different general methods—MDA (multiple displacement amplification) [36], DOP-PCR (degenerate-oligonucleotide-PCR) [37–39], or hybrid methods such as PicoPLEX and multiple annealing and looping-based amplification cycles (MALBAC) [40], which are combinations of the first two. Each method has its own technical challenges and pitfalls [41–43].

MDA begins with random priming and then utilizes a proofreading, strand-displacing polymerase with high processivity to exponentially replicate genomic DNA. The entire process is isothermal with the amount of product yielded determined primarily by the availability of additional nucleotides and incubation time [36]. Although this method yields product with a low mutation rate and covers the majority of the genome (~80–90%), it is prone to genomic distortion. Specifically, successive rounds of amplification result in an overabundance of copies of genomic loci that were replicated first. This makes determining the original copy number difficult, if not impossible [41]. DOP-PCR methods are based on the thermal cycle-regulated PCR amplification of randomized parts of the genomes. Although relatively well suited for evenly amplifying small sections of DNA, total genome coverage is poor (~10%) and replication less accurate at a base pair level.

The hybrid methods, PicoPLEX and MALBAC (Chap. 7), are less susceptible to genome distortion than the MDA protocols and, in the case of MALBAC, the low coverage of the DOP-PCR protocols [40, 41]. Both techniques are based on three steps: lysis, linear pre-amplification, and exponential amplification. During the lysis step, the genomic DNA is freed from its nuclear conformation. The linear pre-amplification step then amplifies the DNA by using quasi-random primers that bind at numerous loci across the genome. In addition to the genomic sequences, these primers are also designed with complementary ends that allow the amplified DNA to form loops after being released from the nuclear template. These loops are unable to be copied in subsequent annealing and elongation cycles, leaving the sample DNA as the primary template and eventually resulting in an evenly amplified library. The final exponential step denatures the looped amplicons and further PCR amplifies the library using the common primer sequences [40].

For our purposes, we have found the PicoPLEX system, specifically applied with the PicoPLEX DNA-seq Kit, to be an efficient and accurate method of WGA for the reasons described above as well as its formulated compatibility with Illumina NGS platforms. As stated above, after enzymatic lysis individual nuclei are subjected to a two-step thermocycle-based protocol. The first amplification step generates copies of genomic elements that accumulate linearly with each thermocycle. The second step attaches Illumina adaptor sequences to the termini of products from the initial step and exponentially amplifies these chimeras. The adaptor elements appended in this step are necessary for hybridization and cluster formation on the Illumina chip surface. These adaptors also contain the i5 and i7 index elements to create 48 unique dual-index barcode combinations to facilitate multiplex sequencing. In short, the genomic fragments derived from individual nuclei are appended with one of the unique i5 and i7 combinations in 2.5 h without tagmentation, fragmentation, or additional thermocycling. The instructions for the PicoPLEX recipe are well described by the manufacturer and will not be discussed further here. However, each successful PicoPLEX reaction should yield approximately 1–5 μg of DNA (primarily consisting of barcoded DNA fragments sized 200–2000 bp) in 50 μL which should be confirmed by agarose gel visualization before proceeding to the library pooling and cleanup.

3.4 Multiplex Pooling of Individual WGA Product Overview

Pooling individual WGA products together for sequencing in multiplex is necessary to maximize the number of cells analyzed per operation of the Illumina platform. Before libraries can be sequenced, individual samples with compatible indexes are combined in mixture, size-selected, and purified. The PicoPLEX DNA-seq Kit contains 48 unique barcode combinations, allowing for the construction of pooled libraries containing 2–48 individual samples. However, to ensure collection of adequate genomic data per sample, the specific Illumina platform utilized should dictate the quantity and exact combination of index-coded nuclei to be sequenced and should be chosen prior to size selection and purification. Usually 10 μL of a 4 nM solution of library is required by the Illumina platform to produce sequence information; thus the quantities and volumes of individual WGA products are chosen accordingly.

For any bioinformatic pipeline, it is crucial to have enough mappable reads per individual genome to detect CNVs with high statistical confidence. We discovered that roughly a minimum of one million mapped reads is necessary to detect copy number across all chromosomal regions. To achieve this minimum on Illumina platforms such as the NextSeq system, with capacities of producing over 400 million reads per sequencing run, pooling and sequencing 48 unique samples is both more cost effective and sufficient to produce the amount of raw data required. However, for more limited systems such as the MiSeq (25 million maximum reads), we have found that exceeding 16 samples per multiplexed library greatly increases the amount of variability within our data pipeline. For pools containing less than 48 samples, Illumina technical manuals provide guidelines for assembling compatible barcode combinations. Since the sequences of the index adaptors have been carefully chosen to consist of balanced ratios of purine and pyrimidine bases, sequencing efficiencies of these terminal elements are considered to be equivalent. The recommended Illumina bar-coding strategies should be followed for the specific sequencing machine and chip chemistry; some of these products are not compatible between machines and must first be verified.

3.4.1 Pooling WGA Products

To combine individual WGA products into one multiplex mixture for size selection, combine 10 μL of each selected crude PicoPLEX amplification product in a single 1.5 mL microfuge tube and briefly vortex. Be sure to label the tubes containing the remaining volumes of the crude WGA products (the leftover 40 μL) and store individually at −20 °C for later analysis if needed. These curated samples may be re-sequenced as necessary.

3.5 DNA Fragment Size Selection and Library Cleanup Overview

For optimal sequencing results using Illumina NGS platforms, it is recommended that libraries be composed of DNA fragments ranging from 200 to 500 bp (not including adaptors which add an additional 100 to 150 bp (depending on the specific adaptors) to each DNA amplicon) to promote efficient bridge-PCR and cluster formation. Although this size range can be exceeded slightly, sequencing accuracy begins to decrease with inserts greater than 600 bp because of reduced bridge-PCR efficiency. The PicoPLEX DNA-seq Kit produces fragments with total lengths ranging from 200 to 2000 bp (insert size ~100–1900 bp) and recommends a magnetic bead-based cleanup to enrich for fragments that fall in the specified size range. However, the control of size selection with magnetic beads is inefficient and often varies from day to day. Fragments above and below the desired size range do contaminate preparations. These fragments display i5 and i7 sequence, and as a result, compete for annealing sites on the Illumina chip surface leading to (1) clusters containing detection cycles with Q-scores below 30, (2) sequencing data of very short reads that map to reference genomes with low confidence, or (3) failed bridge-PCRs. To maximize the amount and quality of sequence data measured per operation, we developed a polyacrylamide-to-agarose gel electrophoresis technique that is used to first size-select and then purify DNA fragments (Fig. 3). Our method, referred to as Gel2Gel, outperforms magnetic bead-based size selection by stringently enriching optimally sized (300–400 bp) DNA fragments from sample pools (Fig. 3). This method increases the number of mappable reads per individual sample and decreases the variability of the MAD scores used to identify CNVs within single genomes (discussed in further detail in the bioinformatics section below) (Fig. 3c).

Gel2Gel is a two-step process (Fig. 3a). First, stoichiometric amounts of crude single-nucleus, dual-indexed libraries are combined into a single mixture providing equivalent representation of

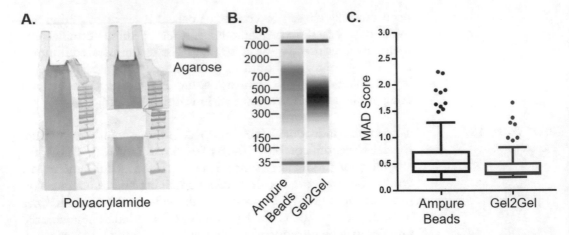

Fig. 3 Gel2Gel is a more effective protocol for cleaning up and size-selecting sequencing libraries, resulting in reduced data variability. (**a**) Pictures of the same representative sequencing library DNA pool as it is processed through Gel2Gel—showing the pool after first separation by polyacrylamide gel, after removal of the gel portion containing the desired DNA fragments, and finally after the fragments have been transferred to agarose gel. (**b**) Bioanalyzer analysis of sequencing library pools cleaned up by either Ampure beads or Gel2Gel. (**c**) Distribution of MAD scores from CNV plots generated from single-neuronal genomes of a neurotypical 26-year-old male from sequencing libraries prepared using Ampure beads ($n = 76$) or Gel2Gel ($n = 73$)

each registered genome. Remember that the combinations of indexes must be compatible for the application and platform. The DNA in this mixture is then fractionated via electrophoresis through the polyacrylamide. Second, fragments of the desired size are excised from the polyacrylamide gel and electrophoresed into a low-melt agarose gel where they are easily purified using any commercially available kit made for this purpose. The polyacrylamide step achieves size selection of library fragments and also purifies library away from contaminating primers, enzymes, and salts. However, DNA is not efficiently purified out of polyacrylamide matrices, and thus electrophoretic mobilization of DNA into low-melt agarose facilitates high-efficiency recovery of library fragments. This approach offers greater resolution than can be achieved through traditional separation by agarose gel alone.

3.5.1 Polyacrylamide-Based Size Selection of DNA Fragments

The volume of the combined libraries may exceed the maximum volume that the well of the polyacrylamide gel will accommodate. In this case, we recommend precipitating the DNA and resuspending in a smaller volume that can easily be loaded into the well of the gel. Here, precipitate total DNA using 0.22 M NaCl and 75% ethanol (final concentration) at -20 °C for 1 h. Pellet the DNA at $16,000 \times g$ for 10 min at 4 °C. Remove the supernatant and dry the DNA pellet briefly at room temperature (do not overdry). Solubilize the pellet in 20 μL of TE buffer (10 mM Tris-Cl pH 8.0, 1 mM EDTA), heat at 65 °C for 5 min, and briefly centrifuge to collect total volume. Add 4 μL of 6× DNA loading buffer (TrackIt Cyan/Orange) and keep on ice until loaded into gel. Prepare a polyacrylamide gel composed of 1 × TAE buffer and 7.5% acrylamide using the BioRad Mini-Protean Tetra Handcast system (10-well, 0.75 mm). Acrylamide polymerization is catalyzed by the addition of ammonium persulfate and TEMED and is well described by BioRad. Assemble the BioRad's Mini-Protean Tetra Handcast Vertical Electrophoresis Cell and fill with the appropriate volume of 1 × TAE buffer. Load 20 μL of pooled library into a single well. For size reference, load 2 μL of 100 bp DNA ladder into an adjacent well, leaving the immediate wells bordering the sample-containing well empty to prevent cross-contamination. Run the gel at 75 mA until the leading visible band, the Orange-G dye within the loading buffer corresponding to the ~50 bp marker, reaches the full length of gel. To visualize the nucleic acids, remove the gel from the cell and glass casting plates and incubate in 25 mL 1 × TAE buffer and 1 × Sybr gold nucleic acid gel stain. Using a UV light, scalpel or razor blade, and the 100 bp ladder, remove the portion of polyacrylamide gel containing DNA fragments from 300 to 700 bp (Fig. 3a). This should yield a small rectangular piece of polyacrylamide.

3.5.2 Polyacrylamide-to-Agarose Electrophoresis

Transfer DNA from the polyacrylamide slice into low-melt agarose by electrophoresis. Insert the slice of polyacrylamide horizontally into a large well of a low-melt agarose gel. In an empty well (not immediately adjacent to the sample) load 2 μL of the DNA ladder. For thorough transfer of DNA, there should be complete contact between the broadest surface area of the polyacrylamide section and the wall of the agarose gel well. This point is critical to the success of this procedure. Any bubbles or liquid-filled gap between the polyacrylamide gel slice and the wall of the low-melt agarose gel could result in loss of sample. Once the polyacrylamide has been inserted and the ladder loaded, run the agarose gel at 10 V/cm length between electrodes until it is obvious that all the visible dye within the DNA ladder has completely entered the gel. This will usually take about 20 min. Because the size selection has already taken place, it is neither necessary nor desirable to run this second gel too far (Fig. 3a). The intent of the agarose gel is to facilitate efficient elution of DNA from the polyacrylamide. Once the DNA fragments have completely entered the low-melt agarose, the electrophoresis may be stopped. Once again, using a UV light and razor or scalpel, extract the portion of gel containing DNA (Fig. 3a).

3.5.3 Purification of Pooled Size-Selected DNA Fragment Libraries

The DNA electrophoresed into the low-melt agarose may be efficiently purified using any commercial product. We successfully used the QIAquick gel extraction kit (Qiagen) and the QIAquick PCR columns (Qiagen) to recover size-selected pools. These

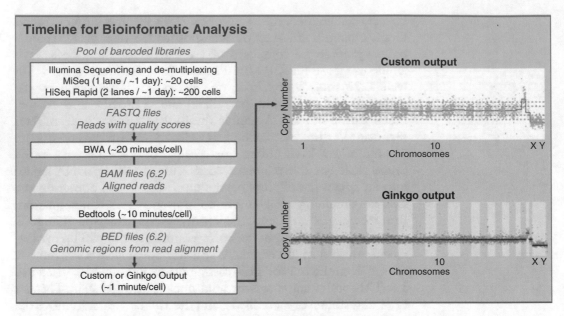

Fig. 4 Flowchart: Bioinformatics Overview. *Right-hand* panels show expected CNV profiles from trisomic male fibroblasts (+Chr21, −X) in a custom pipeline (*top*) or Ginkgo (*bottom*)

Qiagen products are easy to use and reasonably efficient. For the methods detailed herein, we used the QIAquick PCR columns for agarose gel applications according to Qiagen's instructions.

3.6 Bioinformatic Pipeline for the Identification of Large CNV Overview

The wet bench protocols are carried out with the goal of ensuring an accurate, informative final product: a CNV profile summarizing the copy number states across the genome of the single cell from which the DNA was extracted. However, good execution of the previous steps is no guarantee of success; to obtain the desired end result, the sequencing data must be first properly quality-checked and then carefully monitored for any irregularities as it passes through the analyses described below (Fig. 4).

3.6.1 Quality Control and Trimming of Sequence Data

The pool of DNA libraries is sequenced on an Illumina platform sequencer in accordance with the guidelines discussed in Sect. 3.4. After sequencing, de-multiplexing typically follows as an automated step in which all of the sequences read by the sequencer are separated by barcode and written to individual FASTQ files corresponding to each sample in the pool. Either one or two FASTQ files will be generated for each sample depending on whether single-end or paired-end sequencing was carried out. If library preparation, sample pooling, and sequencing were successful, these files will contain upwards of one million unaligned sequence reads per sample.

The first step after receiving FASTQ files is to perform FastQC (http://www.bioinformatics.babraham.ac.uk/projects/fastqc/); see Table 1 for all software versions. This is a tool that checks each file for a variety of possible sequence quality problems, such as low-quality scores (indicating that the sequencer had difficulty making conclusive base calls) or sequence content bias (indicating that the frequencies of A, C, G, and T were not consistent across each base position in the reads). FastQC flags these errors as either a "warning" or a "failure" depending on severity according to preset thresholds. If a FASTQ file grades poorly across several metrics, this typically portends poor mapping to the genome and precludes downstream analysis. While it is not uncommon for one or two libraries to grade poorly, if FastQC consistently flags problems in a single pool or across multiple pools, the issues could be due to a systemic problem. Errors that cause repeated FastQC failures can include everything from adaptor contamination during library preparation to debris inside the flowcell lane.

Next, reads are trimmed to remove amplification primers and other irregularities in relative nucleotide abundance at the ends of reads. Trimming is accomplished using the Trimmer tool from the FASTX-Toolkit (http://hannonlab.cshl.edu/fastx_toolkit/), wherein a specified range of bases is kept while the remainder of each read is removed. FastQC generates a plot of nucleotide frequency for each position across all reads which can serve as a guide for identifying bases for trimming. For PicoPLEX WGA, this plot will show irregular nucleotide frequencies for roughly the first 14 bases unless dark cycles were pre-programmed for these bases during the sequencing run.

3.6.2 CNV Detection Based on Read Depth

Once trimmed, the FASTQ files are ready for our CNV pipeline, a Python wrapper that coordinates a series of genomic operations on the input data, ultimately resulting in detection of CNVs. First, the reads are aligned to the reference human genome (hg19/GRCh37) using the Burrows-Wheeler Aligner (BWA) [44]. For this, we use the BWA "aln" command with standard options. The output SAI files are converted to BAM format using BWA "sampe" and Samtools [45] "view" commands. Next, the BAM file is sorted by reference genome position using Samtools "sort," which allows subsequent operations on the data to be more efficient. Then, all duplicate reads (reads that have the same start and end mapping location as another read) are removed using Picard Tools "MarkDuplicates.jar" (http://broadinstitute.github.io/picard/). At this point, alignment statistics are calculated using Samtools "flagstat," which provide information on the number and percentage of mapped reads. Typically, we see mapping percentages of ~95%; mapping rates that fall below 85–80% are often the result of problems that can be identified by FastQC, as discussed above.

Before the final steps of the pipeline, the BAM file is translated to BED format using Bedtools [46] "bamToBed." At this point, there are two possible routes to generate CNV profiles: our custom protocol or uploading to Ginkgo, an online tool (http://qb.cshl.edu/ginkgo/) [47]. The following sections discuss each option.

**CNV Detection Using
Custom Protocol**

Our protocol carries out CNV detection in a read depth-based fashion by binning mapped reads into custom-generated, large, nonoverlapping genomic windows of uniquely mappable sequence. First, nonunique regions of the genome (which can be determined using the hg19/GRCh37 "wgEncodeCrgMapabilityAlign40mer. bigWig" track, available from UCSC Genome Browser) are masked (i.e., changed to "N") from the full reference genome using Bedtools "maskfasta" to create a modified reference genome FASTA file representing only uniquely mappable sequence. Next, a script can be written to generate sequential bins (typically containing 500 kb) of non-N bases and generate a BED file containing genomic coordinates for the start and end of each bin. There are many ways to script this step; in calculating these bin ranges, our Python script makes use of the pysam module (https://github.com/pysam-developers/pysam) to work quickly with the masked FASTA file, which must first be indexed using Samtools "faidx." The script also records the GC percentage of each genomic bin, which is used in later steps. The resulting bins vary widely in size due to repetitive, "unmappable" features like centromeres and telomeres; thus, the average size is ~665 kb but certain bins can reach megabases in size. Smaller or larger window sizes can be generated if desired.

The BED file from each sequenced cell is then intersected (Bedtools "intersectBed") with the BED file containing bin coordinates. Then, Bedtools "coverageBed" is run to generate a file containing counts of reads mapping to each genomic bin. Using R, read counts are converted to estimated copy number states in a GC content-normalized manner. This process involves separating genomic bins by their GC content in the reference genome. Our script assigns each genomic bin to 1 of 16 groups, designed to contain roughly equal numbers of bins: <34% GC content, 34–35, 35–36, 36–37, 37–38, 38–39, 39–40, 40–41, 41–42, 42–43, 43–44, 44–45, 45–47, 47–49, 49–53, and >53%. For each GC group, the bin with the median read count is assigned a copy number state of 2. Then, the other bins in the same GC group are normalized around the median read count and a corresponding copy number value is assigned. For example, if, among genomic bins with GC content between 42 and 43%, the median number of reads is 500, any 42–43% GC bin with 500 reads will be estimated to have a copy number state of 2. Likewise, the median read count in the 47–49% GC bin may be 600 reads and also represent a copy number state of 2.

CNVs are identified by segmentation using the R package DNAcopy [48]. This tool uses a circular binary segmentation (CBS) algorithm to detect statistically significant "change-points" in copy number state across the genome. Other R packages used for segmentation include HMMcopy [49] and copynumber [50], which use alternative statistical approaches. In our standard approach, the estimated copy number values are run through a "CNA.smooth" step to moderate any outlier values, and then the "segment" command is run using the parameters alpha = 0.001, min.width = 5, and default settings otherwise. These parameters ensure that for a change-point to be called, it must have a p-value <0.001, and each changed segment must be 5 genomic bins in length. Our experience is that these parameters consistently perform well in protecting against false segmentation calls while retaining sensitivity to real CNVs.

Data are visualized by plotting copy number state versus genomic position. Segmentation output from DNAcopy is then superimposed on these data. The median absolute deviation (MAD) is calculated and plotted above and below the median copy number state to provide a general idea of the data's variation. Analytical approaches to distinguish high-confidence calls from low-confidence calls are always improving, but a default threshold that we recommend is 2 ± 2 MADs, as used previously [24]. Once the final CNV calls are made, further analysis of the duplicated or deleted regions can be explored.

CNV Detection Using Ginkgo

CNVs can also be identified using Ginkgo [47]. A minimum of three BED files, each no more than one gigabyte in size, can be uploaded to the Ginkgo website (http://qb.cshl.edu/ginkgo) and analyzed for CNVs. Customizable options are available for various parameters; for direct comparison with our pipeline, the bin type parameter should be set as "variable"; bin size as 500 kb; binning simulation read lengths as 48 bp (the closest option to our own 40 base mappability track); mapping algorithm as "BWA"; segmentation method as "Independent (normalized read counts)"; and all other options left at their default settings.

Following analysis, which can take minutes to hours depending on file number and size, Ginkgo outputs CNV profiles in addition to a clustering tree that tries to identify related samples, heatmaps highlighting any key similarities or differences between the cells analyzed, and several data tables containing bin read counts and CNV calls made, among other statistics.

We have found Ginkgo to be a useful "second opinion" in analyzing our data. An important feature of Ginkgo's analysis is its strict adherence to integer copy number states; that is, unlike DNAcopy, which allows segmentation calls at values like 2.4 and 1.3 to be made, Ginkgo's algorithm forces each segment to conform to a whole-number state. We think both approaches have merit; while it is clear that each genomic locus in a single cell cannot exist partway between two integer copy numbers, it is also conceivable that CNVs smaller than the size of a genomic bin could cause an apparent intermediate copy number state; that is, if 40% of a bin is at copy number 3 while 60% is at copy number 2, then 2.4 becomes an accurate copy number state.

Generally, we prefer our custom protocol for CNV detection given its ease of customizability and modification, but Ginkgo provides a powerful and easy-to-use alternative for copy number analysis. In ideal cases, the techniques will yield similar results, as indeed they have in the past with 99.7% bin-level agreement [47].

4   Conclusions

Brain somatic mosaicism has many unanswered questions, but widespread application of single-cell sequencing approaches is poised to lead to answers. Our protocol has produced hundreds of successful single-cell copy number profiles that have contributed to the field's still-evolving understanding of mosaicism in neurons. In addition to building the base of evidence that many neurons harbor large CNVs, with larger data sets we can begin to analytically explore hypotheses regarding the consequence of mosaic CNVs, such as the existence of CNV "hotspots" or "coldspots." The association of brain somatic mosaicism with neurological diseases also provides several avenues for focused study of brain tissue from affected individuals with the goal of identifying potential mutations that can contribute to disease. Moreover, induced pluripotent stem cells (iPSCs) from individuals with neurological diseases can be differentiated into neurons for the purpose of identifying any differences in CNV characteristics as compared to control iPSCs. Meanwhile, as mechanisms of CNV ontogeny in neurons are revealed, experiments can be performed in cultured NPCs to attempt to replicate, increase, or decrease the rate of CNV incidence during neurogenesis in vitro. In summary, the measurement of brain somatic mosaicism on a cell-by-cell basis is essential for documenting the landscape of brain somatic mosaicism. However, we also envision that accumulated data using single-cell approaches will generate new hypotheses moving forward.

References

1. Stern C (1931) Analyse eines Mosaikindividuums bei Drosophila melanogaster. Bio Zentr 51:194–199

2. Stern C (1936) Somatic crossing over and segregation in Drosophila Melanogaster. Genetics 21(6):625–730

3. Stern C, Doan D (1936) A cytogenetic demonstration of crossing-over between X- and Y-chromosomes in the male Drosophila Melanogaster. Proc Natl Acad Sci U S A 22(11):649–654

4. Stern C, Rentschler V (1936) The effect of temperature on the frequency of somatic crossing-over in Drosophila Melanogaster. Proc Natl Acad Sci U S A 22(7):451–453

5. Cotterman CW (1956) Somatic mosaicism for antigen A2. Acta Genet Stat Med 6(4):520–521

6. Levan A (1956) Chromosome studies on some human tumors and tissues of normal origin, grown in vivo and in vitro at the Sloan-Kettering institute. Cancer 9(4):648–663

7. Campbell IM, Shaw CA, Stankiewicz P, Lupski JR (2015) Somatic mosaicism: implications for disease and transmission genetics. Trends Genet 31(7):382–392. doi:10.1016/j.tig.2015.03.013

8. Campbell IM, Yuan B, Robberecht C, Pfundt R, Szafranski P, McEntagart ME, Nagamani SC, Erez A, Bartnik M, Wisniowiecka-Kowalnik B, Plunkett KS, Pursley AN, Kang SH, Bi W,

Lalani SR, Bacino CA, Vast M, Marks K, Patton M, Olofsson P, Patel A, Veltman JA, Cheung SW, Shaw CA, Vissers LE, Vermeesch JR, Lupski JR, Stankiewicz P (2014) Parental somatic mosaicism is underrecognized and influences recurrence risk of genomic disorders. Am J Hum Genet 95(2):173–182. doi:10.1016/j.ajhg.2014.07.003

9. Jacobs KB, Yeager M, Zhou W, Wacholder S, Wang Z, Rodriguez-Santiago B, Hutchinson A, Deng X, Liu C, Horner MJ, Cullen M, Epstein CG, Burdett L, Dean MC, Chatterjee N, Sampson J, Chung CC, Kovaks J, Gapstur SM, Stevens VL, Teras LT, Gaudet MM, Albanes D, Weinstein SJ, Virtamo J, Taylor PR, Freedman ND, Abnet CC, Goldstein AM, Hu N, Yu K, Yuan JM, Liao L, Ding T, Qiao YL, Gao YT, Koh WP, Xiang YB, Tang ZZ, Fan JH, Aldrich MC, Amos C, Blot WJ, Bock CH, Gillanders EM, Harris CC, Haiman CA, Henderson BE, Kolonel LN, Le Marchand L, McNeill LH, Rybicki BA, Schwartz AG, Signorello LB, Spitz MR, Wiencke JK, Wrensch M, Wu X, Zanetti KA, Ziegler RG, Figueroa JD, Garcia-Closas M, Malats N, Marenne G, Prokunina-Olsson L, Baris D, Schwenn M, Johnson A, Landi MT, Goldin L, Consonni D, Bertazzi PA, Rotunno M, Rajaraman P, Andersson U, Beane Freeman LE, Berg CD, Buring JE, Butler MA, Carreon T, Feychting M, Ahlbom A, Gaziano JM, Giles GG, Hallmans G, Hankinson SE, Hartge P, Henriksson R, Inskip PD, Johansen C, Landgren A, McKean-Cowdin R, Michaud DS, Melin BS, Peters U, Ruder AM, Sesso HD, Severi G, Shu XO, Visvanathan K, White E, Wolk A, Zeleniuch-Jacquotte A, Zheng W, Silverman DT, Kogevinas M, Gonzalez JR, Villa O, Li D, Duell EJ, Risch HA, Olson SH, Kooperberg C, Wolpin BM, Jiao L, Hassan M, Wheeler W, Arslan AA, Bueno-de-Mesquita HB, Fuchs CS, Gallinger S, Gross MD, Holly EA, Klein AP, LaCroix A, Mandelson MT, Petersen G, Boutron-Ruault MC, Bracci PM, Canzian F, Chang K, Cotterchio M, Giovannucci EL, Goggins M, Hoffman Bolton JA, Jenab M, Khaw KT, Krogh V, Kurtz RC, McWilliams RR, Mendelsohn JB, Rabe KG, Riboli E, Tjonneland A, Tobias GS, Trichopoulos D, Elena JW, Yu H, Amundadottir L, Stolzenberg-Solomon RZ, Kraft P, Schumacher F, Stram D, Savage SA, Mirabello L, Andrulis IL, Wunder JS, Patino Garcia A, Sierrasesumaga L, Barkauskas DA, Gorlick RG, Purdue M, Chow WH, Moore LE, Schwartz KL, Davis FG, Hsing AW, Berndt SI, Black A, Wentzensen N, Brinton LA, Lissowska J, Peplonska B, McGlynn KA, Cook MB, Graubard BI, Kratz CP, Greene MH, Erickson

RL, Hunter DJ, Thomas G, Hoover RN, Real FX, Fraumeni JF Jr, Caporaso NE, Tucker M, Rothman N, Perez-Jurado LA, Chanock SJ (2012) Detectable clonal mosaicism and its relationship to aging and cancer. Nat Genet 44(6):651–658. doi:10.1038/ng.2270

10. Lupski JR (2013) Genetics. Genome mosaicism-one human, multiple genomes. Science 341(6144):358–359. doi:10.1126/science.1239503

11. Stern C (1968) Genetic mosaics in animals and man. In: Genetic mosaics and other essays. Harvard University Press, Cambridge, MA

12. De S (2011) Somatic mosaicism in healthy human tissues. Trends Genet 27(6):217–223. doi:10.1016/j.tig.2011.03.002

13. Gottlieb B, Beitel LK, Trifiro MA (2001) Somatic mosaicism and variable expressivity. Trends Genet 17(2):79–82

14. Hall JG (1988) Review and hypotheses: somatic mosaicism: observations related to clinical genetics. Am J Hum Genet 43(4):355–363

15. Mkrtchyan H, Gross M, Hinreiner S, Polytiko A, Manvelyan M, Mrasek K, Kosyakova N, Ewers E, Nelle H, Liehr T, Bhatt S, Thoma K, Gebhart E, Wilhelm S, Fahsold R, Volleth M, Weise A (2010) The human genome puzzle—the role of copy number variation in somatic mosaicism. Curr Genomics 11(6):426–431. doi:10.2174/138920210793176047

16. Taylor TH, Gitlin SA, Patrick JL, Crain JL, Wilson JM, Griffin DK (2014) The origin, mechanisms, incidence and clinical consequences of chromosomal mosaicism in humans. Hum Reprod Update 20(4):571–581. doi:10.1093/humupd/dmu016

17. Thibodeau IL, Xu J, Li Q, Liu G, Lam K, Veinot JP, Birnie DH, Jones DL, Krahn AD, Lemery R, Nicholson BJ, Gollob MH (2010) Paradigm of genetic mosaicism and lone atrial fibrillation: physiological characterization of a connexin 43-deletion mutant identified from atrial tissue. Circulation 122(3):236–244. doi:10.1161/CIRCULATIONAHA.110.961227

18. Vig BK (1978) Somatic mosaicism in plants with special reference to somatic crossing over. Environ Health Perspect 27:27–36

19. Youssoufian H, Pycritz RE (2002) Mechanisms and consequences of somatic mosaicism in humans. Nat Rev Genet 3(10):748–758. doi:10.1038/nrg906

20. Insel TR (2014) Brain somatic mutations: the dark matter of psychiatric genetics? Mol Psychiatry 19(2):156–158. doi:10.1038/mp.2013.168

21. Erwin JA, Paquola AC, Singer T, Gallina I, Novotny M, Quayle C, Bedrosian TA, Alves FI, Butcher CR, Herdy JR, Sarkar A, Lasken RS, Muotri AR, Gage FH (2016) L1-associated genomic regions are deleted in somatic cells of the healthy human brain. Nat Neurosci 19(12):1583–1591. doi:10.1038/nn.4388

22. Lodato MA, Woodworth MB, Lee S, Evrony GD, Mehta BK, Karger A, Lee S, Chittenden TW, D'Gama AM, Cai X, Luquette LJ, Lee E, Park PJ, Walsh CA (2015) Somatic mutation in single human neurons tracks developmental and transcriptional history. Science 350(6256):94–98. doi:10.1126/science. aab1785

23. Cai X, Evrony GD, Lehmann HS, Elhosary PC, Mehta BK, Poduri A, Walsh CA (2014) Single cell, genome-wide sequencing identifies clonal somatic copy-number variation in the human brain. Cell Rep 8(5):1280–1289. doi:10.1016/j.celrep.2014.07.043

24. McConnell MJ, Lindberg MR, Brennand KJ, Piper JC, Voet T, Cowing-Zitron C, Shumilina S, Lasken RS, Vermeesch JR, Hall IM, Gage FH (2013) Mosaic copy number variation in human neurons. Science 342(6158):632–637. doi:10.1126/science.1243472

25. Knouse KA, Wu J, Amon A (2016) Assessment of megabase-scale somatic copy number variation using single cell sequencing. Genome Res 26(3):376–384. doi:10.1101/gr.198937.115

26. Hawkins TL, Detter JC, Richardson PM (2002) Whole genome amplification-applications and advances. Curr Opin Biotechnol 13(1):65–67

27. Hou Y, Wu K, Shi X, Li F, Song L, Wu H, Dean M, Li G, Tsang S, Jiang R, Zhang X, Li B, Liu G, Bedekar N, Lu N, Xie G, Liang H, Chang L, Wang T, Chen J, Li Y, Zhang X, Yang H, Xu X, Wang L, Wang J (2015) Comparison of variations detection between whole genome amplification methods used in single cell resequencing. Gigascience 4:37. doi:10.1186/s13742-015-0068-3

28. Han T, Chang CW, Kwekel JC, Chen Y, Ge Y, Martinez-Murillo F, Roscoe D, Tezak Z, Philip R, Bijwaard K, Fuscoe JC (2012) Characterization of whole genome amplified (WGA) DNA for use in genotyping assay development. BMC Genomics 13:217. doi:10.1186/1471-2164-13-217

29. Pugh TJ, Delaney AD, Farnoud N, Flibotte S, Griffith M, Li HI, Qian H, Farinha P, Gascoyne RD, Marra MA (2008) Impact of whole genome amplification on analysis of copy number variants. Nucleic Acids Res 36(13):e80. doi:10.1093/nar/gkn378

30. Insel TR, Gogtay N (2014) National Institute of Mental Health clinical trials: new opportunities, new expectations. JAMA Psychiat 71(7):745–746. doi:10.1001/jamapsychiatry.2014.426

31. Insel TR (2014) The NIMH research domain criteria (RDoC) project: precision medicine for psychiatry. Am J Psychiatry 171(4):395–397. doi:10.1176/appi.ajp.2014.14020138

32. Chung JY, Insel TR (2014) Mind the gap: neuroscience literacy and the next generation of psychiatrists. Acad Psychiatry 38(2):121–123. doi:10.1007/s40596-014-0054-6

33. Damianov A, Black DL (2010) Autoregulation of fox protein expression to produce dominant negative splicing factors. RNA 16(2):405–416. doi:10.1261/rna.1838210

34. Kim KK, Adelstein RS, Kawamoto S (2009) Identification of neuronal nuclei (NeuN) as fox-3, a new member of the fox-1 gene family of splicing factors. J Biol Chem 284(45):31052–31061. doi:10.1074/jbc.M109.052969

35. Lucas CH, Calvez M, Babu R, Brown A (2014) Altered subcellular localization of the NeuN/Rbfox3 RNA splicing factor in HIV-associated neurocognitive disorders (HAND). Neurosci Lett 558:97–102. doi:10.1016/j.neulet.2013.10.037

36. Dean FB, Hosono S, Fang L, Wu X, Faruqi AF, Bray-Ward P, Sun Z, Zong Q, Du Y, Du J, Driscoll M, Song W, Kingsmore SF, Egholm M, Lasken RS (2002) Comprehensive human genome amplification using multiple displacement amplification. Proc Natl Acad Sci U S A 99(8):5261–5266. doi:10.1073/pnas.082089499

37. Baslan T, Kendall J, Rodgers L, Cox H, Riggs M, Stepansky A, Troge J, Ravi K, Esposito D, Lakshmi B, Wigler M, Navin N, Hicks J (2012) Genome-wide copy number analysis of single cells. Nat Protoc 7(6):1024–1041. doi:10.1038/nprot.2012.039

38. Baslan T, Kendall J, Rodgers L, Cox H, Riggs M, Stepansky A, Troge J, Ravi K, Esposito D, Lakshmi B, Wigler M, Navin N, Hicks J (2016) Corrigendum: genome-wide copy number analysis of single cells. Nat Protoc 11(3):616. doi:10.1038/nprot0316.616b

39. Navin N, Kendall J, Troge J, Andrews P, Rodgers L, McIndoo J, Cook K, Stepansky A, Levy D, Esposito D, Muthuswamy L, Krasnitz A, McCombie WR, Hicks J, Wigler M (2011) Tumour evolution inferred by single cell sequencing. Nature 472(7341):90–94. doi:10.1038/nature09807

40. Zong C, Lu S, Chapman AR, Xie XS (2012) Genome-wide detection of single-nucleotide

and copy-number variations of a single human cell. Science 338(6114):1622–1626. doi:10.1126/science.1229164

41. de Bourcy CF, De Vlaminck I, Kanbar JN, Wang J, Gawad C, Quake SR (2014) A quantitative comparison of single cell whole genome amplification methods. PLoS One 9(8):e105585. doi:10.1371/journal. pone.0105585

42. Gawad C, Koh W, Quake SR (2016) Single cell genome sequencing: current state of the science. Nat Rev Genet 17(3):175–188. doi:10.1038/nrg.2015.16

43. Blainey PC (2013) The future is now: single cell genomics of bacteria and archaea. FEMS Microbiol Rev 37(3):407–427. doi:10.1111/1574-6976.12015

44. Li H, Durbin R (2010) Fast and accurate long-read alignment with burrows-Wheeler transform. Bioinformatics 26(5):589–595. doi:10.1093/bioinformatics/btp698

45. Li H, Handsaker B, Wysoker A, Fennell T, Ruan J, Homer N, Marth G, Abecasis G, Durbin R, 1000 Genome Project Data Processing Subgroup (2009) The sequence alignment/map format and SAMtools.

Bioinformatics 25(16):2078–2079. doi:10.1093/bioinformatics/btp352

46. Quinlan AR, Hall IM (2010) BEDTools: a flexible suite of utilities for comparing genomic features. Bioinformatics 26(6):841–842. doi:10.1093/bioinformatics/btq033

47. Garvin T, Aboukhalil R, Kendall J, Baslan T, Atwal GS, Hicks J, Wigler M, Schatz MC (2015) Interactive analysis and assessment of single cell copy-number variations. Nat Methods 12(11):1058–1060. doi:10.1038/nmeth.3578

48. Seshan VE, Olshen A (2016) DNAcopy: DNA copy number data analysis. (R package). 1.48.0. Edn.

49. Lai D, Ha G, Shah S (2016) HMMcopy: copy number prediction with correction for GC and mappability bias for HTS data (R package). 1.16.0. Edn., Bioconductor

50. Nilsen G, Liestol K, Van Loo P, Moen Vollan HK, Eide MB, Rueda OM, Chin SF, Russell R, Baumbusch LO, Caldas C, Borresen-Dale AL, Lingjaerde OC (2012) Copynumber: efficient algorithms for single- and multi-track copy number segmentation. BMC Genomics 13:591. doi:10.1186/1471-2164-13-591

Chapter 7

Multiple Annealing and Looping-Based Amplification Cycles (MALBAC) for the Analysis of DNA Copy Number Variation

Chenghang Zong

Abstract

The genomes of even close kindred cells are not identical due to various forms of genomic variations. To discover the uniqueness of each genome, we need to examine the genome at single-cell resolution. Here we describe the recent progress in the development of single-cell whole-genome amplification methods and the state of art for analyzing one of the major forms of genomic variations—copy number of variations (CNVs). Robust detection of CNVs in single cells has allowed successful clinical applications such as prenatal genome screening and diagnosis.

Key words Single-cell sequencing, Single-cell WGA, MALBAC, MDA, Copy number variations

1 Introduction

The development of single-cell whole-genome amplification (WGA) methods has made it feasible to examine the genomic variations at single-cell resolution [1–12]. However, WGA methods in general are still prone to amplification bias, which cause some regions of the genome to be overrepresented while some others underrepresented in the final amplified products. Sequencing of the biased products therefore will lead not only to the low genome coverage because that many reads are consumed in covering the overamplified regions, but also to the distorted representation of the original genome of the single cell. Such distortions will make it challenging for the reliable detection of copy number variations as well as single-nucleotide variations in single cells.

The first generation of single-cell whole-genome amplification methods are mainly polymerase chain reaction (PCR)-based reaction. However, limited by the priming efficiency and randomly introduced PCR bias at the initial stage of amplification, significant portion of genomes (over 50%) are not covered. The development

José María Frade and Fred H. Gage (eds.), *Genomic Mosaicism in Neurons and Other Cell Types*, Neuromethods, vol. 131, DOI 10.1007/978-1-4939-7280-7_7, © Springer Science+Business Media LLC 2017

of multiple displacement amplification (MDA) method has resolved the priming issue with φ29 polymerase performing the extension at 30°, which allows the efficient random hybridization between hexamers to the genome [1]. As the result, MDA can robustly cover 80% of the genome using the recently developed alkaline-based scheme for cell lysis and DNA melting before amplification [8]. However, MDA reaction depends on the production of displaced single-stranded DNA, which will randomly hybridize with each other at 30° and the products form DNA "nanoball" with the complex hyper-branched structures. The size of the DNA nano-balls varies and leads to the considerable regional bias in genome coverage, and therefore makes MDA unsuitable for reliable detection of copy number variations [7]. It is worth noting that MDA is essentially a nonlinear amplification with the exponential dependence of DNA yield on the time of reaction, which makes it difficult to accurately detect single-nucleotide variations.

To overcome both priming and amplification bias, we recently developed a new WGA method, multiple annealing and looping-based amplification cycles (MALBAC), which introduces multiple cycles of low-temperature annealing to promote efficient priming and quasilinear preamplification with the looping strategy [7]. With the production of >10 copies of amplicons by the preamplification, the initial-stage PCR bias with single copy of DNA fragments can be significantly reduced.

To allow the efficient annealing at low temperature, we design the primers with G, A, and T nucleotides only for the common 27-nucleotide sequence plus an 8-variable nucleotide (5N3G and 5N3T) at 3′ end. We start the amplification with these primers, which can evenly hybridize to the templates at 0 °C. At an elevated temperature of 65 °C, DNA polymerases with strand displacement activity (Bst polymerase) are used to generate semiamplicons with variable lengths (0.5–1.5 kb), which are then melted off from the template at 94 °C. Amplification of the semiamplicons gives full amplicons that have complementary ends. The temperature is cycled to 58 °C to allow the looping of full amplicons, which prevents the amplification of the amplicons in the following cycle. Five cycles of preamplification are followed by exponential amplification of the full amplicons by PCR to generate micrograms of DNA required for next-generation sequencing (Fig. 1). In the PCR, oligonucleotides with the common 27-nucleotide sequence are used as the primers.

In the recent study, we performed the single-cell amplification with single-SW480 cancer cells using both MDA and MALBAC. With ~25× mean sequencing depth, we consistently achieved ~85% and up to 93% genome coverage at ≥1× depth on either strand (Fig. 2a). As a comparison, we performed MDA on a single cell from the same cancer cell line. At 25× mean sequencing depth, MDA covered 72% of the genome even at low depth (~1×). Substantial

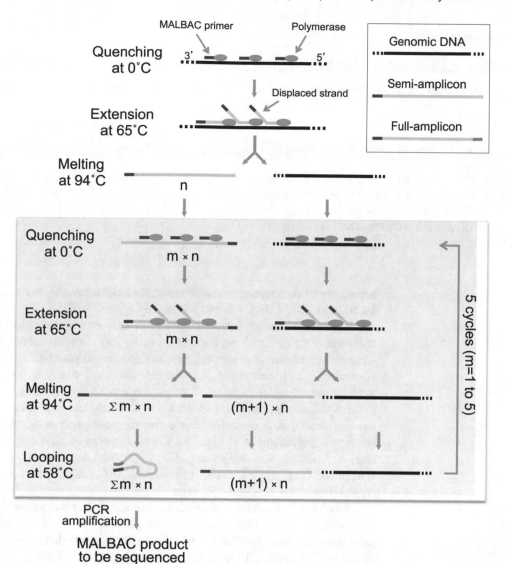

Fig. 1 Overview of MALBAC. Low-bias single-cell whole-genome amplification (WGA). Lysis of a single cell is followed by melting genomic DNA into single-stranded DNA molecules. MALBAC preamplification prior to additional PCR amplification is performed for high-throughput sequencing. First, MALBAC primers anneal randomly to single-stranded DNA molecules and are extended by a polymerase with displacement activity, which creates semi-amplicons. In the next cycle, single-stranded amplicons with complementary sequences on both ends are generated. The 3′ ends are protected by loop formation at intermediate temperature, which prevents the formation of chimeras and further amplification. The above cycles are repeated five times to generate amplicons with overlapping genome coverage that contain universal complementary sequences on both ends for subsequent PCR amplification. Reprinted from Science 338, Zong C, Lu S, Chapman R. A., Xie X. S., "Genome-Wide Detection of Single-Nucleotide and Copy-Number Variations of a Single Human Cell," 1622–1626, 2012, with the permission from Science

Chromosome 1

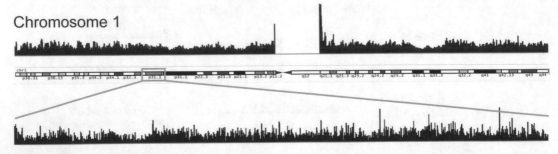

Fig. 2 Profile of sequencing coverage over the entirety of chromosome 1 of a single cell and zooming into a 2 kilobase region (zoomed in region). Reprinted from Science 338, Zong C, Lu S, Chapman R. A., Xie X. S., "Genome-Wide Detection of Single-Nucleotide and Copy-Number Variations of a Single Human Cell," 1622–1626, 2012, with the permission from Science

variations of the coverage have been reported for MDA [5, 12, 13]. In contrast, MALBAC coverage is highly reproducible.

Lorenz curves and power spectrum can be used to evaluate coverage uniformity along the genome. To plot Lorenz curves, the sites of the whole genome are first ranked based on the depth of coverage. The cumulative fraction of the total reads is plotted against the cumulative fraction of genome that the reads covered (Fig. 3). The diagonal line indicates a perfectly uniform distribution of reads, and deviation from the diagonal line indicates an uneven distribution of reads. The Lorenz curves of bulk sequencing, MALBAC, and MDA at ~25× mean sequencing depth are compared (Fig. 3). It is evident that MALBAC outperforms MDA in uniformity of genome coverage.

The power spectrum of read density variations shows the spatial scale at which the variations take place (Fig. 4). For MDA, large amplitudes at low frequencies (inverse genome distance) were observed, indicating that large contiguous regions of the genome are over- or underamplified. In contrast, MALBAC has a power spectrum similar to that of the unamplified bulk.

CNVs due to insertions, deletions, or multiplications of genome segments are frequently observed in almost all categories of human tumors (13, 24, 25). MALBAC's lack of large-scale bias makes it amenable to probing CNVs in single cells. We determined the digitized CNVs across the whole genomes of three individual cells from the SW480 cancer cell line (Fig. 5a–c). CNVs of five cells are included in the supplementary materials. The chromosomes exhibit distinct CNV differences among the three individual cancer cells and in the bulk result (Fig. 5d), which are difficult to resolve by MDA (Fig. 5e). For the MALBAC data, a hidden Markov model is used to quantify CNVs. The gross features of CNVs detected by MALBAC are consistent with karyotyping data (data not shown). Although the majority of copy numbers are consistent

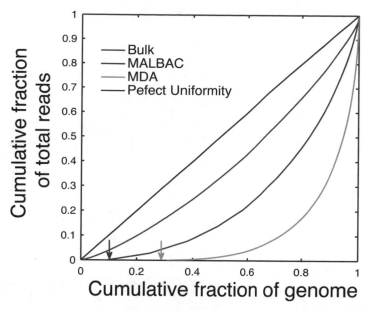

Fig. 3 Lorenz curves of MALBAC, MDA, and bulk sample. A Lorenz curve gives the cumulative fraction of reads as a function of the cumulative fraction of genome. Perfectly uniform coverage would result in a diagonal line, and a large deviation from the diagonal is indicative of biased coverage. The *blue* and *green arrows* indicate the uncovered fractions of the genome for MALBAC and MDA, respectively. All samples are sequenced at 25× depth. Reprinted from Science 338, Zong C, Lu S, Chapman R. A., Xie X. S., "Genome-Wide Detection of Single-Nucleotide and Copy-Number Variations of a Single Human Cell," 1622–1626, 2012, with the permission from Science

between single cells, we also observe cell-to-cell variations as labeled by the dashed boxes in Fig. 3.

For noninvasive prognosis and diagnosis of cancer, it is desirable to monitor genomic alterations through the circulatory system. MALBAC can be used to profile the copy number variations in circulating tumor cells. Interestingly, with the analyses of the whole genome of single CTCs, Ni et al. discovered the reproducibility of the CNV patterns among five patients with lung cancer adenocarcinoma, in contrast to the patients with a mixture of ADC and SCLC (Fig. 6) [14]. This data indicates that copy number variations (CNVs), one of the major genomic variations, can be specific to cancer types, reproducible from cell to cell, and even from patient to patient. The high degree of similarity in CNV pattern has been reported in the recent large scale of Pan-Cancer Analysis [15].

The reliable detection of CNV in single cell by MALBAC has also enabled its application in preimplantation genetic diagnosis or screening (PGD or PGS) in IVF with the goal of selecting a normal

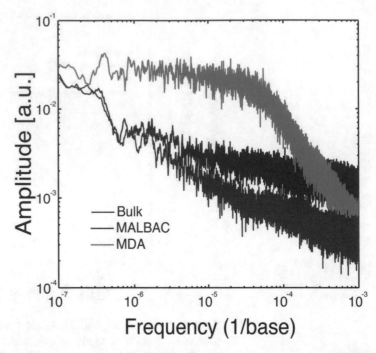

Fig. 4 Power spectrum of read density throughout the genome (as a function of spatial frequency). MALBAC performs similarly to bulk, whereas the MDA spectrum shows high amplitude at low frequency, demonstrating that regions of several megabases suffer from under- and overamplification. Reprinted from Science 338, Zong C, Lu S, Chapman R. A., Xie X. S., "Genome-Wide Detection of Single-Nucleotide and Copy-Number Variations of a Single Human Cell," 1622–1626, 2012, with the permission from Science

fertilized egg [16]. If the genetic disorders are from the father, one can sequence one or a few cells from the blastocyst stage of the embryo, and if the genetic disorders are from the mother, sequencing polar bodies can be used to deduce the genome of the haploid female pronucleus, information regarding aneuploidy, as well as SNVs in disease-associated alleles. With the high coverage of MALBAC, Hou et al. have successfully demonstrated that the aneuploidy of female pronucleus can be accurately deduced from the genomes of two polar bodies. As the result, whole-genome analyses of single-human oocytes based on MALBAC allow accurate and cost-effective embryo selection for in vitro fertilization (Fig. 7). In Fig.7a, the two polar bodies are isolated and MALBAC WGA is performed. The copy number profiles are determined by sequencing and the CNVs of the pronucleus are predicted and compared with the direct experimental measurement. The consistent result proves that MALBAC can provide reliable detection of copy number variations for PGD or PGS.

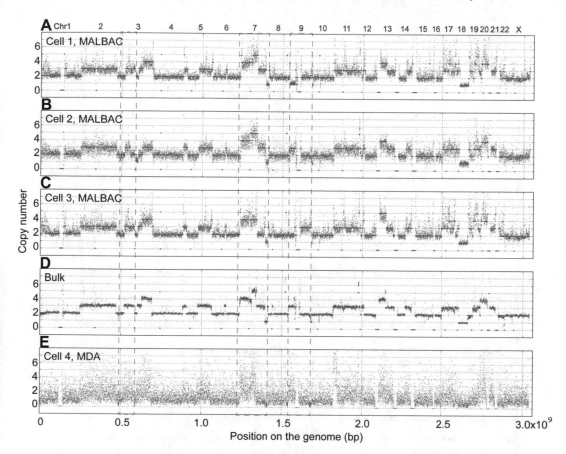

Fig. 5 CNVs of single-cancer cells. Digitized copy numbers across the genome are plotted for three single cells (**a–c**) as well as the bulk sample (**d**) from the SW480 cancer cell line. The *bottom panel* shows the result based on MDA amplification (**e**). *Green lines* are fitted CNV numbers obtained from the hidden Markov model (see supplementary materials). The single cells are sequenced at only 0.8× depth, whereas the bulk and MDA are done at 25×. The regions within the *dashed box* exhibit the CNV differences among single cells and the bulk, which cannot be resolved by MDA. The binning window is 200 kb. Reprinted from Science 338, Zong C, Lu S, Chapman R. A., Xie X. S., "Genome-Wide Detection of Single-Nucleotide and Copy-Number Variations of a Single Human Cell," 1622–1626, 2012, with the permission from Science

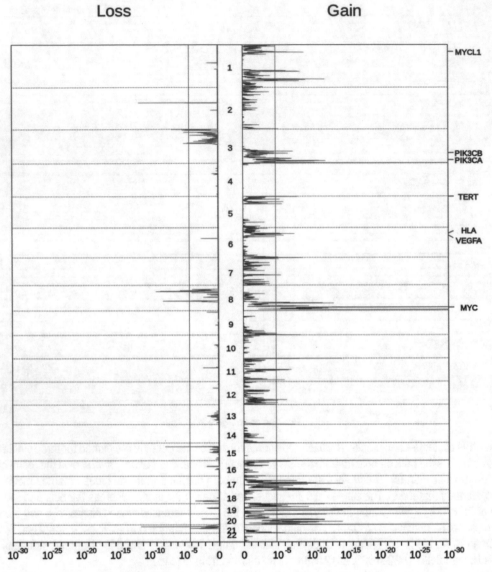

Fig. 6 Statistical significance of gain and loss regions among single circulating tumor cells of six patients with lung cancer adenocarcinoma. Reprinted from PNAS 110(52), Ni X, Zhuo M, Su Z, Duan J, Gao Y, Wang Z, Zong C, Bai H, Chapman AR, Zhao J, Xu L, An T, Ma Q, Wang Y, Wu M, Sun Y, Wang S, Li Z, Yang X, Yong J, Su XD, Lu Y, Bai F, Xie XS, Wang J. "Reproducible copy number variation patterns among single circulating tumor cells of lung cancer patients," 21,083–8, 2013, with the permission from PNAS

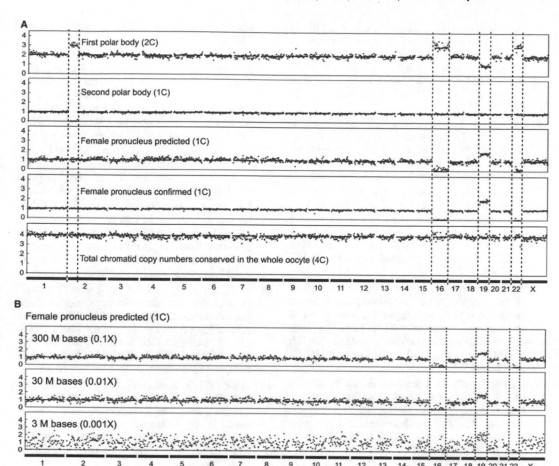

Fig. 7 (a) Number of chromosomes of pronucleus can be deduced by its polar bodies. **(b)** Influence of sequencing depth on aneuploidy deduction. Three levels of sequencing depth (0.1×, 0.01×, and 0.001×) were analyzed, and the results suggest that 0.01× data are sufficient for aneuploidy deduction at megabase resolutions. Reprinted from Cell 155, Hou Y, Fan W, Yan L, Li R, Lian Y, Huang J, Li J, Xu L, Tang F, Xie XS, Qiao J, "Genome Analyses of Single Human Oocytes," 1492–1506, 2013, with the permission from Elsevier"

References

1. Dean FB, Nelson JR, Giesler TL, Lasken RS (2001) Rapid amplification of plasmid and phage DNA using phi 29 DNA polymerase and multiply-primed rolling circle amplification. Genome Res 11(6):1095–1099. doi:10.1101/gr.180501. PubMed PMID: 11381035; PubMed Central PMCID: PMC311129

2. Navin N, Kendall J, Troge J, Andrews P, Rodgers L, McIndoo J et al (2011) Tumour evolution inferred by single-cell sequencing. Nature 472(7341):90–94. doi:10.1038/nature09807. PubMed PMID: 21399628; PubMed Central PMCID: PMC4504184

3. Evrony GD, Cai X, Lee E, Hills LB, Elhosary PC, Lehmann HS et al (2012) Single-neuron sequencing analysis of L1 retrotransposition and somatic mutation in the human brain. Cell 151(3):483–496. doi:10.1016/j.cell.2012.09.035. PubMed PMID: 23101622; PubMed Central PMCID: PMC3567441

4. Lu S, Zong C, Fan W, Yang M, Li J, Chapman AR et al (2012) Probing meiotic recombination and aneuploidy of single sperm cells by whole-genome sequencing. Science 338(6114):1627–1630. doi:10.1126/science.1229112. PubMed PMID: 23258895; PubMed Central PMCID: PMC3590491

5. Wang J, Fan HC, Behr B, Quake SR (2012) Genome-wide single-cell analysis of recombination activity and de novo mutation rates in human sperm. Cell 150(2):402–412. doi:10.1016/j.cell.2012.06.030. PubMed PMID: 22817899; PubMed Central PMCID: PMC3525523

6. Xu X, Hou Y, Yin X, Bao L, Tang A, Song L et al (2012) Single-cell exome sequencing reveals single-nucleotide mutation characteristics of a kidney tumor. Cell 148(5):886–895. doi:10.1016/j.cell.2012.02.025. PubMed

7. Zong C, Lu S, Chapman AR, Xie XS (2012) Genome-wide detection of single-nucleotide and copy-number variations of a single human cell. Science 338(6114):1622–1626. doi:10.1126/science.1229164. PubMed PMID: 23258894; PubMed Central PMCID: PMC3600412

8. Wang Y, Waters J, Leung ML, Unruh A, Roh W, Shi X et al (2014) Clonal evolution in breast cancer revealed by single nucleus genome sequencing. Nature 512(7513):155–160. doi:10.1038/nature13600. PubMed PMID: 25079324; PubMed Central PMCID: PMC4158312

9. Huang L, Ma F, Chapman A, Lu S, Xie XS (2015) Single-cell whole-genome amplification and sequencing: methodology and applications. Annu Rev Genomics Hum Genet 16:79–102. doi:10.1146/annurev-genom-090413-025352. PubMed

10. Baslan T, Kendall J, Ward B, Cox H, Leotta A, Rodgers L et al (2015) Optimizing sparse sequencing of single cells for highly multiplex copy number profiling. Genome Res 25(5):714–724. doi:10.1101/gr.188060.114.

PubMed PMID: 25858951; PubMed Central PMCID: PMC4417119

11. Lodato MA, Woodworth MB, Lee S, Evrony GD, Mehta BK, Karger A et al (2015) Somatic mutation in single human neurons tracks developmental and transcriptional history. Science 350(6256):94–98. doi:10.1126/science.aab1785. PubMed PMID: 26430121; PubMed Central PMCID: PMC4664477

12. Hou Y, Song L, Zhu P, Zhang B, Tao Y, Xu X et al (2012) Single-cell exome sequencing and monoclonal evolution of a JAK2-negative myeloproliferative neoplasm. Cell 148(5):873–885. doi:10.1016/j.cell.2012.02.028. PubMed

13. Zhang K, Martiny AC, Reppas NB, Barry KW, Malek J, Chisholm SW et al (2006) Sequencing genomes from single cells by polymerase cloning. Nat Biotechnol 24(6):680–686. doi:10.1038/nbt1214. PubMed

14. Ni X, Zhuo M, Su Z, Duan J, Gao Y, Wang Z et al (2013) Reproducible copy number variation patterns among single circulating tumor cells of lung cancer patients. Proc Natl Acad Sci U S A 110(52):21083–21088. doi:10.1073/pnas.1320659110. PubMed PMID: 24324171; PubMed Central PMCID: PMC3876226

15. Zack TI, Schumacher SE, Carter SL, Cherniack AD, Saksena G, Tabak B et al (2013) Pan-Cancer patterns of somatic copy number alteration. Nat Genet 45(10):1134–1140. doi:10.1038/ng.2760. PubMed PMID: 24071852; PubMed Central PMCID: PMC3966983

16. Hou Y, Fan W, Yan L, Li R, Lian Y, Huang J et al (2013) Genome analyses of single human oocytes. Cell 155(7):1492–1506. doi:10.1016/j.cell.2013.11.040. PubMed

Chapter 8

Competitive PCR for Copy Number Assessment by Restricting dNTPs

Luming Zhou, Robert A. Palais, Yotam Ardon, and Carl T. Wittwer

Abstract

Copy number variation (CNV) reflects a gain or loss in the number of copies of DNA fragments in a genome. CNV is common in genetic diseases and is known to cause particular neurodegenerative diseases. We developed a dNTP-limited, competitive PCR technique to identify relative copy number differences between a reference and one or more target genes. Suitable fragments with single melting domains, well-separated melting temperatures, and no common homologs were identified by uMelt melting curve prediction software. Relative product amounts were maintained during multiplex PCR into the plateau phase by limiting dNTPs. After PCR, fluorescent melting curve analysis was automatically performed with the saturating DNA dye, LCGreen® Plus. Exponential background was removed, melting curves were plotted as negative derivative melting peaks, and the reference peak was normalized by both position (temperature) and height of each peak. With the reference peak normalized, the height of the target peaks established the copy number order that can be quantified against standards. Using chromosome X variation, the best dNTP concentration to distinguish copy numbers was about 6 μM each and CVs of about 1% were obtained with high-resolution melting analysis. The method was applied to spinal muscular atrophy, trisomies 13, 18, and 21, and cystic fibrosis gene deletions. The method is rapid, economical, and closed tube, and can be used for diagnosis or confirmation of copy number differences identified by high-throughput screening methods.

Key words Copy number variation, CNV, High-resolution melting analysis, HRMA, Melting temperature, Tm, Spinal muscular atrophy

1 Introduction

Copy number variation (CNV) is a common type of genetic variation. About 13% of the genes in the human genome have variation in their copy number [1]. Deletions, duplications, and unbalanced translocations all affect the diploid status of an individual [2]. Many neurogenic diseases are caused by the loss or gain of large segments of DNA sequence. For example, spinal muscular atrophy (SMA) is very common worldwide. SMA is the second most common fatal genetic disease, after cystic fibrosis, in people of European descent. SMA affects voluntary muscle movement through the central

José María Frade and Fred H. Gage (eds.), *Genomic Mosaicism in Neurons and Other Cell Types*, Neuromethods, vol. 131, DOI 10.1007/978-1-4939-7280-7_8, © Springer Science+Business Media LLC 2017

nervous system. Approximately 95–98% of SMA cases are caused by a homozygous deletion of the survival motor neuron 1 (*SMN1*) gene [2]. Duchenne muscular dystrophy (DMD) is another genetic disorder characterized by progressive muscle degeneration and weakness. DMD is caused by an absence of dystrophin, a protein that helps keep muscle cells intact. Approximately 65% of mutations causing DMD are deletions or duplications of one or more exons, clustered in two hotspot regions [3–6]. Charcot-Marie-Tooth neuropathy type 1 (CMT1) is a demyelinating peripheral neuropathy characterized by distal muscle weakness and atrophy, sensory loss, and slow nerve conduction velocity. It is a slowly progressive disorder often associated with deformity and bilateral foot drop. Approximately 70–80% of all CMT1 cases involve duplication of the *PMP22* gene [7]. Neurofibromatosis type I (NF1) causes tumors throughout the nervous system. Approximately 5–20% of all NF1 patients carry a heterozygous deletion in *NF1* of approximately 1.4 Mb [8, 9]. Cystic fibrosis is the most prevalent genetic disease in Western populations. Approximately 1–3% of cystic fibrosis cases are caused by deletions in *CFTR* that may cover either the entire gene or one or more exons [10, 11]. Down syndrome is caused by an additional copy of an entire chromosome (trisomy 21), and trisomies 13 and 18 also occur.

In addition to causing neurologic disease, copy number variants also affect many other disease processes. For example, large deletions in *BRCA1* or BRCA2 may predispose individuals to breast cancer [12–14]. Increased *CCL3L1* copy number is associated with susceptibility to HIV infection [15, 16]. Many somatic copy number changes are correlated to cancer; for example, high *EGFR* copy number is related to colon cancer and non-small-cell lung cancer [17–20].

Clinically, the most commonly used method for CNV detection is fluorescence in situ hybridization (FISH) [21, 22]. This technique, popular among pathologists, visualizes each copy as a colored spot that can be viewed through a fluorescent microscope. FISH adequately identifies the copy number of DNA segments that are 1–100 kb or longer; however, CNVs involving shorter segments are more difficult to detect. In the past decade following the sequencing of the human genome [23], several molecular techniques that can detect shorter CNVs have revealed a remarkable degree of structural variation among normal individuals. The most popular of these techniques are single-nucleotide polymorphism (SNP) arrays, real-time qPCR, multiplex ligation-dependent probe amplification (MLPA), and massively parallel sequencing (MPS). Array-based techniques are the most efficient for high-resolution scans of genome-wide variation [24–27]. The resolution of high-density targeted arrays can now approach tens of base pairs. Massively parallel sequencing can also be used for CNV detection where the resolution is only limited by the complexity of the DNA

[28, 29]. Additional methods for assessing copy number variants, often from single cells, include flow cytometry (Chap. 4, this volume), quantitative reverse transcriptase PCR (Chap. 11, this volume), and others (Chaps. 6 and 7, this volume). Although extraordinary in their genomic coverage, these methods are time consuming and/or require costly equipment and reagents.

Real-time qPCR can calculate relative copy number from the change in quantification cycle (ΔCq) [30, 31]. qPCR is simple, and relatively rapid, and the equipment and disposables are not as expensive as arrays or MPS. However, the throughput and copy number resolution of qPCR are limited. In the human genome, typical germline copy number ratios (sample to reference) range from 1:2 to greater than 5:1. The 3:2 ratio indicating trisomy (e.g., Down syndrome) is especially common. Theoretically, qPCR can detect 1:2 CNVs (the deletion of 1 out of 2 copies), but considerable care is required in practice for reliable results, and it is very difficult to distinguish ratios nearer to 1.0 by qPCR. While qPCR is the gold standard for gene expression, it is less commonly used in copy number assays.

Digital PCR is useful for both relative and absolute quantification. Digital PCR partitions PCR into small droplets so that, on average, there is only about one template molecule per partition, and thousands of partitions are read out as either negative (no template) or positive (one or more templates). The ratio of positive to negative reactions determines how many copies are present [32]. Although elegant in design, digital PCR requires expensive equipment and careful preparation is critical for accurate results.

MLPA is another method that has been widely used to detect CNVs associated with genetic disease. MLPA can detect large deletions, duplications, and complex rearrangements in genes [33, 34]. It is often used to complement sequencing for full gene analysis by assessing the copy number of multiple exons. For each exon tested, two probes ligate when the complementary target is present, forming probes of different lengths and colors that are easily distinguished by capillary electrophoresis. Over 50 exons can be tested simultaneously using MLPA; however, MLPA requires a long ligation time and the customized oligonucleotide probes can be expensive and difficult to design. All of the approaches described above except real-time and digital PCR require more than a day to complete.

High-resolution melting analysis for genotyping and scanning is simple, fast, accurate, and inexpensive [35]. As is the case with sequencing, however, large deletions or duplications that encompass the primers are not identified unless the deletion is X-linked [36]. Recently, high-resolution melting techniques have been applied to targeted copy number assessment. In some protocols, homologous sequences are required to identify a common primer pair. These methods are based on competitive PCR to amplify both a target and a competitor with the same primer set so as to retain their quantitative relationship [37]. In addition, melting analysis requires an internal sequence difference between reference and target that provides distinguishable melting temperatures (Tms). Duplex melting with a single primer set that amplifies segmental duplications can quantify trisomies [38], and a similar method using homologous sequences, can identify microdeletions or microinsertions [39]. The reference and target PCR products must lie on different chromosomes, with distinct Tms so their melting peaks can be distinguished and compared. These methods enable relative quantification because identical primers ensure equal efficiencies of amplification.

We have developed a technique for CNV detection that does not require identical primer sets. Usually, different primers amplify with different efficiencies while approaching the PCR plateau, resulting in a loss of relative quantification. However, the efficiency of different primer sets can be maintained by (1) limiting the number of cycles, (2) limiting the concentration of DNA polymerase, or (3) limiting the amount of dNTPs [40]. Achieving relative quantification by limiting dNTP concentrations is simple, fast, less expensive, and more precise than other current copy number variation detection methods. There is no need to adjust template concentrations or design specialized oligonucleotides. Only a PCR instrument and fluorescent DNA melting acquisition are required. Real-time PCR is not necessary, and melting analysis requires less than 5 min on some dedicated melting instruments. Applications of this targeted technique include not only direct diagnosis, but also confirmation of specific copy number variants suspected by screening techniques such as arrays and MPS.

2   Materials

2.1 Human Whole Blood and Genomic DNA Samples

One and two copies of chromosome X DNA (male and female) were purified from laboratory volunteers. DNA from human cell lines with 3 (NA03623), 4 (NA11226), and 5 (NA 06061) copies of chromosome X were obtained from the Coriell Institute for Medical Research, as well as DNA from a patient affected with spinal muscular atrophy (SMA) (NA00232), an SMA carrier (NA003814), and a cystic fibrosis carrier with a heterozygous deletion of exons 2 and 3 (NA18668). Excess DNA from 50 clinical products of conception including trisomies of chromosomes 13, 18, and 21 was obtained from ARUP Laboratories under IRB 7275.

2.2 Primers

Primers were synthesized on a 40 nmol scale at the University of Utah core facility with cartridge purification. One hundred μL of TE' buffer (10 mM Tris, 0.1 mM EDTA, pH 8.0) was added to

Table 1
Primer sequences and Tms, chromosome numbers and regions, and PCR product sizes, Tms and associated Figures[a]

Chromosome	Region	Primer 1	Tm (°C)[b]	Primer2	Tm (°C)[b]	Size (bp)	Tm (°C)[c]	Figure
7	*CFTR* exon 6	CTCCTCATGGGGCTAATCTG	63	AAGTCCACAGAAGGCAGACG	66	54	82	1,2,4,5
X	*CYBB* exon 10	CCTTCAGGATAGCGGTTGAT	63	CTTGAGAATGGATGCGAAGG	62	121	88	1,2,3,4,5
7	*CFTR* exon 7	TTGTGATTACCTCAGAAATGATTGA	62	CATTGCTTCTTCCCAGCAGT	64	68	78	3,6,7
13	g.76560967-1048[d]	AACGGGAGGGGTGTATGTTT	65	GCAGACTAGGTGCCCAACTT	66	82	85	7
18	*TPGS*2 exon 4	ACATCATTCCACTGGGAAGC	64	CCAGAGTGGGTGCATTAGGA	65	99	84	7
21	g.19181856-964[e]	TCAGACTTGGACAGCCACAC	66	CACTTGGGGAATTGACTCACA	64	109	84	7
5	*SMN*1 exon 7	TTCCTTATTTTCCTTACAGGGTTT	62	CCTTCCTTCTTTTTGATTTTGTCTG[f]	62	58	74	6
5	*SMN*2 exon 7	TTCCTTTATTTTCCTTACAGGGTTT	62	CCTTCCTTCTTTTTGATTTTGTCTA[f]	61	58	73	6
21	*FTCD* exon 6	AGCCAGGTTCTTCTCATCCA	64	GCCAGGACGTCTGAAGAAAG	64	61	82	8
7	*CFTR* exon 2	TCTGTTGATTCTGCTGACAATCT	63	TGAACATACCTTTCCAATTTTTCA	61	50	73	8
7	*CFTR* exon 3	GGGATAGAGAGCTGGCTTCA	65	GCCGAAGGGCATTAATGAGT	64	54	77	8

[a]Reference PCR targets are shown in red. The copy number target(s) are shown in black below the corresponding reference target

[b]Predicted Tms (see text)

[c]Observed Tms

[d]Genomic bounds of the PCR product on chromosome 13, GRCh38 primary assembly, (NC_000013.11)

[e]Genomic bounds of the PCR product on chromosome 21, GRCh38 primary assembly (NC_000021.9). The PCR product is near to the short tandem repeat D21S11

[f]*SMN1* and *SMN2* are differentiated by allele specific amplification determined by the 3'-base

dissolve the oligonucleotides. Two µL was used to measure the oligo concentration, which was adjusted with TE' to 50 µM and stored at −20 °C. Primer sequences and Tms, PCR amplicon lengths, and amplicon Tms are shown in Table 1.

2.3 Equipment

A NanoDrop™ spectrophotometer (Thermo Scientific) was used to measure DNA concentration. A real-time thermocycler with high-resolution melting analysis (LightCycler 480, Roche) was used for PCR amplification and high-resolution melting, unless otherwise specified, to produce melting data from which copy number variation was identified by subsequent analysis.

2.4 Assay Design Software

Primer3 web version 4.0.0 from the University of Massachusetts was used to design primers (http://bioinfo.ut.ee/primer3/). The University of California Santa Cruz (UCSC) Genome Browser (https://genome.ucsc.edu/) and the National Center for Biotechnology Information (NCBI, https://www.ncbi.nlm.nih.gov/) were used to check for variants in the amplicons. DNA uMelt software (https://www.dna.utah.edu/index.html) from the University of Utah was used to predict target and reference amplicon melting curves and Tms.

2.5 Melting Analysis and Quantification Software

Custom melting analysis software (MeltWizard 6, available from RAP for research use) was used to analyze and quantify the reference and target peaks for relative DNA quantification. The algorithms used are described below under Sect. 3.4.

3 Methods

3.1 Genomic DNA Extraction

DNA was extracted using commercial kits specific to the tissue type. The DNA was quantified on a NanoDrop spectrophotometer, using its absorbance at 260 nm. DNA was stored at 4 °C for up to 1 week, with longer storage at −20 or −80 °C.

3.1.1 Genomic DNA Extraction from Whole Blood

Three mL whole blood was drawn into EDTA vacuum tubes. The contents of each tube were mixed by gently inverting five times. The EDTA-treated whole blood was stored at room temperature (15–25 °C) for 1 day, or at 4 °C for up to 5 days. The DNA was extracted using a Gentra Puregene Blood Kit (QIAGEN®).

3.1.2 DNA Extraction from Fetal Villi

Fifty fresh products of conception samples with fetal villi, including trisomies of chromosomes 13, 18, and 21, were macro-dissected away from maternal tissue, and metaphase karyotypes were obtained after growth by ARUP. DNA was extracted from residual fetal tissue by a 5 PRIME™ ArchivePure™ DNA purification cell and tissue kit (Fisher Scientific).

3.2 Primer and Multiplex PCR Amplicon Design

Human genome databases NCBI and UCSC genome browsers were used to check the reference and target sequences for uniqueness and variation. The reference sequence must not have homologs in the genome, and both reference and target sequences should have minimal internal variation.

Primer3 and uMelt were used in combination to design reference and target multiplex PCR primers. The primer design software Primer3 was used to adjust primer Tms and the length of the PCR product. The multiplex PCR primer Tms were designed to be as close as possible to 60 °C. The lengths of amplicons were optimally in the range of 50–120 bp. The reference and target sequences were entered into uMelt to predict amplicon Tms. The ΔTm between reference and target amplicons should be between 2 and 10 °C. If the ΔTm is out of this range, Primer3 may be used to adjust the amplicon length, and GC or AT 5′-primer tails can be added to one or both primers.

Duplex reactions with target genes on chromosomes X, 5, 7, 13, 18, and 21 were paired with reference genes on different chromosomes for relative quantification. *CYBB* exon 10 was used to target chromosome X with *CFTR* exon 6 on chromosome 7 as the reference gene. *CFTR* exon 7 was the reference gene for chromosomes 5, 13, 18, and 21. A fragment of *FTCD* exon 6 on chromosome 21 was used as the reference for *CFTR* exon 2 and exon 3 deletions. The primer sequences of reference and CNV target fragments are shown in Table 1. Tms were calculated using the online "Tm Tool" found at https://www.dna.utah.edu/tm/ using 6.25 µM each dNTP (25 µM total dNTPs) and 2.0 mM total Mg^{++}.

3.3 PCR Conditions

Final PCR concentrations were 0.4 U KlenTaq1™ (Ab Peptides) with 64 ng anti-Taq antibody (eEnzyme), 2 mM $MgCl_2$, 50 mM Tris (pH 8.3), 1× LCGreen Plus dye (BioFire Defense), 500 mg/L bovine serum albumin (Sigma), 0.5 µM each primer, 50 ng DNA, and 6.25 µM each dNTP in10 µL unless otherwise specified. PCR was performed on a LightCycler 480. An initial denaturation step at 95 °C for 2 min was followed by 35 cycles of 10 s denaturation at 95 °C and 30 s annealing/extension at 65 °C. For melting, the samples were programmed to reach 95 °C momentarily (0 s), cooled to 55 °C for 5 s, and then melted from 65 to 95 °C with 15 acquisitions/°C (rate of 0.065 °C/s). Excitation and emission filters on the LC480 were centered at 450 and 500 nm for PCR and melting.

3.4 Analysis for Relative Quantification

Several transformations were applied to the raw high-resolution melting data (fluorescence vs. temperature) to rank and quantify copy number ratios. The fluorescence is comprised of two temperature-dependent components. The desired component is the signal reflecting the quantity of DNA present in the double-stranded state, represented by the helicity, H. The undesired component is the fluorescence background. H is 100% at low temperature and 0% at high temperature, sharply decreasing across a narrow interval around the melting temperature, T_M. As a first step, H was obtained by modeling the background component as an exponential [41] and subtracting the background from the temperature-fluorescence data points (t_j, f_j) to obtain (t_j, h_j) for the temperature vs. helicity graph (the melting curve). In the second step, the negative derivative of each resulting melting curve was calculated. For each point of the melting curve, all points (t_k, h_k) in the window W_j of fixed width w defined by $t_k \leq t_j + w$ were fit by the least squares polynomial of degree $n = 2$, $h(t) = at^2 + bt + c$ also known as the $n = 2$ Savitzky-Golay fit or simply "SG2" fit [42]. Using $-h'(t) = -2at - b$, each point on the negative derivative curve was obtained as $(<t>_j, -2a<t>_j - b)$, where $<t>_j$ is the mean of the t_k in the window bounded by t_j. In the third step, the locations of the peaks of the negative derivative curves corresponding to reference and target amplicons were determined for internal temperature correction and amplitude normalization. For each peak, all points near the peak maximum were fit by a SG2 polynomial of the form $d(t) = at^2 + bt + c$, whose peak occurs at $(-b/(2a), -(b^2 - 4ac)/(4a))$. Then, the arithmetic mean of the temperatures of the reference peaks and of a target peak was computed. For each derivative curve, the temperatures of all points were transformed by the first-degree polynomial that maps its particular two peak temperatures to these means. In the final transformation, the geometric (multiplicative) mean of the amplitudes of reference peaks was computed, and each derivative curve was scaled by a constant to map its amplitude to the mean. The resulting amplitudes of target peaks were then used for CNV quantification. For continuous ratios (e.g., those arising from mixed sources) quantitative copy numbers of unknowns were obtained by weighting copy numbers of known standards or evaluating a polynomial fit to the data. When copy number variants were known to occur in integer ratios with small denominators, e.g., 3:2, these discrete copy number ratios were determined by performing unbiased hierarchal clustering on the transformed target peaks and assigning the ratio of a known standard to all samples in the same cluster.

3.5 Competitive PCR with Different Concentrations of dNTPs

The duplex PCR of *CFTR* exon 6 as a reference and *CYBB* exon 10 on chromosome X as the target can quantify chromosome X copy number. Ten µL reactions were made with 2 µL of 5× PCR working solution, 1 µL of each 5 µM primer, 1 µL of 50 ng/µL female human genomic DNA, 1 µL of 10× dNTP (varying concentrations), and water to 10 µL. The 10× dNTP concentrations were 2000, 1000, 500, 250, 125, 62.5, 31.3, and 15.6 µM.

The copy number ratio of the reference *CFTR* amplicon on chromosome 7 to the target *CYBB* amplicon on chromosome X is 1.0 (2:2). As the dNTP concentration is decreased from 200 to 3.13 µM, the PCR Cq occurs three cycles later and the fluorescence decreases by 90% (Fig. 1a). Decreasing the concentration from 200 to 25 µM decreases the peak height of the higher Tm amplicon disproportionately to the lower Tm amplicon. However, at low dNTP concentrations (3–12 µM), the peak height ratios of both the reference and target are similar (Fig. 1b). Only at high cycle numbers do the relative amounts of PCR products deviate.

The effect of dNTP concentration on the derivative peak height and the relative rank between samples with 1, 2, 3, and 4 copies of chromosome X are shown in Fig. 2. The dNTP concentrations were 200, 50, 6.25, and 1.56 µM. At very low dNTP concentrations (1.56 µM) the fluorescence is very low, reflecting very little amplification (Fig. 2a and magnified inset). The best relative quantification occurred at 6.25 µM dNTPs, where the fluorescence was still limited but each copy number was distinguishable (Fig. 2b). At higher dNTP concentrations (50 µM), the peaks for 3 and 4 copies were no longer separated, as the product quantity began to reflect primer limitation rather than initial template amounts (Fig. 2c). At higher dNTP concentrations (200 µM), no relative quantification was possible (Fig. 2d).

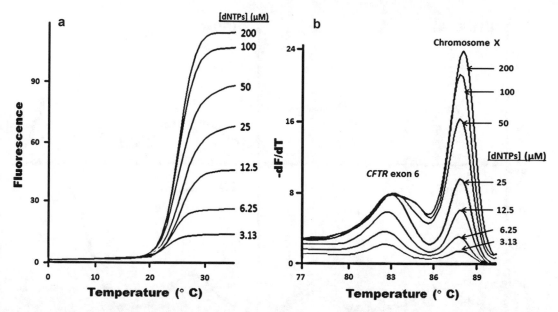

Fig. 1 The effect of varying dNTP concentrations on duplex real-time PCR product fluorescence. (**a**) Decreasing dNTP concentrations in PCR increased Cq and lowered the amount of product produced. (**b**) High-resolution melting analysis revealed distinct melting curves for reference and copy number targets. dNTP concentrations up to 10 μM retained a constant ratio between targets. However, above 10 μM dNTPs the target-to-reference peak ratio increased. Relative quantification at the endpoint is only possible if the ratio of the 2 PCR products does not change. Primer sequences as well as product sizes and Tms are given in the Table 1. Reprinted with permission from Clinical Chemistry, 61, 724–33 (2015)

3.5.1 Precision and Accuracy	The precision of copy number analysis by competitive PCR with limiting dNTPs depends on the resolution of the melting instrument. The precision obtained with the high-resolution LightScanner produces CVs of <1.0%, allowing detection of a 1.11-fold increase (Fig. 3). The copy number accuracy depends on using a standard curve to correct for any nonlinear response of peak height to copy number. This calibration is dependent on the particular target and reference amplicons. In general, the calibration curve is not linear and increasingly diverges at the extremes. Therefore, although very small differences in copy number can be detected and ranked, quantification of these differences requires a standard curve, and standards may not be easily available. Accurate absolute quantification of copy number changes is a challenge for many techniques.
3.6 Effect of Initial Template Concentration	Limiting the dNTP concentration maintains the ratio of target to reference amplicons even as PCR reaches the plateau, thus permitting relative quantification without concern for the initial template concentration. Fig. 4a shows real-time PCR of samples with either 5 or 50 ng of genomic DNA per sample. The Cqs from PCR of this tenfold difference differ by 3–4 cycles and are not related to the number of copies of chromosome X. However, once all the samples

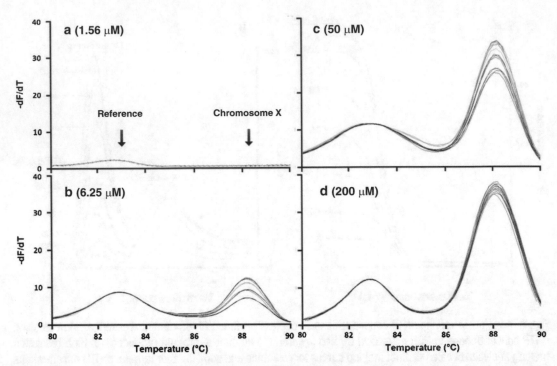

Fig. 2 Using restricted dNTPs to resolve the number of chromosome X copies relative to a reference gene. (**a**) At 1.56 μM dNTPs, only the reference sequence was weakly amplified. The chromosome X sequence showed no amplification, even when magnified in the inset. (**b**) At 6.25 μM, both sequences were amplified and 1, 2, 3, and 4 copies (*bottom to top*) of chromosome X were distinguished when normalized to the reference. (**c**) At 50 μM dNTPs, the relative peak height of the chromosome X target increased, but the 3 and 4 copies of X could not be distinguished. (**d**) No copy number distinctions could be made at 200 μM dNTPs. Thirty-five cycles of PCR were performed to reach plateau before melting analysis was performed. Samples were analyzed in quadruplicate. Sequences and PCR product information are given in Table 8.1. Reprinted with permission from Clinical Chemistry, 61, 724–33 (2015)

reach PCR plateau under conditions of restricted dNTPs, the X chromosome differences are readily apparent, and are not affected by the initial template concentration (Fig. 4b). This comparison highlights one advantage of competitive PCR with limiting dNTPs—its robustness to template concentration uncertainty. CNV detection by dNTP limitation does not require accurate template adjustment in order to obtain a stable, consistent relative copy number. In contrast, limiting PCR by cycle number is highly sensitive to the concentration of initial template DNA.

3.7 Limiting PCR Through Cycle Number Control

It is also possible to detect copy number variation with standard dNTP concentrations (200 μM) by carefully controlling the total number of PCR cycles and the template concentration. DNA samples having 1, 2, 3, or 4 copies of chromosome X were adjusted to a template concentration of 50 ng/μL. At this concentration, PCR plateaus after 35 cycles with a Cq near the 23rd cycle (Fig. 1a).

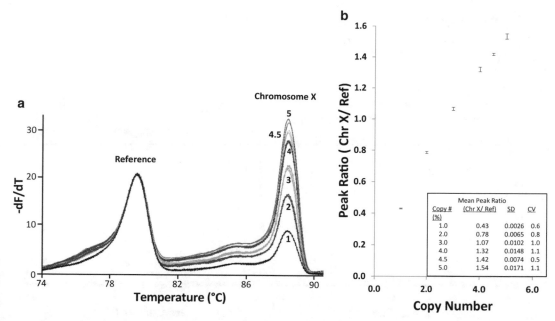

Fig. 3 Resolution of chromosome X copy number using duplex PCR, restricted dNTPs, and normalization against a reference gene. Samples were amplified on the LC480 (Roche) with 3.13 μM dNTPs and melted on a LightScanner (BioFire Defense) at 0.1 °C/s. The 4.5 copy chromosome X sample was obtained by mixing equal volumes of the 4× and 5× standards. (**a**) The normalized melting curve with the copy numbers identified. (**b**) The peak height ratio plotted against the copy number. The nonlinear curve was attenuated at higher copy numbers. Nevertheless, a fold change of 1.11 (4.5 to 5.0 copies) was easily differentiated. Samples were run in quadruplicate. Error bars are ±1 SD. The inset tabulates the peak heights and their variance, revealing CVs between 0.5 and 1.1%. The points fit the second-degree polynomial: $y = -0.0263x^2 + 0.4344x + 0.0092$ with $R^2 = 0.9996$. Modified with permission from Clinical Chemistry, 61, 724–33 (2015)

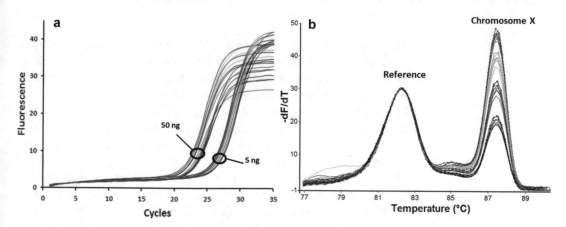

Fig. 4 Duplex PCR and relative quantification using a tenfold difference in starting template concentration. (**a**) Real-time quantification curves are shown for samples with 1, 2, 3, and 4 copies of chromosome X, using either 50 or 5 ng of starting DNA in a 10 μL reaction. The Cqs reflect the difference in starting concentrations. (**b**) Despite the differences in starting concentrations, the relative ratios of target to reference peak heights were preserved, and the ability to detect copy number changes was maintained (lowest to highest target clusters: 1, 2, 3, and 4 copies). Samples were run in quadruplicate. Reprinted with permission from Clinical Chemistry, 61, 724–33 (2015)

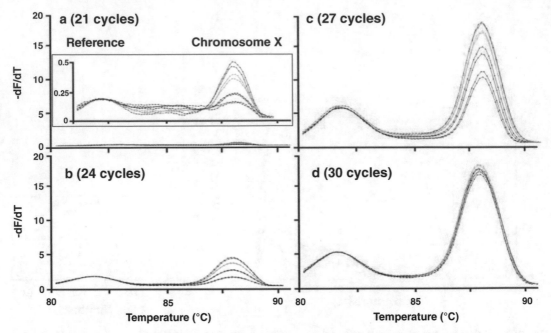

Fig. 5 Establishing chromosome X copy number by limiting the cycle number. (**a**) At 21 cycles, very weak fluorescence signals were observed. The inset shows an expanded view of the melting curves. Copy number information could be determined from these expanded curves, albeit with high noise. (**b**) At 24 cycles, both reference and target signals were increased, and copy number differences were all resolved. (**c**) Both signals were stronger still at 27 cycles, although copy number differences were more difficult to assess. (**d**) By 30 cycles, no copy number information could be extracted. This series was run on 2 replicates of each DNA sample with 200 µM of each dNTP. PCR was performed as described in Methods up through cycle 17. Starting on the 18th cycle, a melting curve was acquired during the transition from 72 °C to 95 °C with 15 acquisitions/°C. Melting acquisition was repeated every 3rd cycle (2 normal PCR cycles followed by 1 melting "cycle") ending at cycle 30. Reprinted with permission from Clinical Chemistry, 61, 724–33 (2015)

PCR was stopped at different cycles before and after the Cq to determine the effect of cycle number on the copy number ratio. When the PCR was stopped at the 18th cycle, no amplicon melting peaks were visible (data not shown). When the PCR was stopped at the 21st cycle (before the Cq), the replicate curves corresponding to samples having each of the four different copies of chromosome X were distinguishable, but the melting peaks of both the target and reference products were very small (Fig. 5a). When the PCR was stopped at the 24th cycle (near the Cq), the curves clustered into the four different copy numbers of chromosome X and melting signals were strong (Fig. 5b). When PCR was stopped after the 27th cycle (after the Cq), the melting peaks for 3 copies and 4 copies of chromosome X were barely distinguishable (Fig. 5c). When the PCR was stopped at the 30th cycle (just before the plateau), no chromosome X copy number variants could be distinguished (Fig. 5d). Limiting PCR amplification at the Cq cycle provided the best precision for CNV detection. This emphasizes the risk of using cycle limitation in contrast to dNTP restriction as a strategy for CNV quantification by melting analysis. If different amounts of DNA are used, reactions must be allowed to progress to their own Cq and not a fixed cycle number, a function not available on real-time instruments. Because the ability to quantify depends on cycle limitation at the Cq, and because Cq depends on the template concentration, concentration variation can degrade the precision of cycle-limited PCR in a way that limited dNTPs do not.

We also investigated other approaches to address the challenge of producing amplicons in proportion to the template ratio at plateau. In duplex PCR, primers, $MgCl_2$, $Tris^+$, BSA, polymerase, and dNTPs are present. Limiting primer concentrations or restricting buffer components ($MgCl_2$, $Tris^+$, BSA) did not maintain the initial template ratio. Polymerase and dNTPs are directly involved in polymerization. Although limiting polymerase was partly successful in maintaining the ratio under some conditions, restricting dNTPs resulted in better precision and dynamic range.

3.8 Applications

3.8.1 Spinal Muscular Atrophy

Spinal muscular atrophy (SMA) is a common disease of newborns worldwide with fatal consequences, associated with copy number variation in the survival motor neuron 1 gene (*SMN1*). Carriers having only 1 copy of *SMN1* (heterozygous deletion) occur in approximately 1 in 40–50 individuals [43, 44]. The incidence of SMA is 1:6000–10,000 births [44–46]. SMA is characterized by the loss of spinal cord motor neurons that control voluntary muscle movement.

SMN1 was identified in 1995 and is located on chromosome 5q13 [2]. The gene consists of nine exons that code for the *SMN* protein. A highly homologous gene, *SMN2*, is located centromeric to *SMN1*. *SMN1* is the functional gene, while transcription of *SMN2* results in alternative splicing and exclusion of exon 7 in 90% of transcripts with only 10% expression of the full-length protein. The diagnosis of SMA is based on molecular genetic testing. Mutations in *SMN1* cause SMA, while higher copy numbers of *SMN2* [3–5] decrease the severity of the phenotype. Most individuals with SMA (95–98%) are homozygous for a complete deletion of *SMN1* while about 2–5% are compound heterozygotes for an *SMN1* deletion/conversion mutation and an *SMN1* intragenic mutation [47].

SMN1 and *SMN2* can be distinguished by a single base difference in exon 7 at c.840C > T. We have used this distinction to combine duplex PCR with allele-specific amplification of *SMN1* or *SMN2* and a highly conserved fragment of *CFTR* from chromosome 7. This provides the relative copy numbers of *SMN1* and *SMN2* against the *CFTR* reference. Controls include DNA from an affected patient (*SMN1* homozygous deletion) and an SMA carrier (*SMN1* heterozygous deletion). Twenty-four randomly obtained DNA samples from healthy individuals were

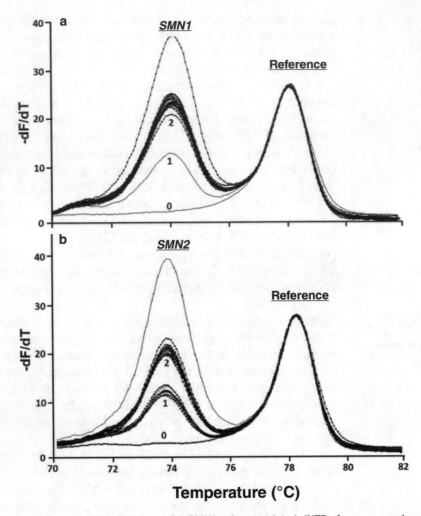

Fig. 6 Duplex amplification of (a) *SMN1* and (b) *SMN2* using restricted dNTPs for copy number assessment. *SMN1* and *SMN2* are on chromosome 5, while the reference gene is on chromosome 7 (Table 8.1). Observed genotype frequencies for 24 normal samples approximated the expected frequencies [56–58]. Common genotypes were two copies of *SMN1* and two copies of *SMN2* (58% observed, 57% expected), 2:1 (33%, 28%), 2:0 (4%, 3%), and 3:2 (4%, 3%). PCR included an initial 1 min denaturation and 35 cycles of 95 °C for 10 s and 65 °C for 20 s. Reprinted with permission from Clinical Chemistry, 61, 724–33 (2015)

tested with 6.25 μM dNTPs. The relative copy numbers clearly clustered into groups of 0, 1, and 2 and more than 2 copies of *SMN1* and *SMN2* (Fig. 6).

3.8.2 Down Syndrome   Down syndrome is the leading genetic cause of intellectual disability and is associated with trisomy of chromosome 21. It affects millions of patients who face a variety of health issues including congenital heart disease, Alzheimer's disease, leukemia, cancers, and Hirschsprung's disease. The incidence of trisomy 21 is influenced by maternal age and population differences (approximately 1 in

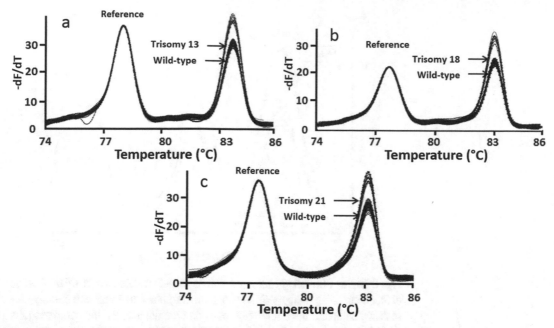

Fig. 7 Duplex PCR using restricted dNTPs to determine trisomy. The copy numbers of (**a**) chromosome 13, (**b**) chromosome 18, and (**c**) chromosome 21 were detected by normalization to a reference peak on chromosome 7. Reprinted with permission from Clinical Chemistry, 61, 724–33 (2015)

1000 live births worldwide) [48–51]. The only other trisomies that are compatible with life are trisomy 13 (Patau syndrome) and trisomy 18 (Edwards syndrome), and both lead to early infant death.

The wild-type copy number ratio of chromosomes 13, 18, and 21 to the reference on chromosome 7 is 1.0. The copy number ratio is 1.5 in trisomy patients. Duplex PCR with 6.25 μM dNTPs was performed on 50 previously typed samples that included trisomies 13, 18, and 21, as well as wild-type samples. The number of samples in each category was blinded to the experimenter and the initial template concentrations ranged from 10 to 200 ng/μL. For each trisomy target, one additional wild-type control and one additional trisomy control were included. PCR products were obtained from all 52 samples in each assay. Target-to-reference ratios were conserved throughout 35 cycles of amplification into the PCR plateau. The trisomy samples were easily distinguishable visually after analysis, as well as automatically by unbiased hierarchical clustering. Nine of the samples were identified as trisomy 13 (Fig. 7a), eight as trisomy 18 (Fig. 7b), and 14 as trisomy 21 (Fig. 7c). The remaining 19 samples were wild-type. All samples were correctly identified for an apparent sensitivity and specificity of 100%.

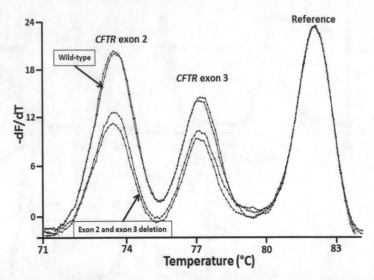

Fig. 8 Triplex PCR using restricted dNTPs to detect deletions in *CFTR*. A large deletion spanning multiple *CFTR* exons on chromosome 7 was detected by normalizing samples with a reference peak on chromosome 21 and comparing the peak heights of exons 2 and 3 to a control wild-type sample. Reprinted with permission from Clinical Chemistry, 61, 724–33 (2015)

3.8.3 Exon Deletions in Cystic Fibrosis

Cystic fibrosis is the most common life-shortening genetic disease in Caucasian populations. The carrier frequency in Caucasians is 1 in 22–25 with an incidence of 1 in 2000–2500 newborns. About 1–3% of cystic fibrosis cases have one or more large deletions [10, 52]. Triplex competitive PCR with 6.25 μM dNTPs was used to detect deletions of *CFTR* exons 2 and 3. Small amplicons within exons 2 and 3 were chosen to avoid *CFTR* variants. Figure 8 shows the melting data after background removal and normalization. The amplicon Tms of exons 2 (73 °C) and 3 (77 °C) are lower than the Tm of the reference amplicon on chromosome 21 (82 °C). When peak heights of the reference gene are normalized, the heterozygous deletion is confirmed by the lower peak heights of exons 2 and 3 compared to the wild-type control without any deletion (Fig. 8).

4   Notes

4.1   The Importance of Avoiding Sequence Variants in Competitive Amplicons

The utility of the reference segment in competitive multiplex PCR depends on its conservation and uniqueness in the human genome. Sequence variants can occur in most regions, and when they occur in reference or target amplicons, they distort melting curve shape and compromise copy number assessment. Although such distortion is very useful in detecting and genotyping variants [53], it confounds melting peak analysis for relative quantification. To avoid variants in the design phase, it is important to check the reference and target amplicons in NCBI, UCSC, or other human genome databases to eliminate or minimize the frequency of variation in the region and confirm the uniqueness of the sequences. The amplicon lengths should be short (between 50 and 120 bp) to reduce the likelihood of including variation. Short amplicons have the additional benefit of yielding products without internal melting domains, an essential property for accurate quantification.

4.2 ΔTm of Amplicons

The primer Tm and amplicon size can be easily adjusted using Primer3 or other design tools. However amplicon length is less important than amplicon Tm. The web software "uMelt" accurately predicts reference and target amplicon melting temperatures and confirms the absence of any internal melting domains that would alter derivative peak shapes. Both reference and target should consist of just one melting domain and have a ΔTm between adjacent peaks of about 5 °C. If the ΔTm is less than 2 °C, the peaks will not separate and CNVs may not be detected. If the ΔTm is more than 10 °C, PCR amplification efficiency of the different amplicons may vary, biasing product yield and degrading CNV information.

The reference amplicon Tm can be either higher or lower than that of the target(s), depending on amplicon length and GC content. If the target has high GC content, then the reference amplicon should be designed to have a lower Tm; otherwise the reference Tm is typically higher than the target.

4.3 Multiplex PCR Efficiency

During multiplex PCR, different amplification efficiencies may arise from variation in primer annealing, polymerase extension, and/or product denaturation. Increasing the annealing time can equalize primer annealing efficiencies, but may also generate nonspecific products. Annealing temperatures 1–5 °C higher than the primer Tms and 30 s annealing times are recommended with amplicons between 50 and 120 bp. If both reference and target amplicons are very short (40–60 bp), 10 s annealing is usually sufficient. A high annealing temperature reduces the probability of nonspecific amplification, while the long annealing time equalizes primer annealing efficiency, leading to better precision.

5 Conclusions

Competitive PCR with limited dNTPs is a simple and rapid method to detect relative copy number changes between specific targets. A saturating DNA dye, eg. LCGreen Plus, is used so that melting analysis can provide a convenient readout after background removal and normalization. Both visual and automatic clustering can detect copy number changes, but quantification requires analysis using a standard curve.

Assay precision is excellent with CVs of about 1%. This is better than most implementations of digital PCR. For example, in droplet digital PCR, 20,000 partitions are usually formed and counted with CVs of 3–4% [54, 55]. The precision of digital PCR can be improved by increasing the number of partitions, but this increases the analysis time, and only some systems can count more than 20,000 events.

Competitive PCR can be performed in 96- or 384-well plates, but has low throughput compared to arrays or massively parallel sequencing. Each locus requires a unique primer pair and a reference amplicon that melts at a sufficiently different temperature than the target amplicon. Nevertheless, competitive PCR with restricted dNTPs simplifies workflow, is amenable to clinical diagnostics, and can be used to confirm copy number changes identified by high-throughput techniques such as arrays and massively parallel sequencing.

References

1. Stankiewicz P, Lupski JR (2010) Structural variation in the human genome and its role in disease. Annu Rev Med 61:437–455

2. Lefebvre S, Burglen L, Reboullet S et al (1995) Identification and characterization of a spinal muscular atrophy-determining gene. Cell 80(1): 155–165

3. Koenig M, Hoffman EP, Bertelson CJ et al (1987) Complete cloning of the Duchenne muscular-dystrophy (DMD) cDNA and preliminary genomic organization of the DMD gene in normal and affected individuals. Cell 50(3):509–517

4. Dendunnen JT, Grootscholten PM, Bakker E et al (1989) Topography of the Duchenne muscular-dystrophy (DMD) gene—FIGE and cDNA analysis of 194 cases reveals 115 deletions and 13 duplications. Am J Hum Genet 45(6):835–847

5. Gillard EF, Chamberlain JS, Murphy EG et al (1989) Molecular and phenotypic analysis of patients with deletions within the deletion-rich region of the Duchenne muscular-dystrophy (DMD) gene. Am J Hum Genet 45(4): 507–520

6. Forrest SM, Smith TJ, Cross GS et al (1987) Effective strategy for prenatal prediction of Duchenne and Becker muscular-dystrophy. Lancet 2(8571):1294–1297

7. Taioli F, Cabrini I, Cavallaro T et al (2011) Inherited demyelinating neuropathies with micromutations of peripheral myelin protein 22 gene. Brain 134(Pt 2):608–617

8. Riva P, Corrado L, Natacci F et al (2000) NF1 microdeletion syndrome: refined FISH characterization of sporadic and familial deletions with locus-specific probes. Am J Hum Genet 66(1):100–109

9. Jenne DE, Tinschert S, Reimann H et al (2001) Molecular characterization and gene content of breakpoint boundaries in patients with neurofibromatosis type 1 with 17q11.2 microdeletions. Am J Hum Genet 69(3): 516–527

10. Schneider M, Hirt C, Casaulta C et al (2007) Large deletions in the CFTR gene: clinics and genetics in Swiss patients with CF. Clin Genet 72(1):30–38

11. Quemener S, Chen JM, Chuzhanova N et al (2010) Complete ascertainment of intragenic copy number mutations (CNMs) in the CFTR gene and its implications for CNM formation at other autosomal loci. Hum Mutat 31(4): 421–428

12. Hartmann C, John AL, Klaes R et al (2004) Large BRCA1 gene deletions are found in 3% of German high-risk breast cancer families. Hum Mutat 24(6):534

13. Agata S, Dalla Palma M, Callegaro M et al (2005) Large genomic deletions inactivate the BRCA2 gene in breast cancer families. J Med Genet 42(10):e64

14. Petrij-Bosch A, Peelen T, van Vliet M et al (1997) BRCA1 genomic deletions are major founder mutations in Dutch breast cancer patients. Nat Genet 17(3):341–345

15. Gonzalez E, Kulkarni H, Bolivar H et al (2005) The influence of CCL3L1 gene-containing segmental duplications on HIV-1/AIDS susceptibility. Science 307(5714):1434–1440

16. Huik K, Sadam M, Karki T et al (2010) CCL3L1 copy number is a strong genetic determinant of HIV seropositivity in Caucasian intravenous drug users. J Infect Dis 201(5): 730–739

17. Cunningham D, Humblet Y, Siena S et al (2004) Cetuximab monotherapy and cetuximab plus irinotecan in irinotecan-refractory metastatic colorectal cancer. N Engl J Med 351(4):337–345

18. Hirsch FR, Varella-Garcia M, Cappuzzo F et al (2007) Combination of EGFR gene copy number and protein expression predicts outcome for advanced non-small-cell lung cancer patients treated with gefitinib. Ann Oncol 18(4):752–760

19. Algars A, Lintunen M, Carpen O et al (2011) EGFR gene copy number assessment from areas with highest EGFR expression predicts response to anti-EGFR therapy in colorectal cancer. Br J Cancer 105(2):255–262

20. Dahabreh IJ, Linardou H, Kosmidis P et al (2011) EGFR gene copy number as a predictive biomarker for patients receiving tyrosine kinase inhibitor treatment: a systematic review and meta-analysis in non-small-cell lung cancer. Ann Oncol 22(3):545–552

21. Visakorpi T, Hyytinen E, Kallioniemi A et al (1994) Sensitive detection of chromosome copy number aberrations in prostate cancer by fluorescence in situ hybridization. Am J Pathol 145(3):624–630

22. Wang F, Fu S, Shao Q et al (2013) High EGFR copy number predicts benefits from tyrosine kinase inhibitor treatment for non-small cell lung cancer patients with wild-type EGFR. J Transl Med 11(1):90

23. International Human Genome Sequencing Consortium (2004) Finishing the euchromatic sequence of the human genome. Nature 431(7011):931–945

24. Sharp AJ, Locke DP, McGrath SD et al (2005) Segmental duplications and copy-number variation in the human genome. Am J Hum Genet 77(1):78–88

25. Urban AE, Korbel JO, Selzer R et al (2006) High-resolution mapping of DNA copy alterations in human chromosome 22 using high-density tiling oligonucleotide arrays. Proc Natl Acad Sci U S A 103(12):4534–4539

26. Komura D, Shen F, Ishikawa S et al (2006) Genome-wide detection of human copy number variations using high-density DNA oligonucleotide arrays. Genome Res 16(12):1575–1584

27. Lai WR, Johnson MD, Kucherlapati R et al (2005) Comparative analysis of algorithms for identifying amplifications and deletions in array CGH data. Bioinformatics 21(19):3763–3770

28. Xi R, Hadjipanayis AG, Luquette LJ et al (2011) Copy number variation detection in whole-genome sequencing data using the Bayesian information criterion. Proc Natl Acad Sci U S A 108(46):E1128–E1136

29. Sepulveda N, Campino SG, Assefa SA et al (2013) A Poisson hierarchical modelling approach to detecting copy number variation in sequence coverage data. BMC Genomics 14(1):128

30. D'Haene B, Vandesompele J, Hellemans J (2010) Accurate and objective copy number profiling using real-time quantitative PCR. Methods 50(4):262–270

31. Ingham DJ, Beer S, Money S et al (2001) Quantitative real-time PCR assay for determining transgene copy number in transformed plants. BioTechniques 31(1):132–134, 136–140

32. Huggett JF, Foy CA, Benes V et al (2013) The digital MIQE guidelines: minimum information for publication of quantitative digital PCR experiments. Clin Chem 59(6):892–902

33. Janssen B, Hartmann C, Scholz V et al (2005) MLPA analysis for the detection of deletions, duplications and complex rearrangements in the dystrophin gene: potential and pitfalls. Neurogenetics 6(1):29–35

34. Schouten JP, McElgunn CJ, Waaijer R et al (2002) Relative quantification of 40 nucleic acid sequences by multiplex ligation-dependent probe amplification. Nucleic Acids Res 30(12): e57

35. Vossen RHAM, Aten E, Roos A et al (2009) High-resolution melting analysis (HRMA)-more than just sequence variant screening. Hum Mutat 30(6):860–866

36. Hill HR, Augustine NH, Pryor RJ et al (2010) Rapid genetic analysis of X-linked chronic granulomatous disease by high-resolution melting. J Mol Diagn 12(3):368–376

37. Zimmermann K, Mannhalter JW (1996) Technical aspects of quantitative competitive PCR. BioTechniques 21(2):268

38. Guo QW, Xiao L, Zhou YL (2012) Rapid diagnosis of aneuploidy by high-resolution melting analysis of segmental duplications. Clin Chem 58(6):1019–1025

39. Stofanko M, Goncalves-Dornelas H, Cunha PS et al (2013) Simple, rapid and inexpensive quantitative fluorescent PCR method for detection of microdeletion and microduplication syndromes. PLoS One 8(4):e61328

40. Zhou LM, Palais RA, Paxton CN et al (2015) Copy number assessment by competitive PCR with limiting deoxynucleotide triphosphates and high-resolution melting. Clin Chem 61(5): 724–733

41. Palais R, Wittwer CT (2009) Mathematical algorithms for high-resolution DNA melting analysis. Methods Enzymol 454:323–343

42. Press WH, Teukolsky SA, Vetterling WT et al (2007) Savitzky-Golay smoothing filters. In: Numerical recipes—the art of Scientific computing, 3rd edn. Cambridge University Press, Cambridge, UK, pp 766–772

43. YN S, Hung CC, Lin SY et al (2011) Carrier screening for spinal muscular atrophy (SMA) in 107,611 pregnant women during the period 2005-2009: a prospective population-based cohort study. PLoS One 6(2):e17067

44. Sugarman EA, Nagan N, Zhu H et al (2012) Pan-ethnic carrier screening and prenatal diagnosis for spinal muscular atrophy: clinical laboratory analysis of >72,400 specimens. Eur J Hum Genet 20(1):27–32

45. Pearn J (1978) Incidence, prevalence, and gene frequency studies of chronic childhood spinal muscular-atrophy. J Med Genet 15(6): 409–413

46. Cusco I, Barcelo MJ, Soler C et al (2002) Prenatal diagnosis for risk of spinal muscular atrophy. BJOG 109(11):1244–1249

47. Parsons DW, McAndrew PE, Iannaccone ST et al (1998) Intragenic telSMN mutations: frequency, distribution, evidence of a founder effect, and modification of the spinal muscular atrophy phenotype by cenSMN copy number. Am J Hum Genet 63(6):1712–1723

48. Carothers AD, Hecht CA, Hook EB (1999) International variation in reported livebirth prevalence rates of down syndrome, adjusted for maternal age. J Med Genet 36(5):386–393

49. Canfield MA, Honein MA, Yuskiv N et al (2006) National estimates and race/ethnic-specific variation of selected birth defects in the United States, 1999–2001. Birth Defects Res A Clin Mol Teratol 76(11):747–756

50. Murthy SK, Malhotra AK, Mani S et al (2007) Incidence of down syndrome in Dubai, UAE. Med Princ Pract 16(1):25–28

51. Wahab AA, Bener A, Teebi AS (2006) The incidence patterns of down syndrome in Qatar. Clin Genet 69(4):360–362

52. Dork T, Macek M, Mekus F et al (2000) Characterization of a novel 21-kb deletion, CFTRdele2,3(21 kb), in the CFTR gene: a cystic fibrosis mutation of Slavic origin common in central and East Europe. Hum Genet 106(3): 259–268

53. Farrar JS, Wittwer CT (2016) High-resolution melting curve analysis for molecular diagnostics. In: Patrinos GP, Ansorge WJ, Danielson PB (eds) Molecular diagnostics, 3rd edn. Academic Press, New York, pp 79–102

54. Sedlak RH, Cook L, Huang ML et al (2014) Identification of chromosomally integrated human herpesvirus 6 by droplet digital PCR. Clin Chem 60(5):765–772

55. Hindson BJ, Ness KD, Masquelier DA et al (2011) High-throughput droplet digital PCR system for absolute quantitation of DNA copy number. Anal Chem 83(22):8604–8610

56. Ogino S, Gao SZ, Leonard DGB et al (2003) Inverse correlation between SMN1 and SMN2 copy numbers: evidence for gene conversion from SMN2 to SMN1 (vol 11, pg 275, 2003). Eur J Hum Genet 11(9):723–723

57. Kao HY, YN S, Liao HK et al (2006) Determination of SMN1/SMN2 gene dosage by a quantitative genotyping platform combining capillary electrophoresis and MALDI-TOF mass spectrometry. Clin Chem 52(3):361–369

58. Sheng-Yuan Z, Xiong F, Chen YJ et al (2010) Molecular characterization of SMN copy number derived from carrier screening and from core families with SMA in a Chinese population. Eur J Hum Genet 18(9):978–984

Chapter 9

Using Cloning to Amplify Neuronal Genomes for Whole-Genome Sequencing and Comprehensive Mutation Detection and Validation

Jennifer L. Hazen, Michael A. Duran, Ryan P. Smith, Alberto R. Rodriguez, Greg S. Martin, Sergey Kupriyanov, Ira M. Hall, and Kristin K. Baldwin

Abstract

Recent studies of somatic mutation in neurons and other cell types suggest that somatic cells can acquire hundreds to thousands of new mutations over their lifetimes. Each individual mutation can have extremely low prevalence, with many mutations restricted to a single cell. Because of their rarity, somatic mutations can be challenging to detect and reliably distinguish from false-positive calls arising from amplification, sequencing, or bioinformatic methods. In these scenarios, a variety of methods are required to compensate for the limited applicability and technical artifacts inherent in any single approach. In the method we describe, somatic cell nuclear transfer (SCNT, also known as cloning) is used to reprogram single neurons to blastocysts from which we derive embryonic stem cells. Division of these cells faithfully amplifies the neuronal genome for next-generation sequencing and genome-wide mutation detection. This approach allows the detection of false positives due to amplification artifacts and is applicable to all classes of mutations. While it is both sensitive and reliable, our method is lower throughput than single-cell sequencing-based approaches and may also fail to amplify the most severely compromised neuronal genomes. In this chapter, we outline current methods for generating neuron-derived SCNT embryonic cell lines, discuss best practices for genome-wide mutation detection, and address the advantages and limitations of this approach.

Key words Somatic mutation, Postmitotic neuron, Somatic cell nuclear transfer, Whole-genome sequencing, Mobile element insertion, Structural variant mutation, Copy number variants, Indel mutation, Single-nucleotide variant mutation

---

1 Introduction

1.1 Somatic Mutation Discovery and Cloning

A series of pioneering studies, many performed using the techniques outlined in this book, demonstrated that neuronal genomes undergo changes ranging from aneuploidies and copy number variations to single-nucleotide variants [1–15]. Clinical studies suggest that somatic mutations harbored by as few as 20% of brain cells can cause severe functional disturbances [16], and a number

José María Frade and Fred H. Gage (eds.), *Genomic Mosaicism in Neurons and Other Cell Types*, Neuromethods, vol. 131, DOI 10.1007/978-1-4939-7280-7_9, © Springer Science+Business Media LLC 2017

of neurologic diseases are associated either with certain types of somatic mutation [17–26] or with impaired DNA repair pathways [27–29]. The functional impact of lower prevalence mutations is less clear but depending on their total load many have a significant cumulative effect on cellular function [30–32].

A key challenge in studying somatic mutation is producing a high-confidence picture of genome-wide mutations within a single cell. DNA content assays reveal changes in the size of the genome but not where changes have occurred. FISH, sequence capture, and other targeted sequencing-based approaches provide more information about the types of mutations observed but are limited to predefined categories of mutations.

One important advance towards producing genome-wide profiles of somatic mutation is single-cell sequencing (SCS). In these experiments, the genomes of single cells are amplified by DNA polymerases in vitro until sufficient DNA is generated for whole-genome sequencing [33]. Using SCS, researchers are able to generate genome-wide sequencing data for many cells easily and rapidly. However, SCS approaches remain vulnerable to amplification artifacts [34]. DNA polymerases working in isolation in vitro without the full collection of cellular DNA repair machinery produce far more mutations than those occur during replication. Also, unlike genome replication linked to cell division, in vitro amplification is not required to make a full genomic copy before starting the next round of replication. As a result SCS often produces uneven genomic coverage [17]. This can produce additional false-positive mutations that obscure one's ability to distinguish artifacts from bona fide somatic mutations (see Sect. 1.3). Finally, because SCS destroys the original DNA, it is not possible to independently validate candidate mutations using molecular approaches such as PCR.

As a result, SCS-based techniques are either limited to large copy number variant detection, which is robust to amplification artifacts [33], or must account for high false-positive and -negative rates statistically. Statistical approaches improve estimates of the absolute number of true somatic mutations but cannot specifically remove artifactual mutations. These residual mutations can mask critical mutational signatures and hinder attempts to identify the molecular mechanisms responsible for true somatic mutations.

To expand genome-wide somatic mutation studies to reliably detect smaller and more complex types of mutations, we aimed to develop an approach that incorporates more faithful whole-genome amplification along with an unequivocal means to validate candidate mutations. The most faithful method of genome amplification is through cell division. However, once neurons exit the cell cycle during development, they cannot, in general, be stimulated to resume cell division [35]. Some exceptions have been reported, but these approaches require suppression of key genome maintenance pathways [36–39]. To overcome this issue, we turned to

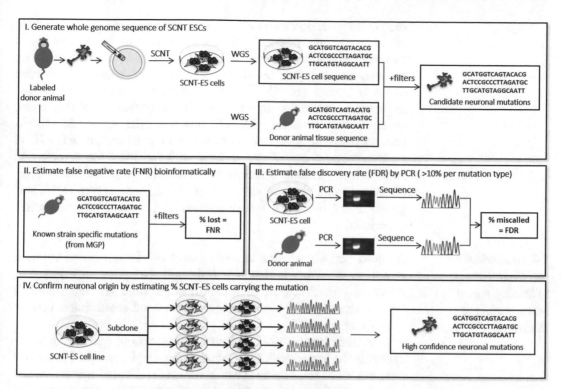

Fig. 1 Schematic outline of SCNT-based mutation detection. Labeled neurons are taken from a donor animal and are subjected to somatic cell nuclear transfer (SCNT) with the aim of deriving embryonic stem (ES) cells. ES cells and tissue from the donor animal are sequenced to determine neuron-specific mutations. The false-negative rate (FNR) is determined by calculating how many known strain-specific mutations survive our somatic mutation filters [14]. The false-positive rate is assayed by PCR validation of subset of mutation calls. Finally, we ensure that candidate somatic mutations did not arise during reprogramming or ES cell culture by confirming the presence of putative somatic mutations in all early-passage subclones

developmental reprogramming via somatic cell nuclear transfer (SCNT or cloning) as a way of converting the cellular identity of a postmitotic neuron to that of a dividing cell. The basic steps of this procedure are selection of appropriate neuronal subtype and transgenic donor animal (Sect. 2.1), donor neuron isolation (Sects. 2.2 and 2.3), somatic cell nuclear transfer (Sect. 2.4), SCNT-ES cell derivation (Sects. 2.5 and 2.6), and whole-genome sequencing and analysis (Sects. 2.7–2.15). For diagram of the experimental overview, see Fig. 1.

1.2 Somatic Cell Nuclear Transfer (SCNT)

SCNT, one of the oldest methods of developmental reprogramming, relies on the natural ability of factors in the oocyte cytoplasm to reset the developmental state of the egg and sperm nuclei [40–45]. In SCNT, the oocyte chromosomes are removed with a micropipette, and replaced with the neuronal genome. The "reconstructed" embryo is then activated to stimulate the beginning of early embryonic development and neuronal genome reprogramming. Within a day, the neuronal genome is copied and the embryo starts to undergo cell division. After several days of in vitro culture, the embryo develops to the blastocyst stage and contains a mature inner cell mass (ICM). In order to harvest enough DNA for sequencing and downstream validation experiments, we generate an embryonic stem cell (ES cell) line from the ICM cells. To do this, we place the blastocyst on a feeder cell layer and allow the ICM cells to migrate out of the embryo and onto the feeder cells. As soon as the neuronal genome is sufficiently amplified for whole-genome sequencing (usually 3–5 passages) we harvest DNA and generate cell stocks for future analysis and downstream validation experiments.

1.3 Advantages and Limitations of Cloning-Based Somatic Mutation Discovery

As described above, one key advantage of cloning-based somatic mutation discovery is that it is unbiased. No prior knowledge of expected mutation type or location is required and all mutation types including single-nucleotide variants and small indels can be reliably detected.

Another key feature of cloning-based whole-genome amplification is the superior fidelity of genome amplification. Lower per base mutation rates means that fewer method-based mutations are introduced. In addition, the uniform amplification produced by cell division allows us to eliminate most of the limited amplification artifacts resulting from errors during S phase. This is because heterozygous mutations present in the neuronal genome must be present in 100% of copies of the genome (or 50% of copies of each chromosome). In contrast, most types of amplification-based non-somatic mutations generated during or after the first round of genome replication will be present in only the copied strand of the original DNA, leading to a true frequency of 25% in the first round of amplification and lower frequencies thereafter. For small mutations such as SNVs and indels located in regions with sufficient sequencing depth, unique sequencing reads can be used to estimate the frequency or the mutation (variant allele frequency—VAF). For mutations lacking sufficient sequencing depth including large and complex mutations, a collection of single-cell subclones can be generated from early-passage SCNT-ES cells. These subclones can be assayed by PCR-based techniques to determine the frequency of candidate mutations in the original SCNT-ES cell line. This subcloning-based method can also be used to independently validate candidate somatic mutations from regions with low overall coverage or nucleotide sequences that are difficult to sequence.

Finally, cloning-based whole-genome amplification produces pluripotent embryonic stem cells, which can be used to assay for the functional impacts of the observed somatic mutations. ES cells can be differentiated to many cell types in vitro or can be used to generate chimeric mice and/or entire animals via tetraploid embryo complementation [46, 47]. Studies of the resulting animals/cells can help reveal whether somatic mutations are likely to impact neuronal function.

As a result of the technical difficulty and low efficiency of SCNT, one important limitation of cloning-based somatic mutation discovery is that it is very low throughput. A successful SCNT experiment using neurons will generate ~70 reconstructed embryos, around 1% of which will go on to generate SCNT-ES cell lines [14]. Many experiments are unsuccessful for unknown reasons.

An additional important consideration when using SCNT and when comparing SCNT-based mutational estimates to figures produced by other methods is the possibility for selection bias. In contrast to in vitro amplification, in vivo genome amplification requires the genome be sufficiently intact to be copied and to support the cellular function of the early embryo and embryonic stem cells. However, we note that the frequency of failed development of reconstructed embryos generated from neurons is approximately the same as those reported from other cell types. Therefore it may be that the frequency of cells with genomes that are incompatible with reprogramming, perhaps due to genomic mutations, is approximately the same for neurons as for other cell types.

A final limitation of SCNT-based whole-genome amplification is that it has not been shown to be possible for human neurons. Furthermore, while successful SCNT has recently been reported using human somatic cells and oocytes [48], the cost and difficulty of obtaining human oocytes are at present likely prohibitive for studies aiming to look directly at human neurons. Encouragingly, however, many of the results reported in our mouse studies are concordant with more recent human single-cell sequencing-based studies, indicating that the mechanisms and types of neuronal mutations may be generally shared between the two species.

2   Materials and Methods

2.1   Donor Animal Choice

As the goal of the experiment is to study the genomes of neurons in particular, and potential differences between different neuronal subtypes, it is important to be certain of the identity of the original reprogrammed cell. During reprogramming, key transcriptional and epigenetic aspects of neuronal cell identity are reset to an early embryonic state. Therefore, without taking specific precautions, it is impossible to be certain the originally reprogrammed cell was a neuron at the end of the experiment.

To ensure donor cell identity, when possible we recommend use of an irreversible genetic marking scheme, or use of a

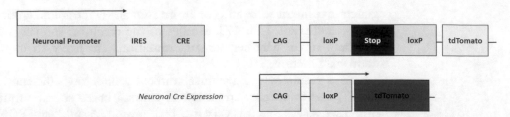

Fig. 2 Cre/loxP genetic and fluorescent labeling of donor neurons. Cre recombinase expression is driven by a promoter specific to the neuronal subtype of interest. Cre recombinase is thus specifically expressed in the neurons of interest and recombines the loxP sites to excise a stop cassette before the fluorescent reporter (in this case tdTomato). This results in a permanent genetic change and fluorescent protein expression specifically in the neuronal cell type of interest

particularly well-characterized and bright fluorescent marker of neuronal subtype identity. One such method is to use Cre-loxp labeling [49, 50], in which a stop cassette preceding a fluorescent protein is excised by Cre recombinase exclusively in the cell type of interest. In this way, the target cells are permanently fluorescently labeled and also harbor an irreversible genetic change that marks their original identity (Fig. 2). When such systems are not available, one can use a transgenic line in which a cell type-specific promoter directly drives fluorescent protein expression. The disadvantage of this approach is that there can be no retrospective confirmation of target cell identity, as the cell type-specific promoter will be reset during reprogramming. However, generating multiple lines from these cells can ensure that any rare cases of cloning from an incorrect cell type are unlikely to alter the overall conclusions. For rare cell types, purification of fluorescent cells (for example by FACS) can be used so that there is limited opportunity to choose the wrong cell as a nuclear donor during SCNT.

Another important consideration when selecting a transgenic mouse line is the brightness of the fluorescent label. Neurons must be visibly fluorescent on an inverted microscope to facilitate picking the correct cell type during nuclear transfer. In our experience, dim fluorescence greatly increases the time it takes to find donor neurons and decreases the number of nuclear transfers that can be completed per experiment.

To ensure specific labeling, we rigorously characterize our transgenic mouse lines by immunostaining with markers for the cell type of interest and for non-neuronal cell types (see **Note 2**). In the olfactory bulb we have had success using Ki67 to label dividing cells, Iba1 to label microglia, S100b to label astrocytes and olfactory ensheathing cells, and Olig2 to label oligodendrocytes [14].

2.2 Considerations in Neuronal Dissociation

Dissociation of adult neurons can be challenging and may require specific tailoring to the neuronal subtype of interest. Below we provide our protocol for isolating healthy, morphologically intact mitral and tufted neurons from the olfactory bulb (modified from Brewer et al. [51]) (see **Note 3**). We suggest optimizing dissociation time, dissociation enzyme, composition of the holding media (media used to store cells following dissociation and before nuclear transfer), and rigorousness of trituration (see **Note 4**).

We strongly recommend preparing practice cell preparations before the scheduled day of nuclear transfer as part of the optimization process. We find that the best way to judge the quality of a cell preparation is to attempt to pick up donor cells and isolate their nuclei using the injection micropipette. In the ideal final cell suspension, the target neurons should be dissociated to single cells to allow neurons to be picked up efficiently in the micropipette (see **Note 5**). Target neurons should remain phase bright with sharp refractive edges and ideally retain some visible evidence of subcellular structure before lysis. They should retain membrane and nuclear integrity when manipulated with micropipette and the cell membrane should only break when a pulse is applied with the piezo drill. The nucleus should not break during this piezo pulse and should remain resistant to subsequent micromanipulation.

Limited cell debris within the cell preparation is not a problem as long as it does not obscure visibility and access to target neurons. However, the final cell solution must not contain free DNA from lysed cells. Free DNA will stick to the micropipette and cause other cells and debris to stick to the pipette, making further manipulation impossible. If stickiness is observed during optimization attempts, free DNA can be removed by treatment with a small amount of DNase I, ideally during papain treatment or shortly thereafter (see below). If DNase I is required, care must be taken to wash away DNase I before micromanipulation so that target neuron DNA is not damaged.

Practice nuclear isolation sessions are also important, as finding the correct diameter injection micropipette is crucial to success. The pipette opening should be narrow enough to stretch and deform the neuron as it enters the injection pipette, which aids in breaking the cell membrane. However, it should not be so small that it radically deforms or damages the nuclear membrane. For mitral and tufted neurons, which are quite large, we use 7 or 8 μ diameter injection pipettes.

A final logistical consideration: Advanced planning is required to save tissue from the donor animal as a control to remove inherited mutations from the dataset. During the neuronal dissociation, we generally arrange for a second person to perform dissections to harvest donor animal tissues. We save at -80 °C tissues from a range of germ layers including brain, thymus (in young animals), lungs, heart, liver, spleen, stomach, kidney, and tail. When possible, we choose to sequence thymus as our control tissue because T cells within the thymus have known mutations in the T cell receptor genes that can be used as a positive control for structural variant detection. For older animals due to thymic involution with aging, we use the spleen as an alternative control tissue.

2.3 Neuronal Dissociation for Nuclear Transfer

Brain tissue is dissected (under a fluorescent dissecting microscope if necessary to isolate desired brain regions) and placed immediately in ice-cold HAGB media, which is composed of Hibernate-A basal media supplemented with 1× B27 without vitamin A and 500 μM Glutamax. For aged mice (>1 year old) we observe a slight increase in viability replacing standard Hibernate-A with Ca^{2+}-free Hibernate-A (BrainBits) supplemented with 0.5 mM $CaCl_2$ and 6 mM $MgCl_2$ in all steps of the protocol (see **Note 6**).

To facilitate enzymatic digestion, the tissue is chopped with a sharp razor blade taking care to slice rather than crush the tissue (see **Note 7**). Tissue chunks are transferred to 3 mL of a pre-activated solution of 10 units/mL papain (PAP2, Worthington Biochemical) dissolved in HAG (Hibernate-A plus 500 μM Glutamax). To activate the papain solution, incubate for at least 30 min at 37 °C. To increase enzyme activity, we keep the papain solution warm by attempting to coordinate the end of the activation period with the start of tissue digestion. To facilitate this, we routinely leave the enzyme at 37 °C for as much as 30 min longer than needed for activation and observe no decrease in enzyme activity.

For papain digestion, the tissue is placed in a 30 °C water bath for 10 min with constant vigorous shaking to increase tissue penetration and slough off dead/dying surface tissue. If evidence of free DNA appears (long spindly, sticky debris and/or previously unattached tissue chunks adhering to one another) either during digestion or in subsequent trituration steps add a small amount of DNase I solution (no more than 6 μg total, Roche 10,104,159,001). It is especially important to eliminate free DNA prior to trituration if tissue chunks are adhering to one another, as large collections of tissue can clog the trituration pipette and cause compression damage to the target neurons within the tissue. At the end of digestion, tissue chunks are allowed to settle and papain is gently removed, taking care not to remove chunks of tissue.

To disperse neurons into a single-cell solution 2 mL of ice-cold HAGB is added and tissue is triturated ten times with fire-polished Pasteur pipettes (9″ borosilicate glass with cotton plug) in three rounds. Trituration should be forceful but care should be taken not to introduce bubbles into the cell suspension. We make three sizes of fire-polished Pasteur pipette so that in each round, the diameter of the fire-polished Pasteur pipettes decreases slightly to compensate for the decreasing size of tissue chunks (see **Note 8**). At the end of each round of trituration, the remaining tissue chunks are allowed to settle and overlaying cell suspension is removed and stored on ice.

At the end of trituration, we purify the resulting 6 mL of cell suspension either with a 70 μ nylon cell strainer or with an OptiPrep density gradient. We find that the OptiPrep density gradient is more effective at eliminating debris and dead or dying cells. The OptiPrep density gradient is prepared and centrifuged as in Brewer et al. [51]. During optimization experiments, we strongly suggest examining each of the four layers of the density gradient and the cell pellet to see where your target cells are located.

After density column or cell strainer, the resulting cell solution is diluted in 10 mL of ice-cold HAGB and centrifuged for 5 min at 200 r.c.f. and 4 °C to eliminate residual papain and DNase I. The resulting cell pellet is resuspended in ~100 μL ice-cold HAGB and stored on ice until the time of nuclear transfer.

2.4 Somatic Cell Nuclear Transfer

Our protocol for SCNT is based largely on the work of Kishigami et al. [52], and Eggan et al. [53] which we strongly recommend reviewing. Here, we briefly summarize their protocol with modifications to reflect advances in the literature and adaptations to performing nuclear transfer on adult neurons (see **Note 9**).

In preparation for nuclear transfer, we pre-equilibrate our working stocks of CO_2-buffered medias at 37 °C, 5% CO_2, to establish optimal pH. These include KSOMaa (LifeGlobal Group, formerly Zenith Biotech, ZEKS-050) and Ca-free CZB medium (recipe as in Kishigami et al. [52]). Because these media are made and stored in small quantities, they tend to degas at 4 °C quite quickly as evidenced by media color change. We find that the efficiency of NT is significantly improved when the medias are pre-equilibrated.

Oocytes are harvested from 8–10-week-old B6D21F females (C57BL/6 × DBA/2). To superovulate, oocyte donor animals are injected with 5 IU of PMSG at 5:30 pm 3 days prior to nuclear transfer, and 5 IU of hCG at 5:30 pm the evening before nuclear transfer. At 9 am the morning of nuclear transfer, oocyte-cumulus cell complexes are collected from oviduct ampullae in M2 media (Cytospring, M2102). Cumulus cells are removed from oocytes using 0.1% hyaluronidase in M2 medium supplemented with BSA (Cytospring M2102HB).

The following embryo micromanipulations are performed on a Nikon Eclipse microscope equipped with a Hoffman condenser, Narishige NT-88-V3 micromanipulators, and a Primetech piezo drill. Importantly, the microscope is mounted on an air-pressured bench to minimize external vibrations.

In order to make the metaphase II spindle visible for enucleation, oocytes are cultured for 30 min (37 °C, 5% CO_2) in KSOMaa medium supplemented with 4 mg/mL BSA (Sigma A3311). Enucleation is performed using 15°-angled piezo drill micropipettes with 7 μ diameter (Origio, PIEZO-7-15) in a drop of M2 supplemented with cytochalasin B 5 μg/mL (Sigma, C6762). Importantly, we perform enucleation on a 37 °C-heated stage, which helps keep the spindles easily visible for enucleation and allows larger groups of oocytes to be enucleated together (20–30 oocytes per group) (see **Note 10**).

In preparation for nuclear transfer, neuronal cells are mixed thoroughly in a drop of 10% PVP solution made by reconstituting a vial of lyophilized PVP (Irvine Scientific, 99,219) with 1 mL of M2 (see **Note 11**). Fluorescent cells are selected as nuclear donors and collected in groups of 10–15 in the injection pipette (7 or 8 μ diameter, 15°-angled piezo drill micropipettes, Origio). No attempt is made at this point to break the cell membrane or isolate the nucleus. We prefer to do this immediately before nuclear transfer within the same drop as enucleated oocytes both to keep the cell alive longer and because the cell suspension drop is quite crowded. Once the cell membrane is broken, fluorescence dissipates, and nuclei are often less brightly fluorescent than intact cells. Therefore, it can be easy to lose the neuron/nucleus or confuse it with another cell.

After nuclear transfer, embryos are activated in Ca-free CZB medium supplemented with 10 mM $SrCl_2$, 5 μg/mL cytochalasin B, and 5 nM trichostatin A for 6 h (see **Note 12**). Following activation, embryos are cultured in KSOMaa medium supplemented with 5 nM TSA for an additional 10 h. Trichostatin A is a histone deacetylase inhibitor thought to aid in epigenetic remodeling of somatic cell nuclei. It has been used extensively to enhance the rate of cloned blastocyst generation (up to fivefold [54]) and ES cell derivation from cloned blastocysts (approximately double, [34]). Following TSA incubation, embryos are cultured in KSOMaa at 37 °C until blastocysts develop (see **Note 13**).

2.5 SCNT-ES Cell Derivation and Culture

ES cell lines are derived based on a protocol from Meissner et al. [55] with modifications. Briefly, zona pellucida is removed from blastocyst with a piezo-actuated drill needle [14] and the zona-free embryos are placed on a MEF feeder layer (see **Note 14**). To aid in successful attachment, we use a short glass capillary attached to a mouth pipette to settle the embryo directly over the feeder layer. The inner cell mass cells grow out of the embryo and on to the feeder layer over the following 7–9 days in ES cell derivation media containing 500 mL knockout DMEM (Gibco), 80 mL knockout serum replacement (Gibco), 6 mL MEM nonessential amino acids (Gibco), 6 mL Glutamax (Gibco), 6 mL Pen/Step (Gibco), 6 μL B-mercaptoethanol (Sigma M7522), 50 μm final concentration MEK1 inhibitor PD98059 (Cell Signaling Technology 9900), and 2000 units/mL LIF (Chemicon ESG1107).

Once large enough, outgrowths of the inner cell mass are manually picked with a plastic p20 pipette tip set to 5 μL. We judge outgrowths to be large enough when they are clearly visible with a stereo dissecting scope set at 20× magnification. To pick colonies, we use the pipette tip to dislodge the colony in one piece, generally with some surrounding feeder cells. To dissociate the colony, we pick up the colony in 5 μL of media and transfer it to an empty 95-well U-bottom tissue culture plate. To this we add 30 μL 0.25% trypsin-EDTA and incubate at 37 °C for 5 min. After incubation, 70 μL of fresh ES cell derivation media is added and the cell suspension is triturated at least five times (see **Note 15**). All 105 μL of the cell suspension is transferred to one well of a 96-well flat-bottom plate containing a MEF feeder layer and 100 μL of fresh ES cell derivation media. After 2–3 days in culture, colonies with ES cell morphology should be visible. Once colonies are large enough to be passaged they are expanded in ES cell maintenance media (same as ES cell derivation media except without MEK1 inhibitor and with 1000 units/mL LIF).

2.6 Purification of DNA for Whole-Genome Sequencing

Prior to harvesting DNA for whole-genome sequencing, we adapt ES cells to feeder-free conditions to eliminate contaminating feeder DNA [56]. We usually aim to do this around passage 3 to 5. Genomic DNA from SCNT-ES cells and donor tissue is purified using standard phenol chloroform extraction, ethanol precipitation, and RNase A treatment. We recommend phenol chloroform extraction over commercial kits because it results in minimally sheared, high-molecular-weight genomic DNA.

2.7 Sequence Analysis General Principles

Sequencing technology and analysis techniques continue to evolve rapidly. We expect that the methods we currently use will be out of date shortly. So, we believe the most valuable information we can provide is a general overview and rationale for our work so that the reader can easily adapt our approach to state-of-the-art technology as needed. It is important to note that specific parameters and thresholds may need to be adjusted to achieve high-quality results depending on sequencing depth, sequencing technology, read-length, data quality, software versions, and the species/strain under study. The thresholds that we include are examples that have worked well for us in our specific experimental paradigm. Specific details of previously published analyses not included here include somatic variant false-negative rate estimations, structural variant and MEI breakpoint determination, and detection and validation of shared mutations. For extensive detail and rationale for these methods, please see the supplemental experimental procedures of our recent publication [14].

One important conceptual point within our methodology is that somatic mutations are defined relative to control tissues saved from the original donor animal (see Sect. 2.2). In the strictest sense, somatic mutations are defined as those that are not inherited from the parental germ cells. Obviously whole-genome sequences from the original sperm and egg are not accessible, so we use tissue from the donor animal as a stand in. Because of this, and our low tolerance for evidence of the mutant allele in the control sample (see Sect. 2.8), the earliest somatic mutations and other somatic mutations present at detectable levels in the saved donor tissues are treated as inherited. Therefore, for our purposes, neuronal somatic mutations are those present in the neuron-derived SCNT-ES cell line, and absent in the sequenced donor tissue. These will include mutations that occur during embryonic and neuronal development as well as during postmitotic function.

With respect to sequencing technology, we strongly recommend using PCR-free methods to generate sequencing libraries (for example the TruSeq DNA PCR-Free Library Preparation Kit from Illumina). As discussed above, in vitro amplification is a significant source of false positives and eliminating PCR during library preparations can help further decrease false positives. In addition, sequencing should be paired-end for optimal structural variant detection and structure determination.

For the analysis pipeline, we highly recommend multiple SCNT-ES cell lines and control samples be analyzed in parallel, ideally from similar strain backgrounds. This allows one to compare genotype and sequence quality information for each candidate mutation across all samples (a.k.a. joint multi-sample genotyping). This practice, now becoming common, helps rule out false positives. Areas prone to false positives due to alignment errors become more evident as more samples are considered. Similarly, genotypes are incorrectly assigned with a certain frequency, particularly in regions of low coverage. Having data from additional samples can function similarly to increasing coverage (see preliminary filter 6 for how this information is used).

2.8 Generalized Preliminary Filters for Somatic Variant Discovery

Our generalized preliminary filters for generating a high-confidence list of somatic mutations require that candidate somatic mutations must:

1. **Be absent from any published database of germline mutations from inbred laboratory strains.** Given the size of the genome, the chance of having a somatic mutation overlap an inherited mutation is very low. The chance of missing an inherited mutation in the control tissue sample is much higher.

2. **Be present on one of the 19 autosomes or on the X and Y chromosome**. At present, we find that calls present on the "random" or "unknown" scaffolds are highly susceptible to alignment errors. It is also important to exclude calls made on the mitochondrial genome, as very little if any mitochondria come from the original neuron. Most come from the oocyte.

3. **Have sufficient sequencing depth to be confident in the predicted genotype.** A lack of information can lead to incorrect calls.

4. **Be predicted to have a non-reference genotype in the SCNT-ES cell with high confidence.**

5. **Be predicted to have a reference genotype in the control tissue sample with high confidence.** Again because of the size of the mouse genome, somatic mutations are highly unlikely to occur at a place where the inherited genotype differs from reference. We find that cases where the control and SCNT-ES cell sample genotypes differ, and the control sample genotype is not reference, usually result from alignment errors. If one is expressly interested in somatic mutations where the true genotype of the control sample is not reference, it is relatively simple to generate custom scripts to search for these mutations separately.

6. **Be predicted to have a reference genotype in all samples not from the same donor animal with high confidence.** Because of the complex and delicate nature of SCNT, many experiments will fail, while successful experiments will result in the production of several SCNT-ES cells. When multiple SCNT-ES cell lines are generated from the same animal, bona fide somatic mutations that occur prior to neuronal cell cycle exit may be shared between one or more ES cell lines. These mutations are of interest because they can help reconstruct lineage relationships [11, 12], and should be left in the dataset. However, assuming that the distribution of somatic mutations has a large random component, true somatic mutations are highly unlikely to recur in neurons from different animals. Indeed, in our experience, calls made in multiple neurons from different animals are generally false positives. To eliminate these false positives, we require strong evidence that all samples from different donor animals (both SCNT-ES cells and control samples) have the reference genotype. However, mutations that recur across multiple neurons may be quite interesting and could suggest fragile regions or programmed genomic change. We generally write a custom script to look for these mutations in a separate analysis. For an example, see the section labeled "Detection and validation of shared mutations" in our recent publication [14].

2.9 Initial Alignment and Post-Processing

For all bioinformatic methods described here forward, default parameters are used in software packages except as explicitly noted. SCNT-ES cell lines and control samples for each mouse are sequenced using Illumina whole-genome paired-end sequencing with reads 150 bp in length from templates of approximately 400 bp. As our sequencing facility produces unaligned BAMs containing readgroup metadata, each lane is then separately aligned to the mm10 reference genome (NCBI GRCm38_68) using SpeedSeq "realign" (v0.1.0) [57]. If reads are provided as FASTQ files, SpeedSeq "align" should be used with user-specified meta-data for each pair of FASTQs.

SpeedSeq automatically marks duplicates, position sorts, and extracts discordant and split reads for downstream SV detection [58]. For comprehensive variant discovery, we recommend a median per-sample read depth of no less than 30× in the final BAM.

2.10   SNV and Indel Detection

Single-nucleotide variant (SNV) and indel calling is performed using GATK v3.4 [59, 60]. HaplotypeCaller is run on each sample individually in parallel using the GVCF mode (—emitRefConfidence GVCF). Genotypes are then calculated across all samples with GATK GenotypeGVCF, yielding a single VCF file with raw SNV and indel calls for all samples. GATK VariantRecalibrator and ApplyRecalibration steps are then run on SNPs (—mode SNP), and indels (—mode INDEL) independently, to assign the calls into four sensitivity tranches. To generate the required "truth" sets for SNPs, we start with the high-confidence 129S1 and C57BL6 SNP calls from the MGP v5 [61], intersect these with our own autosomal GATK SNV calls from above, and select the highest ranked calls by QUAL, GQ, MQ0F, and MQ scores (228, 127, 0, 60). To generate an indel "truth" set, we us all high-confidence 129S1 and C57BL6 indel calls from the MGP.

From this call set, we select putative somatic variants using custom scripts modeled after the approach taken in Kong et al. [62]. For each mouse, we partition samples into three sets:

1. The control sample (donor animal tissue sample) for that mouse
2. The SCNT-ES cell line(s) for that mouse
3. The "other" samples, comprised of all other samples from all other mice

In order to be called a putative somatic mutation in a particular SCNT-ES cell line, a variant locus/allele pair must meet all of the following criteria:

1. The same alternate allele may not be reported at the same locus in any inbred mouse strain by the MGP. For indels, which are known to have higher false-positive rates than SNVs, we are slightly more rigorous, and require that the variant must also not overlap any indel reported by the MGP regardless of the type and size of the indel.
2. The call must appear in one of the 19 autosomes or the X or Y chromosome.
3. The control sample and the SCNT-ES cell line(s) from the mouse of interest must each have a total read depth between 10 and 250.

4. One or more SCNT-ES cell lines for the mouse of interest must have an AAG/RR ratio of phred likelihood scores $\geq 10^{10}$ and a VAF $\geq 30\%$. For the X and Y chromosomes in male mice, the VAF must be greater than or equal to 95%.

5. The control RR/AAG ratio of phred likelihood scores must be $\geq 10^5$ and the control VAF must be at most 5% for SNVs and must be 0% for indels.

6. For all "other" samples, the RR/AAG ratio of phred likelihood scores must be ≥ 1, and the VAF must be $\leq 5\%$ for SNVs and 0% for indels.

Where "RR," "AR," and "AA" refer to homozygous reference (R), heterozygous, or homozygous alternate allele (A) genotypes, respectively. "AAG" refers to the alternate allele genotype, which depends on the chromosome and sex of the mouse. For the autosomes and X chromosome of female mice we use AR, and for the X/Y chromosomes of male mice we use AA.

2.11 Structural Variation Breakpoint Detection

Structural variant breakpoints are detected and genotyped using Lumpy and SVTyper via SpeedSeq [63, 64]. The "sv" module of SpeedSeq is run on the discordant and split read BAMs from all samples, requiring at least four confirming reads for a call, using a minimum alignment mapping quality of 10, and excluding all genomic regions in which any cell line has an aligned read depth > 500.

The resulting SV calls are considered putative somatic structural variants if they meet all of the following criteria:

1. The SV must not be previously reported as a germline polymorphism by MGP in any mouse strain. We define a SV call as previously reported if it shares the same variant type (e.g., deletion) and is at the same genomic location (50% reciprocal overlap, bedtools intersect -r -f 0.5) as a previously reported SV.

2. The call appears in one of the 19 autosomes or the X or Y chromosome.

3. The call must have at least five supporting discordant read pairs and/or split reads from one SCNT-ES cell line and a minimum VAF of 15% as reported by SVTyper. We choose a lower VAF threshold for SVs than for SNVs as the alternate allele is often underrepresented during SV discovery and genotyping relative to SNP/indel calls.

4. The call may not have any supporting reads in any other SCNT-ES cell or control sample from any mouse.

2.12 Copy Number Variation Detection by Read Depth Analysis

Our current copy number variation (CNVs) detection strategy is extensively described in our recent publication [14]. The only difference is that we now use the mm10 reference genome, which does not have publicly available mappability tracks. To generate mappability tracks de novo we recommend the gem-mappability tool [65] with parameter "–l150" followed by gem-2-wig which produces a mappability track for 150mers.

2.13 Mutation Validation via Sanger Sequencing

For SNVs and indels, we find it very useful to take advantage of quality score or ranking functions built into many mutation detection pipelines. Our strategy is generally to validate a handful of the highest and lowest ranked candidate mutations, as well as a randomly selected group of candidate mutations totaling around 10% of the complete call set.

If we discover many false positives, we adjust our filters or the parameters of mutation detection pipeline and repeat validation experiments. During our initial optimization of GATK for example, we found several rounds of validation necessary. In our final high-confidence call sets, close to 100% of candidate calls were positively validated.

For SVs, MEIs, and CNVs, we use intentionally lenient parameters that produce many false positives to ensure that we do not overlook at bona fide somatic mutations. Because of this, all putative mutations are validated by PCR and Sanger sequencing.

The PCR and Sanger sequencing techniques we use are extensively described in the text and methods of our previous publication [14].

2.14 Validation of Somatic Origin of Candidate Mutations

As described in the introduction, one key advantage of our approach is that we are better able to eliminate amplification artifacts that may arise from cell division and reprogramming. We do this by requiring that mutations are present in all cells within the SCNT-ES cell population. The rationale is that mutations present in the original neuron must also be present in all resulting SCNT-ES cells and copies of the genome. Artifactual amplification and reprogramming-associated mutations that arise later will be present in smaller proportion of the ES cell population and amplified genomes.

To ensure that SNVs and indels are present in all SCNT-ES cells, we use a minimum variant allele frequency cutoff. Variant allele frequency (VAF) is defined as the fraction of total reads that cover a given locus bearing the alternate (non-reference) allele. In our case, we use a 0.30 VAF cutoff, below which we assume that mutations are heterozygous and mosaic within the SCNT-ES cell population. We established our 0.30 cutoff based on previous experience; however, one could generate a dataset-specific cutoff empirically by examining the VAF distribution of a "gold standard set" of inherited heterozygous mutations (see the section "Single Nucleotide Variant and Indel False Negative Rate Estimation" in the Supplemental Experimental Procedures section of our recent publication [14] for details of the gold standard set).

For structural variants, mobile element insertions, and copy number variants, an allele frequency-based approach is not as robust, as the alternative allele can be underrepresented in sequencing data due to uneven coverage or less robust alignment. To overcome this, rather than sampling the combined SCNT-ES cell population, we generate and expand single-cell subclones from early-passage SCNT-ES cells (see **Note 16**). The fraction of subclones harboring the variant allele is then taken as a reflection of the prevalence of the mutation within the larger SCNT-ES cell population. To assay for the presence/absence of a candidate somatic mutation in subclone DNA, we perform PCR using diagnostic validation primers [14].

2.15 Generation of Early-Passage SCNT-ES Cell Subclones and DNA Purification

To eliminate contamination from feeder cell DNA, subclones must be generated from feeder-free ES cells. To do this, early-passage SCNT ES cells (prior to passage 6, if possible) are seeded at low density to allow physical space for picking. To pick, scrape a colony gently off the surface using a plastic pipette tip. Pick up the colony and transfer it directly to one well of a 96-well flat-bottom tissue culture dish coated with gelatin. Allow the colony to adhere, recover, and grow in ES cell maintenance media until the colony is large enough to be passaged.

DNA from SCNT-ES cell subclones is purified in 96-well format. In this format, it is important not to disturb or dislodge the attached cells. DNA will remain intertwined with cellular components adherent to the plate until the final TE resuspension step. Subclones are grown to confluency washed with PBS and incubated in 50 µL lysis buffer (100 mM Tris pH 8.0, 5 mM EDTA, 0.2% SDS, 200 mM NaCl, 100 µg/mL proteinase K) for 1 h at 55 °C in a humidified hybridization oven. To remove soluble impurities, 80 µL of cold 100% ethanol is added to the lysed cells. After addition of ethanol, the solution becomes viscous. To reduce the viscosity and dissolve impurities, pipette several times, again being careful not to dislodge attached cells. After pipetting, incubate the solution for 1 h at −80 °C and then 5 min at room temperature. Remove the supernatant, wash the wells twice with 70% ethanol, and air-dry the empty wells at room temperature for 20 min. Resuspend the resulting DNA in 35–80 µL of TE by scraping the bottom of the well with a pipette tip while pipetting up and down. It is important to ensure that pipette tips are firmly attached to the pipetman during this step, especially when using a multichannel pipette, as tips can easily become dislodged and cross contaminate other wells. To improve resuspension of DNA, samples are incubated overnight at 37 °C. This step can also be performed at 55 °C for 1 h with slightly less efficiency. PCR is performed on 1ul of the final DNA solution.

3 Notes

1. Although FACS can yield a pure population of labeled cells for nuclear transfer, we find that FACS can decrease the viability of neurons, leading to less development in subsequent NT.

2. In addition to tissue staining, we have also validated our neuronal preps by cytospin (protocol will vary depending on the instrument available) or by letting the neurons adhere to cell culture plates, and then fixing and staining the wells. This is particularly useful to demonstrate an absence of contaminating cells if the transgenic mouse line labels cells beyond the neuron of interest in closely associated brain regions.

3. Anecdotally, we find that the faster the dissociation protocol and the shorter the time between euthanasia of the animal and nuclear transfer, the higher the success rate. Therefore we suggest preparing all reagents ahead of time and using practice preparations as an opportunity to streamline the dissociation protocol. We have also observed that the viability of cell suspensions is increased by maximizing time spent on ice. For all steps where a temperature is not explicitly stated, samples are stored on ice.

4. We have observed significant differences in the quantity and quality of donor neurons based on age and cell type. Adjusting dissociation conditions can significantly improve the quality of donor neurons.

5. Note that for neuronal cell types with limited viability as single cells, we have had success leaving target neurons in small clumps of cells by shortening enzyme digestion time. Removing neurons from these clumps requires practice. Using the piezo drill can help.

6. We have the greatest success when Hibernate-A-based media solutions are prepared fresh from stocks no greater than 24 h ahead of use. All media must be stored at 4 °C or on ice except when specifically noted otherwise.

7. The chopping step is not necessary with small or easily dissociated tissues. For example we routinely dissociated rod photoreceptors from retina without chopping.

8. Take care that fire-polished pipettes do not have sharp edges and that the openings are not so small that tissue impedes the flow of media through the opening which can damage the neurons.

9. Please note that nuclear transfer is a highly complex, delicate procedure that can fail for unknown reasons. For this reason, we strongly recommend using specific vendor and part numbers for the reagents we list here when possible. We also strictly

follow the manufacturer's preparation instructions and strictly adhere to their indicated expiration dates.

10. It is important to remember to turn off warming stage and allow it to cool prior to the NT step. The heated stage is helpful during enucleation but can cause oocytes to lyse during nuclear transfer.

11. Because of the high percentage of PVP, the M2-PVP solution is quite viscous and it can increase the time it takes for neurons to settle to the bottom of the dish. This is undesirable as it takes much more time to find fluorescent cells and the experiment needs to proceed quickly at this point. We find that premixing the desired amount of cell suspension with M2-PVP immediately prior to placing it in the drop on the NT dish greatly accelerates the rate at which the cells settle to the bottom.

12. When we make a new batch of CZB or activation reagent stock solutions, before use in SCNT experiments, we always run a test activation on parthenote or in vitro-fertilized embryo development.

13. We strongly recommend that prior to enucleation, 10–20 high-quality oocytes be reserved for a parthenote development control. Unfertilized oocytes can be activated and cultured in parallel with reconstructed SCNT embryos. If in vitro culture conditions are optimal all or nearly all activated oocytes will develop to the blastocyst stage. We include this control in all experiments as a check on oocyte and reagent quality.

14. NT-blastocysts can often look irregular in comparison to normal or blastocyst and parthenote controls. We find that anything with a blastocoel cavity (small or large) can be placed on feeders.

15. In addition to the ICM cells, various differentiated cell types will expand from the blastocyst and surround the ICM cells. The central goal of trypsinization is to separate the ICM cells from these cell types. It is not completely necessary to obtain a single-cell suspension.

16. It is very important to generate subclones from SCNT-ES cells as early as possible. The longer ES cells are in culture, the greater the potential for bottlenecks and other population-wide changes that may obscure the true prevalence of a putative somatic mutation in the original SCNT-ES cell population.

4   Conclusions

The cloning-based somatic mutation methodology described above enables high-confidence genome-wide detection of all categories of mutations. It can be easily adapted to new bioinformatic approaches to mutation discovery and to new advances in sequencing technology. One major improvement, which would greatly expand the utility and scope of this method, would be the discovery of a method to reprogram neurons with higher throughput, perhaps via transcription factor-mediated reprogramming. However, at present no reliable high-throughput methods to induce neurons to reenter the cell cycle or reprogram to pluripotency have been reported.

In the future, we anticipate that cloning-based somatic mutation discovery can help guide the interpretation of data from complementary methods, and that together the field will begin to produce a clear picture of somatic mutation across many different neuronal subtypes and neuronal ages. Results of these studies should help establish how somatic mutation impacts neurons across the brain and across the animal's life span, both in healthy animals and in models of neurodegenerative disorders and of disorders of genome maintenance.

References

1. Rehen SK, McConnell MJ, Kaushal D, Kingsbury MA, Yang AH, Chun J (2001) Chromosomal variation in neurons of the developing and adult mammalian nervous system. Proc Natl Acad Sci 98(23):13361–13366

2. Muotri AR, Chu VT, Marchetto MCN, Deng W, Moran JV, Gage FH (2005) Somatic mosaicism in neuronal precursor cells mediated by L1 retrotransposition. Nature 435(7044):903–910

3. Rehen SK, Yung YC, McCreight MP, Kaushal D, Yang AH, Almeida BSV, Kingsbury MA, Cabral KMS, McConnell MJ, Anliker B, Fontanoz M, Chun J (2005) Constitutional aneuploidy in the normal human brain. J Neurosci 25(9):2176–2180

4. Coufal NG, Garcia-Perez JL, Peng GE, Yeo GW, Mu Y, Lovci MT, Morell M, O'Shea KS, Moran JV, Gage FH (2009) L1 retrotransposition in human neural progenitor cells. Nature 460(7259):1127–1131

5. Baillie JK, Barnett MW, Upton KR, Gerhardt DJ, Richmond TA, De Sapio F, Brennan PM, Rizzu P, Smith S, Fell M, Talbot RT, Gustincich S, Freeman TC, Mattick JS, Hume DA, Heutink P, Carninci P, Jeddeloh JA, Faulkner GJ (2011) Somatic retrotransposition alters the genetic landscape of the human brain. Nature 479(7374):534–537

6. Evrony GD, Cai X, Lee E, Hills LB, Elhosary PC, Lehmann HS, Parker JJ, Atabay KD, Gilmore EC, Poduri A, Park PJ, Walsh CA (2012) Single-neuron sequencing analysis of L1 Retrotransposition and somatic mutation in the human brain. Cell 151(3):483–496

7. Poduri A, Evrony Gilad D, Cai X, Elhosary Princess C, Beroukhim R, Lehtinen Maria K, Hills LB, Heinzen Erin L, Hill A, Hill RS, Barry Brenda J, Bourgeois Blaise FD, Riviello James J, Barkovich AJ, Black Peter M, Ligon Keith L, Walsh Christopher A (2012) Somatic activation of AKT3 causes hemispheric developmental brain malformations. Neuron 74(1):41–48. doi:10.1016/j.neuron.2012.03.010

8. McConnell MJ, Lindberg MR, Brennand KJ, Piper JC, Voet T, Cowing-Zitron C, Shumilina S, Lasken RS, Vermeesch JR, Hall IM, Gage FH (2013) Mosaic copy number variation in human neurons. Science 342(6158):632

9. Suberbielle E, Sanchez PE, Kravitz AV, Wang X, Ho K, Eilertson K, Devidze N, Kreitzer AC, Mucke L (2013) Physiologic brain activity causes DNA double-strand breaks in neurons, with exacerbation by amyloid-[beta]. Nat Neurosci 16(5):613–621. doi:10.1038/nn.3356; http://www.nature.com/neuro/journal/v16/n5/abs/nn.3356.html#supplementary-information

10. Gole J, Gore A, Richards A, Chiu Y-J, Fung H-L, Bushman D, Chiang H-I, Chun J, Lo Y-H, Zhang K (2013) Massively parallel polymerase cloning and genome sequencing of single cells using nanoliter microwells. Nat Biotechnol 31(12):1126–1132

11. Evrony Gilad D, Lee E, Mehta Bhaven K, Benjamini Y, Johnson Robert M, Cai X, Yang L, Haseley P, Lehmann Hillel S, Park Peter J, Walsh Christopher A (2014) Cell lineage analysis in human brain using endogenous

Retroelements. Neuron 85(1):49–59. doi:10.1016/j.neuron.2014.12.028

12. Lodato MA, Woodworth MB, Lee S, Evrony GD, Mehta BK, Karger A, Lee S, Chittenden TW, D'Gama AM, Cai X, Luquette LJ, Lee E, Park PJ, Walsh CA (2015) Somatic mutation in single human neurons tracks developmental and transcriptional history. Science 350(6256):94

13. Upton Kyle R, Gerhardt Daniel J, Jesuadian JS, Richardson Sandra R, Sánchez-Luque Francisco J, Bodea Gabriela O, Ewing Adam D, Salvador-Palomeque C, van der Knaap MS, Brennan Paul M, Vanderver A, Faulkner Geoffrey J (2015) Ubiquitous L1 mosaicism in hippocampal neurons. Cell 161(2):228–239. doi:10.1016/j.cell.2015.03.026

14. Hazen Jennifer L, Faust Gregory G, Rodriguez Alberto R, Ferguson William C, Shumilina S, Clark Royden A, Boland Michael J, Martin G, Chubukov P, Tsunemoto Rachel K, Torkamani A, Kupriyanov S, Hall Ira M, Baldwin Kristin K (2016) The complete genome sequences, unique mutational spectra, and developmental potency of adult neurons revealed by cloning. Neuron 89(6):1223–1236. doi:10.1016/j. Neuron.2016.02.004

15. Erwin JA, Paquola ACM, Singer T, Gallina I, Novotny M, Quayle C, Bedrosian TA, Alves FIA, Butcher CR, Herdy JR, Sarkar A, Lasken RS, Muotri AR, Gage FH (2016) L1-associated genomic regions are deleted in somatic cells of the healthy human brain. Nat Neurosci 19(12):1583–1591. doi:10.1038/nn.4388; http://www.nature.com/neuro/journal/v19/n12/abs/nn.4388.html#supplementary-information

16. Cai X, Evrony Gilad D, Lehmann Hillel S, Elhosary Princess C, Mehta Bhaven K, Poduri A, Walsh Christopher A (2014) Single-cell, genome-wide sequencing identifies clonal somatic copy-number variation in the human brain. Cell Rep 8(5):1280–1289. doi:10.1016/j.celrep.2014.07.043

17. Lathe R, Harris A (2009) Differential display detects host nucleic acid motifs altered in scrapie-infected brain. J Mol Biol 392(3):813–822. doi:10.1016/j.jmb.2009.07.045

18. Jeong B-H, Lee Y-J, Carp RI, Kim Y-S (2010) The prevalence of human endogenous retroviruses in cerebrospinal fluids from patients with sporadic Creutzfeldt–Jakob disease. J Clin Virol 47(2):136–142

19. Muotri AR, Marchetto MC, Coufal NG, Oefner R, Yeo G, Nakashima K, Gage FH (2010) L1 retrotransposition in neurons is modulated by MeCP2. Nature 468(7322):443–446

20. Douville R, Liu J, Rothstein J, Nath A (2011) Identification of active loci of a human endogenous retrovirus in neurons of patients with amyotrophic lateral sclerosis. Ann Neurol 69(1):141–151. doi:10.1002/ana.22149

21. Kaneko H, Dridi S, Tarallo V, Gelfand BD, Fowler BJ, Cho WG, Kleinman ME, Ponicsan SL, Hauswirth WW, Chiodo VA, Kariko K, Yoo JW, D-k L, Hadziahmetovic M, Song Y, Misra S, Chaudhuri G, Buaas FW, Braun RE, Hinton DR, Zhang Q, Grossniklaus HE, Provis JM, Madigan MC, Milam AH, Justice NL, Albuquerque RJC, Blandford AD, Bogdanovich S, Hirano Y, Witta J, Fuchs E, Littman DR, Ambati BK, Rudin CM, Chong MMW, Provost P, Kugel JF, Goodrich JA, Dunaief JL, Baffi JZ, Ambati J (2011) DICER1 deficit induces Alu RNA toxicity in age-related macular degeneration. Nature 471(7338):325–330

22. Coufal NG, Garcia-Perez JL, Peng GE, Marchetto MCN, Muotri AR, Mu Y, Carson CT, Macia A, Moran JV, Gage FH (2011) Ataxia telangiectasia mutated (ATM) modulates long interspersed element-1 (L1) retrotransposition in human neural stem cells. Proc Natl Acad Sci 108(51):20382–20387

23. Tan H, Qurashi A, Poidevin M, Nelson DL, Li H, Jin P (2012) Retrotransposon activation contributes to fragile X premutation rCGG-mediated neurodegeneration. Hum Mol Genet 21(1):57–65

24. Li W, Jin Y, Prazak L, Hammell M, Dubnau J (2012) Transposable elements in TDP-43-mediated neurodegenerative disorders. PLoS One 7(9):e44099. doi:10.1371/journal.pone.0044099

25. Li W, Prazak L, Chatterjee N, Gruninger S, Krug L, Theodorou D, Dubnau J (2013) Activation of transposable elements during aging and neuronal decline in drosophila. Nat Neurosci 16(5):529–531

26. Bundo M, Toyoshima M, Okada Y, Akamatsu W, Ueda J, Nemoto-Miyauchi T, Sunaga F, Toritsuka M, Ikawa D, Kakita A, Kato M, Kasai K, Kishimoto T, Nawa H, Okano H, Yoshikawa T, Kato T, Iwamoto K (2014) Increased L1 Retrotransposition in the neuronal genome in schizophrenia. Neuron 81(2):306–313. doi:10.1016/j.neuron.2013.10.053

27. Borgesius NZ, de Waard MC, van der Pluijm I, Omrani A, Zondag GCM, van der Horst GTJ, Melton DW, Hoeijmakers JHJ, Jaarsma D, Elgersma Y (2011) Accelerated age-related cognitive decline and neurodegeneration, caused by deficient DNA repair. J Neurosci 31(35):12543

28. Madabhushi R, Pan L, Tsai L-H (2014) DNA damage and its links to neurodegeneration. Neuron 83(2):266–282. doi:10.1016/j.neuron.2014.06.034

29. Jeppesen DK, Bohr VA, Stevnsner T (2011) DNA repair deficiency in neurodegeneration. Prog Neurobiol 94(2):166–200. doi:10.1016/j.pneurobio.2011.04.013

30. Poduri A, Evrony GD, Cai X, Walsh CA (2013) Somatic mutation, genomic variation, and neurological disease. Science 341(6141):1237758

31. Shendure J, Akey JM (2015) The origins, determinants, and consequences of human mutations. Science 349(6255):1478–1483

32. McKinnon PJ (2013) Maintaining genome stability in the nervous system. Nat Neurosci 16(11):1523–1529. doi:10.1038/nn.3537

33. Gawad C, Koh W, Quake SR (2016) Single-cell genome sequencing: current state of the science. Nat Rev Genet 17(3):175–188. doi:10.1038/nrg.2015.16

34. Macaulay IC, Voet T (2014) Single cell genomics: advances and future perspectives. PLoS Genet 10(1):e1004126. doi:10.1371/journal.pgen.1004126

35. Herrup K, Neve R, Ackerman SL, Copani A (2004) Divide and die: cell cycle events as triggers of nerve cell death. J Neurosci 24(42):9232

36. Hiler D, Chen X, Hazen J, Kupriyanov S, Carroll Patrick A, Qu C, Xu B, Johnson D, Griffiths L, Frase S, Rodriguez Alberto R, Martin G, Zhang J, Jeon J, Fan Y, Finkelstein D, Eisenman Robert N, Baldwin K, Dyer Michael A (2015) Quantification of Retinogenesis in 3D cultures reveals epigenetic memory and higher efficiency in iPSCs derived from rod photoreceptors. Cell Stem Cell 17(1):101–115. doi:10.1016/j.stem.2015.05.015

37. Ajioka I, Martins RAP, Bayazitov IT, Donovan S, Johnson DA, Frase S, Cicero SA, Boyd K, Zakharenko SS, Dyer MA (2007) Differentiated horizontal interneurons clonally expand to form metastatic retinoblastoma in mice. Cell 131(2):378–390. doi:10.1016/j.cell.2007.09.036

38. Friedmann-Morvinski D, Bushong EA, Ke E, Soda Y, Marumoto T, Singer O, Ellisman MH, Verma IM (2012) Dedifferentiation of neurons and astrocytes by oncogenes can induce gliomas in mice. Science 338(6110):1080–1084

39. Kim J, Lengner CJ, Kirak O, Hanna J, Cassady JP, Lodato MA, Wu S, Faddah DA, Steine EJ, Gao Q, Fu D, Dawlaty M, Jaenisch R (2011) Reprogramming of postnatal neurons into induced pluripotent stem cells by defined factors. Stem Cells 29(6):992–1000

40. Wilmut I, Schnieke AE, McWhir J, Kind AJ, Campbell KHS (1997) Viable offspring derived from fetal and adult mammalian cells. Nature 385(6619):810–813

41. Wakayama T, Perry ACF, Zuccotti M, Johnson KR, Yanagimachi R (1998) Full-term development of mice from enucleated oocytes injected with cumulus cell nuclei. Nature 394(6691):369–374

42. Eggan K, Baldwin K, Tackett M, Osborne J, Gogos J, Chess A, Axel R, Jaenisch R (2004) Mice cloned from olfactory sensory neurons. Nature 428(6978):44–49

43. Eggan K, Jaenisch R (2006) Generation of embryonic stem (ES) cell-derived embryos and mice by tetraploid–embryo complementation. Mammalian and avian Transgenesis—new approaches. Springer, Heidelberg

44. Hochedlinger K, Jaenisch R (2002) Monoclonal mice generated by nuclear transfer from mature B and T donor cells. Nature 415(6875):1035–1038

45. Makino H, Yamazaki Y, Hirabayashi T, Kaneko R, Hamada S, Kawamura Y, Osada T, Yanagimachi R, Yagi T (2005) Mouse embryos and chimera cloned from neural cells in the postnatal cerebral cortex. Cloning Stem Cells 7(1):45–61

46. Nagy A, Gocza E, Diaz EM, Prideaux VR, Ivanyi E, Markkula M, Rossant J (1990) Embryonic stem cells alone are able to support fetal development in the mouse. Development 110(3):815–821

47. Nagy A, Rossant J, Nagy R, Abramow-Newerly W, Roder JC (1993) Derivation of completely cell culture-derived mice from early-passage embryonic stem cells. Proc Natl Acad Sci 90(18):8424–8428

48. Tachibana M, Amato P, Sparman M, Gutierrez Nuria M, Tippner-Hedges R, Ma H, Kang E, Fulati A, Lee H-S, Sritanaudomchai H, Masterson K, Larson J, Eaton D, Sadler-Fredd K, Battaglia D, Lee D, Wu D, Jensen J, Patton P, Gokhale S, Stouffer Richard L, Wolf D, Mitalipov S (2013) Human embryonic stem cells derived by somatic cell nuclear transfer. Cell 153(6):1228–1238. doi:10.1016/j.cell.2013.05.006

49. Lewandoski M (2001) Conditional control of gene expression in the mouse. Nat Rev Genet 2(10):743–755

50. Madisen L, Zwingman TA, Sunkin SM, SW O, Zariwala HA, Gu H, Ng LL, Palmiter RD, Hawrylycz MJ, Jones AR, Lein ES, Zeng H (2010) A robust and high-throughput Cre reporting and characterization system for the whole mouse brain. Nat Neurosci 13(1):133–140; http://www.nature.com/neuro/journal/v13/n1/suppinfo/nn.2467_S1.html

51. Brewer GJ, Torricelli JR (2007) Isolation and culture of adult neurons and neurospheres. Nat Protoc 2(6):1490–1498

52. Kishigami S, Wakayama S, Van Thuan N, Ohta H, Mizutani E, Hikichi T, Bui H-T, Balbach S, Ogura A, Boiani M, Wakayama T (2006) Production of cloned mice by somatic cellnuclear transfer. Nat Protoc 1(1):125–138; http://www.nature.com/nprot/journal/v1/n1/suppinfo/nprot.2006.21_S1.html

53. Eggan K, Jaenisch R (2006) Cloning the laboratory mouse by nuclear transfer. In: Pease S, Lois C (eds) Mammalian and Avian transgenesis—new approaches. Springer, Berlin, pp 69–96. doi:10.1007/978-3-540-28489-5_4

54. Kishigami S, Mizutani E, Ohta H, Hikichi T, Thuan NV, Wakayama S, Bui H-T, Wakayama T (2006) Significant improvement of mouse cloning technique by treatment with trichostatin a after somatic nuclear transfer. Biochem Biophys Res Commun 340(1):183–189. doi:10.1016/j.bbrc.2005.11.164

55. Meissner A, Eminli S, Jaenisch R (2009) Derivation and manipulation of murine embryonic stem cells. In: Audet J, Stanford W (eds) Stem cells in regenerative medicine, Methods in molecular biology, vol 482. Humana Press, New York, pp 3–19. doi:10.1007/978-1-59745-060-7_1

56. Tamm C, Pijuan Galitó S, Annerén C (2013) A comparative study of Protocols for mouse embryonic stem cell culturing. PLoS One 8(12):e81156. doi:10.1371/journal.pone.0081156

57. Chiang C, Layer RM, Faust GG, Lindberg MR, Rose DB, Garrison EP, Marth GT, Quinlan AR, Hall IM (2015) SpeedSeq: ultra-fast personal genome analysis and interpretation. Nat Methods 12:966–968. doi:10.1038/nmeth.3505

58. Tarasov A, Vilella AJ, Cuppen E, Nijman IJ, Prins P (2015) Sambamba: fast processing of NGS alignment formats. Bioinformatics 31:2032–2034. doi:10.1093/bioinformatics/btv098

59. DePristo M, Banks E, Poplin R, Garimella K, Maguire J, Hartl C, Philippakis A, del Angel G, Rivas MA, Hanna M, McKenna A, Fennell T, Kernytsky A, Sivachenko A, Cibulskis K, Gabriel S, Altshuler D, Daly M (2011) A framework for variation discovery and genotyping using next-generation DNA sequencing data. Nature Genetics 43:491–498. doi:10.1038/ng.806

60. Van der Auwera GA, Carneiro MO, Hartl C, Poplin R, Del Angel G, Levy-Moonshine A, Jordan T, Shakir K, Roazen D, Thibault J, Banks E, Garimella KV, Altshuler D, Gabriel S, DePristo MA (2013) From FastQ data to high confidence variant calls: the genome analysis toolkit best practices pipeline. Curr Protoc Bioinformatics 43:11.10–11.33. doi:10.1002/0471250953.bi1110s43

61. Keane TM, Goodstadt L, Danecek P, White MA, Wong K, Yalcin B, Heger A, Agam A, Slater G, Goodson M, Furlotte NA, Eskin E, Nellåker C, Whitley H, Cleak J, Janowitz D, Hernandez-Pliego P, Edwards A, Belgard TG, Oliver PL, McIntyre RE, Bhomra A, Nicod J, Gan X, Yuan W, van der Weyden L, Steward CA, Balasubramaniam S, Stalker J, Mott R, Durbin R, Jackson IJ, Czechanski A, Assunção JAG, Donahue LR, Reinholdt LG, Payseur BA, Ponting CP, Birney E, Flint J, Adams DJ (2011) Mouse genomic variation and its effect on phenotypes and gene regulation. Nature 477:289–294. doi:10.1038/nature10413

62. Kong A, Frigge ML, Masson G, Besenbacher S, Sulem P, Magnusson G, Gudjonsson SA, Sigurdsson A, Jonasdottir A, Jonasdottir A, Wong WSW, Sigurdsson G, Walters GB, Steinberg S, Helgason H, Thorleifsson G, Gudbjartsson DF, Helgason A, Magnusson OT, Thorsteinsdottir U, Stefansson K (2012) Rate of de novo mutations and the importance of father/'s age to disease risk. Nature 488:471–475. doi:10.1038/nature11396

63. Layer RM, Chiang C, Quinlan AR, Hall IM (2014) LUMPY: a probabilistic framework for structural variant discovery. Genome Biol 15:R84. doi:10.1186/gb-2014-15-6-r84

64. Hall IM, et. al., SV Typer. https://github.com/hall-lab/sv-pipeline

65. Derrien T, Estellé J, Sola SM, Knowles DG, Raineri E, Guigó R, Ribeca P (2012) Fast computation and applications of genome Mappability. PLoS One 7:e30377. doi:10.1371/journal.pone.0030377

Part IV

LINE-1 Retrotransposition

Chapter 10

Analysis of LINE-1 Retrotransposition in Neural Progenitor Cells and Neurons

Angela Macia and Alysson R. Muotri

Abstract

Long interspersed nuclear element-1 (LINE-1 or L1) is a type of retrotransposon that comprise around 17% of the human genome. Increasing evidence has suggested that L1 activity, termed as L1 retrotransposition, may occur in somatic cells such as neural progenitor cells (NPC) in higher rate than other non-brain tissues. Indeed, L1 retrotransposition has been found to be associated with several types of neurological disorders. Thus, L1 activity may contribute to the mosaicism in brain tissues, suggesting an intriguing, and also important role of L1 in the central nervous system.

Key words Line-1, Retrotransposition, Brain, Stem cell, NPCs, Neurons

1 Introduction

The human genome project has shown that the approximately 21,000 protein-coding genes of the genome represent only a small portion of our DNA and nearly 99% of the human genome does not encode proteins. Noncoding DNA is highly repetitive DNA that contains numerous introns, pseudogenes, and transposable elements (TEs) [1]. TEs are repetitive DNA sequences with the ability to move or transpose within the genome. TEs can be found in all living forms, from bacteria to mammals and the percentage occupied by TEs and their ongoing activity can vary widely between organisms, accounting for approximately 45% of the human genome [2].

TEs found in genomes can be divided into two broad classes: DNA transposons and retrotransposons [3]. DNA transposons are sequences able to move by a cut-and-paste mechanism in which the transposon is excised from one location and reintegrated elsewhere using an specialized enzyme termed transposase [4]. Retrotransposons are DNA sequences that move through an RNA intermediate by a copy-and-paste mechanism using a Reverse Transcriptase (RT) activity [5]. Retrotransposons can be

José María Frade and Fred H. Gage (eds.), *Genomic Mosaicism in Neurons and Other Cell Types*, Neuromethods, vol. 131, DOI 10.1007/978-1-4939-7280-7_10, © Springer Science+Business Media LLC 2017

subdivided in two major groups: LTR and Non-LTR retrotransposons. LTR retrotransposons only comprise about 8% of the human
genome. By contrast, the majority of human TEs are non-LTR
retrotransposons, typified by L1, Alu, and SVA elements, which
account for about one-third of the human genome [1].

L1 elements are the only active autonomous retrotransposon
class in our genome [6]. There are more than 500,000 L1 copies
per human genome but the vast majority is unable to move and
only few L1 elements remain active [3, 7–9]. Indeed, the average
human genome contains approximately 80–100 L1 full-length
copies that are able to mobilize, termed retrotransposition-
competent L1s (or RC-L1s) [10, 11]. New L1 insertions can
impact our genome function/regulation and can have mutagenic,
neutral, or beneficial effects. Because the mutagenic effects are
easier to detect, de novo L1 insertions are occasionally associated
with the generation of a human disorder. In general, L1 insertion
occurs randomly in the genome, so any human gene is susceptible
of being disrupted by this element. This implies that the range of
diseases caused by de novo L1 insertions is very broad, including
hemophilia, muscular dystrophy, or lung cancer [12, 13]. Indeed,
estimates of L1 retrotransposition based on the frequency of
disease-causing L1 mutations suggested a rate of 1 insertion per
100–150 births [14, 15]. From an evolutionary point of view, new
L1 insertions have the biological meaning of perpetuating L1s in
the human genome. Thus, the generation of a human genetic disorder by retrotransposition is likely a product of their inherent
capability to generate new transmissible L1 insertions.

2   Retroelements' Biology and Regulation

RC-L1s are approximately 6 Kb in length [8] and contain a
5′UnTranslated Region (UTR), up to three open reading frames
(ORFs), followed by a 3′UTR, and ends in a poly(A) tail (Fig. 1).
The 5′UTR presents both sense and antisense RNA polymerase II
promoter activity and a recently described antisense ORF0 [16–
20]. The ORF0 encodes a very short peptide thought to be
primate-specific. Indeed, ORF0 is only present in human and
chimpanzee genomes and leads to an increase of L1 retrotransposition using an engineered based assay, although its function remains
to be determined [19]. RC-L1 also contains other two ORFs:
ORF1 that encodes for a 40 kDa protein (ORF1p) with RNA
binding and nucleic acid chaperone activity, and ORF2, which
encodes a 150 kDa protein (ORF2p) with both endonuclease
(EN) and reverse transcriptase (RT) activities (Fig. 1) [21–26].
The sense promoter activity of the L1 5′UTR will generate the L1
RNA transcript that is used to translate ORF1p and ORF2p

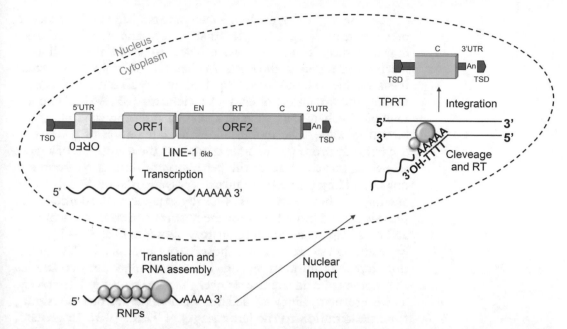

Fig. 1 The retrotransposition cycle. The first step in L1 retrotransposition involves RNA polymerase II-mediated transcription of a genomic L1 locus from an internal promoter. The L1 RNA is exported to the cytoplasm where ORF1 (RNA-binding protein) and ORF2 (endonuclease and reverse transcriptase protein activities) are translated. Both proteins preferentially associate with the L1 RNA transcript to produce a ribonucleoprotein (RNP) particle. The RNP is then transported back into the nucleus. The integration of the L1 element into the genome occurs by a process termed target-primed reverse transcription (TPRT). During TPRT, the L1 endonuclease cleaves the first strand of target DNA, and the L1 RNA is used as a template by the L1 reverse transcriptase. How the second strand is synthesized and integrated is a poorly understood mechanism. Hallmark of the integration process include 2–20 bp-long duplications of the target site (TSD)

necessary for their mobilization. These proteins preferentially assemble with L1 RNA and form a ribonucleoprotein particle (RNP) (Fig. 1) [27–29]. Once the RNP enters the nucleus, L1 will generate a new L1 insertion, potentially using a mechanism termed target-primed reverse transcription (TPRT) [30, 31]. During TPRT, the L1-encoded EN can cleave the DNA at the consensus genomic sequence (5'TTTT/A) [32], exposing a free 3'OH that function as a primer for reverse transcription [33] (Fig. 1). L1 RNA will serve as a template generating the first cDNA linked to the genome. A similar process is thought to occur on the top strand of DNA, giving rise to a newly inserted L1 elsewhere in the genome [30]. Some of the hallmarks of L1 integration by TPRT include target site duplications (TSDs) and an L1 poly-A tail. However, most of the L1 insertions will not result in the generation of a new full-length L1 copy, being 5' truncated copies incapable of forming functional RNPs, and therefore to mobilize again [34]. It is unclear whether the reverse transcriptase is restricted to the nucleus or also occurs in the cytoplasm; thus, alternative mechanisms to the TPRT model might also exist.

During evolution, the human genome has been under selective pressure generating strategies that reduce the activity of L1s. Similarly these elements have evolved to avoid the inhibitory mechanisms raised by the hosts. Ancient L1 subfamilies, molecular fossils unable to mobilize are replaced by new L1 subfamilies, promoting the expansion of active L1 elements [35, 36]. There are multiple mechanisms by which a new L1 insertion can impact the human genome. The mobilization and integration of a L1 copy can not only promote genomic changes as the disruption of a gene sequence, but also lead to the generation of distinct transcription units by adding promoter sequences, polyadenylation signals, or by altering the chromatin status of nearby sequences. In addition, L1s provide novel binding sites for the host transcriptional machinery, and thus help create novel regulatory networks [13, 37–41].

Although the role of most host factors identified on L1 regulation/retrotransposition remains to be determined, several studies shed some light on the matter over the years to study L1 biology. There are many lines of defence against L1 retrotransposition, from transcription to the latest stages of TPRT [42]. In general, DNA methylation, chromatin remodeling, and posttranscriptional regulation of L1 mRNAs are the main processes to control L1 retrotransposition [43]. Several authors have extensively reviewed the different strategies that the cell has developed to coexist with jumping DNA, as some known viral host factors may have evolved from the restriction of ancient endogenous retroelements [34, 44–47]. Although we know many of these cellular factors, we are just starting to uncover new host factors that control and regulate L1 activity during the retrotransposition cycle.

3   Somatic L1 Retrotransposition

L1s and their host genomes are in a situation comparable with a "host-parasite" coevolution, where L1s' function is restricted to its replication and its genetic transmission. Thus, new L1 insertions might accumulate in cells that can be spread over generations (germ cells and early embryo) [40, 48]. Inherited insertions are present in the parent and in all tissues of the new individual. L1 is able to mobilize during early embryogenesis although L1 insertions in germ cells seem to be less frequent [49–51]. More recently, L1 retrotransposition has been associated with viability of fetal oocytes and impairment of preimplantation development in mice [52, 53]. Several in vitro studies had demonstrated that both human embryonic stem cells (hESCs) and human induced pluripotent stem cells (iPSCs), cellular models for early human development, overexpress a constellation of L1 RNA derived sequences and can support L1 retrotransposition using an engineered L1 retrotransposition assay

[18, 51, 54–56]. Nevertheless, the frequency and specific timing when L1 retrotransposition events take place in early human embryogenesis remains to be determined. Pluripotent states are associated with an epigenetic de-repression and transcriptional activation of a myriad of genes; thus, the surrounding environment could contribute to the mobilization of L1 elements. Increasing evidence shows the importance of de novo mutations (present in the offspring but not detected in the parents) in several human disorders. Although the role of somatic retrotransposition is currently unknown, the genomics revolution has demonstrated that there is a load of ongoing retrotransposition in the brain. Intriguingly, somatic L1 insertions have been described in both healthy and pathological human brain samples.

The first evidence demonstrating that L1s are able to mobilize in mammalian brains used neural progenitor cells (NPCs) isolated from adult rat hippocampus [57]. NPCs are multipotent cells found in neurogenic regions of the mammalian brain. Indeed, cortical neurogenesis begins from embryonic neuroepithelial progenitors that give rise to intermediate progenitors and subsequently divide to form neurons and glial cells [58]. During development, neurons migrate from the proliferative zones as hippocampus and subventricular zone toward the surface of the brain to form six distinct histological layers and establish new neuronal networks [59]. Thus, L1-associated mutations occurring in progenitor cells could potentially change the cellular phenotypes in the nascent neurons. In addition, although most of the neurons in the brain are generated before birth, new neurons are continuously generated by progenitor stem cells. Thus, in vitro models that recapitulate neural development are essential to understand L1 behavior in the human brain.

This initial study analyzing L1 mobilization in neural progenitor cells, found L1 retrotransposition in the brains of both male and female transgenic animals, in regions as the striatum, cortex, hypothalamus, hilus, cerebellum, ventricles, amygdala, and hippocampus. In order to study L1 mobilization, authors used the engineered human L1-EGFP reporter assay in both cultured rodent cells and animal models [56, 57]. The L1-EGFP retrotransposition assay relies in the use of an engineered L1 tagged with a reporter cassette that can only be activated after a round of retrotransposition (Fig. 2). This chapter will describe the optimized methodology used to study and analyze L1 retrotransposition in vitro. Muotri et al. not only shows that the mammalian brain is able to accommodate human L1 retrotransposition at a high rate, but also demonstrate that some of these engineered L1 insertions occurred into neuronal expressed genes, altering its expression and, in turn, possibly influencing neuronal cell fate [57]. This study was promptly extended to humans, where the L1-EGFP reporter can

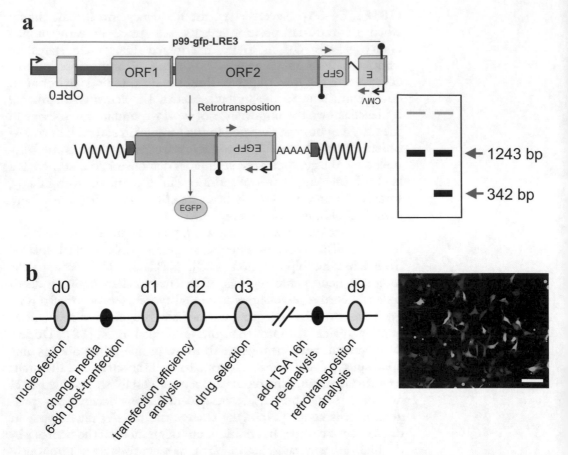

Fig. 2 L1 retrotransposition assay. (**a**) A retrotransposition-competent human L1 contains an EGFP retrotransposition indicator cassette in the 3′UTR. The EGFP gene is cloned backwards the L1 promoter containing its own promoter and polyadenylation signal. In addition, the gene is truncated by an intron, thus, ensuring that EGFP expression will only become activated upon L1 retrotransposition. The *blue arrows* indicate PCR primers flanking the intron present in the EGFP gene. In the figure, the plasmid depicted is p99-gfp-LRE3 (*left panel*). The 1243-bp product amplifies the EGFP cassette containing the intron (vector). The 343-bp PCR product indicates the spliced tagged L1 (insertion), indicative of a retrotransposition event (*right panel*). (**b**) Timeline if the retrotransposition assay described in the methods. The *right panel* shows a representative image of transfected NPC with the p99-gfp-LRE3 vector. L1 retrotransposition using the wild-type construct will result in green-expressing cells. Scale bar: 100 μm

retrotranspose in NPCs isolated from the human fetal brain and in NPCs derived from human embryonic stem cells (hESCs) [60]. Researchers evaluated the endogenous L1 copy-number variation (CNV) in human tissue by a multiplex TaqMan quantitative polymerase chain reaction (qPCR). The estimation was approximately of 80 more endogenous L1 copies per cell in human brain than heart or liver. Notably, the adult hippocampus, a major neurogenic niche in the brain, exhibited elevated L1 copies compared with other brain regions.

The finding that neuronal cell types are permissive for L1 retrotransposition in both humans and rodent raises the question of whether L1 has a function in neurogenesis. Although the functional impact of L1 retrotransposition in the brain is less clear, enhanced environments could stimulate its mobilization. Indeed, mice in running wheels had threefold more L1 retrotransposition than mice in sedentary environments [61]. In human, the expression of L1 retroelements has been linked to several psychopathological conditions. Environmental factors as drug consumption or conditions as post-traumatic stress disorder (PTSD) or major depressive disorder (MDD) have been described to present changes in the heterochromatin status in brain cells, which could result in increased L1 expression and misregulation ([62–65] and reviewed in Macia et al. 2017). Nevertheless, further studies are needed to establish if L1 can indeed participate in the disease onset and/or progression.

The presence of L1 mobilization occurring preferentially in brain compared with other somatic tissues raises another important question: is L1 retrotransposition lineage-dependent in which all ectoderm-derived cells as NPCs are able to accommodate L1 activity? In other words, are cells derived from the same germ layer able to accommodate similar levels of mobilization through the myriads of tissues and organs developed after gastrulation? Researchers reported that human tissues derived from mesoderm as adrenal gland, kidney, spleen, and endoderm-derived organs as esophagus or stomach, express low levels of L1 mRNA [66]. More recently, Macia et al. compared endogenous L1 expression and retrotransposition from several human somatic tissues as well as different populations of somatic stem cells derived from hESC [67]. Authors found that L1 expression and engineered retrotransposition is lower in mesodermal-derived cells and in keratinocytes (ectoderm-derived cells) when compared to NPCs and neurons [67]. Thus, these studies strengthen previous research, which support the idea that L1 retrotransposition is occurring frequently during embryogenesis and later in brain.

The use of next generation sequencing has provided additional insights into the L1 role in the mammalian brain, which demonstrate that is indeed made of a mosaic of genomes. In 2011, Baillie et al., identified numerous potential somatic L1 insertions from the hippocampus and caudate nucleus of three elderly postmortem brain samples by a high-throughput approach termed retrotransposon capture sequencing (RC-seq), described in Chap. 12 of this book. This capture method relies on a low number of PCR cycles and thus is less prone to artifacts [68]. However, only a small fraction of L1 insertions with retrotransposition structural hallmarks were validated by PCR and characterized by Sanger sequencing [69]. The development of single-cell genomic analyses has been fundamental to the field in order to determine that the generation of genomic variability in the brain by L1 occurs not only in dividing progenitor neural cells, but also in fully differentiated post-mitotic neurons. Evrony et al. amplified genomic 300 single neurons isolated from cerebral cortex and caudate nucleus utilizing a modified L1-seq approach [14]. Using this new methodology, authors identified and validated the first somatic full-length L1 insertion found in the human brain and estimated approximately 0.6 unique L1 insertions per cell. Nevertheless, only few insertions were validated, as many were false positives or chimeric sequences [70]. Next, Upton et al. adapted RC-seq to single cells isolated from hippocampus, and estimated 13.7 and 6.5 somatic L1 insertions per neuron and glia, respectively [71]. More recently, Erwin et al. estimated the L1 insertion rate in 0.58–1 events per cell, in both neurons and glia from hippocampus and frontal cortex of three healthy individuals. In addition, authors described the presence of somatic L1-associated variants (SLAVs) composed of both de novo L1 insertions and retrotransposition-independent structural variants mediated by L1 EN [72]. This study proposes that genomic regions near fixed L1 sequences are prone to dsDNA damage and therefore repaired by homology-mediated mechanism. These heritable hotspots in the genome can generate differences between individuals, where SLAVs have the potential to impact gene expression and thus contribute to somatic mosaicism [72]. Chapter 13 of this book will describe in detail the methodology used for quantification and analysis of neural mosaicism identified by SLAV-seq.

Single cell genomic analysis contributed to precisely determine timing and load of L1 retrotransposition events in the brain. While some L1 insertions occur in embryonic or intermediate neural progenitors, new L1 events can take place late in the development, which identification it has been challenging. Although the discovery of somatic retrotransposition in the healthy brain is stimulating, little is known about the role of this element in any biological process of a specialized tissue as brain, as well as the host factors contributing to neuronal diversity. What seems to be certainly clear is that neuronal cells are more permissive for L1 retrotransposition than other cell types in the human body.

The use of hESCs, successfully generated from early stage human embryos, has allowed a better understanding of the L1 biology as well being source of differentiation into a myriad of cell types [73]. However, to develop cellular models of human disease, it is necessary to generate cell lines with a genetic background able to recapitulate the human pathology. Using iPSCs from patients' fibroblasts, researchers have been able to mimic early stages of a human neurodevelopmental disease (Fig. 3). Consistent with a potential role of L1 activity in the pathophysiology of the brain,

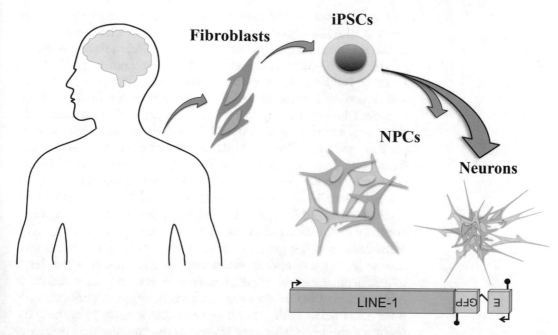

Fig. 3 Neural Progenitor Cells and neurons derived from iPSCs can accommodate L1 retrotransposition. Using NPCs and neurons derived from patients' fibroblasts, we are able to recapitulate early stages of a human neurodevelopmental disease as well as study the possible deregulation of L1 elements. Indeed, L1 elements are de-repressed in a variety of neurodegenerative disorders, although it is unknown if this could alter the brain circuitries or even contribute to neuronal decline in humans and other organisms

several neurodegenerative disorders have shown misregulation of L1, suggesting a contribution to the molecular basis of human genetic disorders. Specifically, Methyl-CpG-binding protein 2 (*MECP2*), a protein involved in global DNA methylation, along with the transcriptional factors Sox2 and the histone deacetylase 1 protein (HDAC1), is known to form up a repressor complex on the L1 promoter region, controlling L1 neuronal transcription and thus retrotransposition [74–76]. Indeed, mutation of *MECP2* in humans cause Rett syndrome (RTT), a progressive neurological disorder being considered part of the autism spectrum disorders (ASD). In the absence of MECP2, L1 is misregulated in both human and mice models. Indeed, there is an increase of endogenous L1 copy number in both RTT postmortem human brain samples and iPSCs-derived NPCs from RTT patients [76], suggesting a contribution of L1 to the genetic disorder [77].

DNA repair pathways also control the regulation of L1 elements in the brain. Ataxia telangiectasia mutated (ATM), a serine/threonine protein kinase, is activated by DNA double-strand breaks. Mutations of ATM cause the disorder Ataxia telangiectasia, characterized by progressive neuronal degeneration, immunodeficiency, and predisposition to cancer. In this study, Coufal et al. found an increase of L1 copies in ATM-deficient hESCs, NPCs and human fetal neural progenitor cells, and in mice model. The absence of ATM allowed the integration of more L1 events as well as longer insertions, suggesting a role of ATM in the recognition of the DNA breaks created during L1 integration [78]. More recently, Bundo et al. proposed a relationship between L1 activity and a major mental disorder as schizophrenia [79]. Authors observed higher L1 copy number in neurons from postmortem prefrontal cortex as well as in iPSCs-derived neurons from schizophrenic patients. The methods for isolation of neuronal nuclei from brain tissue and cultured cells, and the estimation of LINE-1 copy numbers are described in detail in Chap. 11 of this book.

Additionally, enzymes like SAMHD1 and TREX1 dNTP phosphohydrolase and 3′ repair exonuclease, respectively, are known to control the mobilization of L1 retrotransposons [80, 81]. Mutations in either gene are related with Aicardi–Goutières syndrome (AGS), a rare, early onset inflammatory disorder that leads to intellectual and physical problems in humans. The defect of AGS-related genes might cause an accumulation of endogenously produced nucleic acids (as endogenous retroelements) in the cytoplasm of the cells. This may trigger autoimmunity and therefore cause AGS. Although AGS mice do not display any neurological deficit, clinical trials for children with AGS using RT inhibitors to reduce the accumulation of L1 copies have recently began [34, 82–86]. This data suggest that L1 activity could exacerbate some aspects of these diseases or even have a major causative role.

4   Methods to Study and Analyze L1 Retrotransposition in NPCs and Neurons

The development of the cell-based L1 retrotransposition assay has been fundamental for the field. This assay detects retrotransposition events from an exogenous engineered transfected construct, and a reporter gene that is activated as a result of the retrotransposition process [56] (Fig. 2a). Briefly, the reporter cassette (EGFP) is sub-cloned in the 3′UTR of the active L1 element, on the opposite direction of an active L1 element, containing its own promoter and polyadenylation sequences. In addition, the EGFP gene is disrupted by an intron (IVS 2 of the γ-globin gene). Thus, EGFP-positive cells will be detected only after the mRNA is spliced, reverse transcribed, reintegrated into the chromosome and expressed from the reporter's promoter. In sum, only after L1 retrotransposition occurs, the EGFP can be activated and measured (Fig. 2). Over the years, different reporters have been designed to follow L1 retrotransposition using drug selection, microscopy, or luciferase; reviewed in [34]. The assay has been very useful in identifying conserved domains required for L1 retrotransposition as well as the study of restriction factors that control retrotransposition; thus, the L1 retrotransposition assay is widely used across laboratories as an essential tool to decipher aspects of L1 biology [87]. In order to understand the mechanism of L1 integration in the genome, de novo insertions can be isolated and characterized by different methodologies: inverse PCR, recovery assay, biotin capture, etc. [54, 88–91]. In this chapter we show the optimized protocol to test L1 retrotransposition in NPC and in differentiated neurons using an L1-EGFP indicator cassette including analysis and genomic identification of the insertions by inverse PCR.

4.1 Materials

4.1.1 Cell Differentiation and L1 Retrotransposition Analyses

1. 1× Phosphate-buffer saline (PBS), pH 7.4, sterilized (Corning).

2. 5-Bromo-2'-deoxyuridine (BrdU), (Sigma).

3. Ethyl alcohol (Sigma).

4. AD1 Primary Cell 4D–Nucleofector Y Kit (Lonza).

5. Agarose (Sigma).

6. Anti-NeuN mouse monoclonal antibody (Millipore).

7. Cell counter (BioRad) or hemocytometer (Bright-Line™).

8. Cell scraper (Corning).

9. Cell Viability Solution (BD Biosciences).

10. Centrifuge (Eppendorf).

11. Culture media:

    (a) (N2): DMEM/F12 supplemented with 1× N2 and 1% Penicillin/Streptomycin (P/S) (Thermo Fisher).

    (b) (NB): DMEM/F12 supplemented with 0.5× N2, 0.5× B-27, 20 ng/mL of FGF-2 and 1% P/S (Thermo Fisher).

12. Dorsomorphin (Sigma).

13. Endo-free Plasmid Maxi Kit (Qiagen).

14. EVOS® Digital Microscope (Thermo Fisher).

15. FGF-2 (Thermo Fisher).

16. Flow Cytometer and a Fluorescence-activated cell sorter (FACS), (BD Biosciences).

17. Goat serum (Corning).

18. Human/Rat Neuronal Stem Cell Nucleofection kit (Lonza).

19. iRock (Y-27632, Sigma).

20. Laminin (Sigma).

21. Matrigel (Corning).

22. Nanodrop device (Thermo Fisher).

23. Nucleofector® 2b Device (for cells in suspension) and 4D Nucleofector® Device, Unit Y (for adherent cells) (Lonza).

24. Plasmid constructs: p99-gfp-LRE3, p99-gfp-JM111, p99-gfp-LRE3-UB*, p99-gfp-JM111-UB*, and pCEP-EGFP as

described previously in [25, 60, 67, 92]. For more details, see plasmid preparation section of this chapter.

25. Poly-L-ornithine (Sigma).

26. Portable platform shaker (Eppendorf).

27. Puromycin dihydrochloride (Sigma).

28. SB-431542 (Sigma).

29. StemPro Accutase Cell Dissociation Reagent (Thermo Fisher).

30. SYBR® Green (Thermo Fisher).

31. Tissue culture plates.

32. Trichostatin A (Sigma).

33. Triton X-100.

4.1.2   PCR and iPCR

1. DNeasy Blood & Tissue Mini Kit (Qiagen).

2. dNTPs (Invitrogen).

3. Expand Long Template Taq (Roche)

4. MgCl$_2$ (Invitrogen).

5. PCR primers (see "characterization of genomic L1 insertions" in methods' section).

6. Restriction enzymes (NEB). See iPCR methods' section for more information.

7. T4-DNA ligase (NEB).

8. Taq polymerase (Invitrogen).

9. Thermocycler (BioRad).

10. TOPO XL (Invitrogen).

4.2   Methods

4.2.1   Pre Nucleofection

Derivation of NPCs and Neurons from hESCs/ iPSCs

1. In order to generate NPCs from hESCs or iPSCs, previously established methodology is followed [60, 76]. Briefly, pluripotent cells grown on matrigel are cultured during 2 days in N2 media (See materials) containing 1 μm of Dorsomorphin and 10 μm of SB-431542 (Sigma).

2. Undifferentiated cells are detached using a cell scraper and transferred to low-attachment plates or leave at 37 °C under agitation (95 rpm) in the portable platform shaker to allow embryonic body (EB) formation.

3. Change N2 media 2 days after scraping the cells.

4. After 7 days, EBs are plated in a 60 mm matrigel-coated plate, and cultured for 5–7 days using NB media (see materials), changing the media every other day.

5. Rosettes arised in the culture are collected under EVOs microscope, dissociated, and plated on 10 mm plate coated with 10 μg/mL poly-L-ornithine and 2.5 μg/mL Laminin (Sigma) using the NB medium.

6. NPCs are expanded when confluent using StemPro Accutase Cell Dissociation Reagent (Thermo Fisher).

7. To induce neural differentiation, 1×10^5 NPCs are cultured in 24-well plates coated with 20 µg/mL poly-L-ornithine and 5 µg/mL Laminin (Sigma), FGF-2 is withdrawn from the culture, and iNB media is used for 48 h. Next, the media is replaced by NB without FGF-2, and changed every other day. A fully differentiation process is achieved in 30 days.

Plasmid Preparation

All plasmids are purified using Endo-Free Plasmid Maxi kit from Qiagen and analyzed by electrophoresis (1% agarose-SYBR® Green gel). Only highly supercoiled DNA preparations are used in transfection experiments.

– **p99-gfp-LRE3:** contains a full-length retrotransposition-competent L1 element (LRE3) [93] tagged with a *mEGFPI* retrotransposition indicator cassette [92] and contains a puromycin-resistant gene [39] (Fig. 2a).

– **p99-gfp-JM111:** contains a full-length retrotransposition-defective L1 element (L1RP [94]) that contains two engineered missense mutations in L1-ORF1p (RR261/62AA) tagged with a *mEGFPI* retrotransposition indicator cassette and includes a puromycin selection gene.

– **p99-gfp-LRE3-UB*:** is a derivative of plasmid p99-gfp-LRE3 where the CMV promoter that drives expression of EGFP in the mEGFPI cassette has been replaced by the human UBC promoter [67].

– **p99-gfp-JM111-UB*:** is a derivative of plasmid p99-gfp-JM111 where the CMV promoter that drives expression of EGFP in the megfpI cassette has been replaced by the human UBC promoter [67].

– **pCEP-EGFP:** contains the coding sequence of the humanized Enhanced Green Fluorescent Protein (EGFP) cloned in pCEP4 (Invitrogen). Cell are transfected with this expression vector to determine the transfection efficiency of each cell type (Fig. 2b, day2).

4.2.2 NPCs Nucleofection

1. hESC/iPSCs-derived NPCs are cultured with NB and 10 µM iRock (NBi media) for 1 h prior transfection.

2. Cells are detached using StemPro Accutase Cell Dissociation Reagent (Thermo Fisher), centrifuged at 1250 rpm for 5 min, washed with 1× PBS, and filtered with pre-warmed NB media.

3. Cells are nucleofected using the Nucleofector® 2b Device (Lonza). 4×10^6-5×10^6 NPCs are transfer to an Amaxa cuvette with 2–10 µg of the indicated plasmid and the Human or the Rat Neuronal Stem Cell Nucleofection kit (Lonza), selecting the program A-33 for high transfection efficiency.

4. 500 μL of pre-warmed NBi media is inmidiately used to resuspend the NPCs from the Amaxa cuvette and equilibrated in the 37° incubator.

5. 10 min after transfection, NPCs are seeded on 6-well plates coated with 10 μg/mL poly-L-ornithine and 2.5 μg/mL Laminin (Sigma).

6. Replace for fresh NBi media 6–8 h after nucleofection.

7. Cells are fed every other day during the course of the retrotransposition assay.

4.2.3 Post-Nucleofection and Analysis of L1 Retrotransposition in NPCs

1. To determine the transfection efficiency, cells transfected with pCEP-EGFP are analyzed by the percentage of EGFP-expressing cells 48 h post-transfection (d2) (Fig. 2b).

2. Cells are washed with 1 × PBS and detached using StemPro Accutase Cell Dissociation Reagent (Thermo Fisher).

3. Cells are collected from each well in separate microcentrifuge tubes and spin at 1250 rpm for 5 min.

4. Media is removed and NPCs are resuspended in 250 μL 1 × PBS.

5. Add 3 μL of Cell Viability Solution (BD Biosciences) in order to exclude nonviable cells in flow cytometric assays.

6. Determine the percentage of EGFP-expressing cell, gating for live cells on a flow cytometer. The number of live cells that express EGFP serves as an indication of the percentage of cells successfully transfected with plasmids.

7. Transfected NPCs with the engineer L1 constructs are selected for retrotransposition events: Begin drug selection (1–2 μg/mL puromycin selection in NB culture media) 3 days post-transfection (d3) and continue until 9 days post-transfection (d9) (Fig. 2b). Change the puromycin-containing NB medium every other day.

8. Treat a fraction of the NPCs-transfected cells with 250 nM Trichostatin A (TSA) for 16 h prior to flow cytometer analyses as described (Garcia-Perez et al. 2010) to determine whether L1 retrotransposition events are subjected to epigenetic silencing by histone modifications.

9. In order to quantify the percentage of EGFP-expressing cells result of a retrotransposition event; analyze a fraction of transfected cells by flow cytometer. The cells are prepared as previously described in the transfection efficiency assay (d3).

10. In addition, to characterize L1 genomic insertions, genomic DNA can be isolated from a fraction of the NPC-transfected and puromycin-selected cells (d9) (see "characterization of L1 genomic insertions" section of this chapter).

4.2.4 L1 Retrotransposition in Neurons

1. After approximately 30 days of neural differentiation, NB-conditioned media is collected from the culture before transfection.

2. Adhered mature neuronal cells are transfected with 15 μg of the indicated plasmid by using a 4D–Nucleofector Y Unit (Lonza). AD1 Primary Cell 4D–Nucleofector Y Kit (Lonza) is utilized for mature neural cells with the program EH-158.

3. After transfection, hESC-derived neurons are fed with a 1:1 mixture of the collected NB-conditioned media and fresh NB media supplemented with 10 μM iRock (Y-27632, SIGMA).

4. In order to detect somatic retrotransposition in mature neurons and discriminate from progenitor cells, cultures can be incubated with 4 μM 5-Bromo-2′-deoxyuridine (BrdU) during 5 days prior to analyze L1-EGFP-expression by immunostaining. BrdU is used in the detection of proliferating cells, thus mature neurons would be BrdU negative.

5. Alternatively, NeuN-positive cells can be analyzed and collected by FACS. NeuN is used as a marker for mature neurons. 5 days after transfection cells are gently collected and fixed with 70% ice-cold ethanol.

6. Then, samples are permeabilized in 1 × PBS containing 0.1% Triton X-100 and incubated with blocking solution (1 × PBS containing 1% normal goat serum) during 30 min.

7. Samples are then incubated with anti-NeuN mouse monoclonal antibody (1:100, Millipore) during 1 h at 4 °C. Samples are next washed (1 × PBS) and incubated with a fluorescence secondary antibody for 30 min.

8. Cells are washed twice and sorted.

9. After sorting, genomic DNA is isolated from collected NeuN-expressing cells.

4.2.5 Characterization of L1 Genomic Insertions

Intron-Flanking PCR

1. To confirm intron removal and L1 insertion events [92], genomic DNA (gDNA) is isolated form the transfected cells by DNeasy Blood & Tissue Mini Kit (Qiagen) following the manufacturer's instructions.

2. Measure DNA concentration by a Nanodrop device.

3. To assay for intron removal from the EGFP-based retrotransposition assay, 50–100 ng of gDNA are used as a template in 50 μL PCR reactions with the following primers flanking the intron of the EGFP indicator cassette.

   GFP968Fwd (5′-GCACCATCTTCTTCAAGGACGAC);

   GFP1013Rev (5′-TCTTTGCTCAGGGCGGACTG).

4. PCRs are carried out using 1.5 units of Taq polymerase (Invitrogen), 1.5 mM MgCl$_2$, 0.8 mM dNTPs (Invitrogen), and 0.2 μM of each primer. A negative control (no sample)

should be included in the assays. The PCR conditions for the thermocycler (BioRad) are as follows: 95 °C for 2 min, 35 amplification cycles (94 °C for 15 s, 54 °C for 15 s, 72 °C for 30s), and a final step of 72 °C for 7 min.

5. PCR products are resolved on 1% agarose-SYBR® Green gel. The 343-bp PCR product indicates the spliced tagged L1 (insertion); the 1243-bp product contains the intron (vector) (Fig. 2a, right panel).

6. If needed, amplified bands can be excised, purified, and sequenced to ensure identity of amplified products.

Insertion Characterization by Inverse PCR (iPCR)

1. iPCR is carried out as described [51, 87]. In order to characterize L1 genomic insertions, 8 μg of gDNA is digested overnight with a molar excess of *Xba* I, *Ssp* I, or *Bgl* II (NEB).

2. Digested DNAs is ligated at 16 °C for 14 h in a final volume of 500 μL using T4-DNA ligase (NEB).

3. Ligated DNA is precipitated and dissolved in 20 μL of DNase/RNase-Free Distilled Water.

4. Ligated DNA is used to set up a PCR reaction using the primers targeting the EGFP cassette.

iEGFP-1F (5′-CTTGAAGAAGATGGTGCG).
iEGFP-1R (5′-ACAACCACTACCTGAGCACC).

5. PCR reactions are carried out with Expand Long Template Taq (Roche) and the following conditions. Initial step 94 °C for 5 min, 30 amplification cycles (94 °C for 15 s, 64 °C for 30 s, 68 °C for 15 min), and final extension at 72 °C for 15 min. 5 μL of the PCR reaction product is used as a template in a second PCR with the following primers:

iEGFP-2F (5′-TTGAAGAAGTCGTGCTGC).
iEGFP-2R (5′-AAAGACCCCAACGAGAAGCG).

6. Amplified products are resolved in 0.8% agarose-SYBR® Green gel, bands excised and purified.

7. Purified bands are cloned in TOPO-XL (Invitrogen) following manufacturer's instructions and sent to Sanger sequencing with either M13 forward or M13 reverse primers.

8. In order to identify the L1 integration sequence, Blast (http://www.ncbi.nlm.nih.gov/BLAST/) and the Celera databases (http://www.celeradiscoverysystem.com/) can be used.

5   Notes

1. Poly-L-ornithine/Laminin (Sigma) coated-plates must be washed twice in order to avoid toxicity to the cells.

2. In this chapter we show the optimized protocol to test L1 retrotransposition by using a L1-EGFP indicator cassette;

however, other reporter cassettes or drug selection genes can be used.

3. 99-gfp-LRE3-UB* and p99-gfp-JM111-UB* plasmids are recommended to transfect neuronal cultures. Several studies suggest that the CMV promoter is not pan-active in neuronal cells; thus, the CMV promoter that drives EGFP expression was replaced by a ubiquitously expressed human UBC promoter.

4. GFP expression from the retrotransposed mEGFPI reporter cassette is analyzed in cells transfected without pCEP-EGFP, as these transfections are done in parallel.

5. The programs described in the methods for Amaxa nucleofection were previously optimized. Some optimization might be needed with different human cell lines.

6. The nucleofection solution used to transfect neurons (AD1 Primary Cell 4D–Nucleofector Y Kit) is less toxic, thus the media does not need to be changed.

7. Alternatively, other reagents such as Lipofectamine (Invitrogen) or calcium-based transfection can be used for NPC and neural transfection, respectively; however, the best transfection efficiencies were achieved by nucleofection.

8. After nucleofection, resuspend inmidiately the NPCs with the pre-warmed and equilibrated at 37° NBi media, as the nucleofector solution is toxic to the cells.

9. To calculate the adjusted retrotransposition mean = (average number of EGFP-expressing cell)/(pCEP-EGFP-positive cells).

10. In order to calculate a standard deviation, each assay is repeated at least three independent times and each performed with three technical replicates.

11. For the intron-flanking PCR, gDNA can be digested overnight with a molar excess of SwaI (NEB) and then used as a template in the PCR reactions. SwaI cuts within the engineered GFP intron and is used to favor amplification of the spliced product.

Acknowledgments

This work was supported by grants from the National Institutes of Health through the NIH R01MH094753. The work was also supported by the California Institute for Regenerative Medicine (CIRM) award DISC1-08825 and a UCSD CTRI pilot grant to Dr. Muotri.

References

1. Lander ES et al (2001) Initial sequencing and analysis of the human genome. Nature 409(6822):860–921

2. Kazazian HH Jr (2004) Mobile elements: drivers of genome evolution. Science 303(5664):1626–1632

3. Goodier JL, Kazazian HH (2008) Retrotransposons revisited: the restraint and rehabilitation of parasites. Cell 135(1):23–35

4. Munoz-Lopez M, Garcia-Perez JL (2010) DNA transposons: nature and applications in genomics. Curr Genomics 11(2):115–128

5. Moran JV, Gilbert N (2002) Mammalian LINE-1 retrotransposons and related elements. In: Craig N et al (eds) Mobile DNA II. ASM Press, Washington, DC

6. Mills RE et al (2007) Which transposable elements are active in the human genome? Trends Genet 23(4):183–191

7. Venter JC et al (2001) The sequence of the human genome. Science 291(5507):1304–1351

8. Scott AF et al (1987) Origin of the human L1 elements: proposed progenitor genes deduced from a consensus DNA sequence. Genomics 1(2):113–125

9. Dombroski BA et al (1991) Isolation of an active human transposable element. Science 254(5039):1805–1808

10. Brouha B et al (2003) Hot L1s account for the bulk of retrotransposition in the human population. Proc Natl Acad Sci U S A 100(9):5280–5285

11. Sassaman DM et al (1997) Many human L1 elements are capable of retrotransposition. Nat Genet 16(1):37–43

12. Goodier JL (2014) Retrotransposition in tumors and brains. Mob DNA 5(1):11

13. Kazazian HH Jr et al (1988) Haemophilia a resulting from de novo insertion of L1 sequences represents a novel mechanism for mutation in man. Nature 332(6160):164–166

14. Ewing AD, Kazazian HH Jr (2010) High-throughput sequencing reveals extensive variation in human-specific L1 content in individual human genomes. Genome Res 20(9):1262–1270

15. Huang CR et al (2010) Mobile interspersed repeats are major structural variants in the human genome. Cell 141(7):1171–1182

16. Swergold GD (1990) Identification, characterization, and cell specificity of a human LINE-1 promoter. Mol Cell Biol 10(12):6718–6729

17. Speek M (2001) Antisense promoter of human L1 retrotransposon drives transcription of adjacent cellular genes. Mol Cell Biol 21(6):1973–1985

18. Macia A et al (2011) Epigenetic control of retrotransposon expression in human embryonic stem cells. Mol Cell Biol 31(2):300–316

19. Denli AM et al (2015) Primate-specific ORF0 contributes to retrotransposon-mediated diversity. Cell 163(3):583–593

20. Criscione SW et al (2016) Genome-wide characterization of human L1 antisense promoter-driven transcripts. BMC Genomics 17:463

21. Hohjoh H, Singer MF (1997) Sequence-specific single-strand RNA binding protein encoded by the human LINE-1 retrotransposon. EMBO J 16(19):6034–6043

22. Martin SL, Bushman FD (2001) Nucleic acid chaperone activity of the ORF1 protein from the mouse LINE-1 retrotransposon. Mol Cell Biol 21(2):467–475

23. Mathias SL et al (1991) Reverse transcriptase encoded by a human transposable element. Science 254(5039):1808–1810

24. Khazina E et al (2011) Trimeric structure and flexibility of the L1ORF1 protein in human L1 retrotransposition. Nat Struct Mol Biol 18(9):1006–1014

25. Alisch RS et al (2006) Unconventional translation of mammalian LINE-1 retrotransposons. Genes Dev 20(2):210–224

26. Feng Q et al (1996) Human L1 retrotransposon encodes a conserved endonuclease required for retrotransposition. Cell 87:905–916

27. Doucet AJ et al (2010) Characterization of LINE-1 ribonucleoprotein particles. PLoS Genet 6(10)

28. Hohjoh H, Singer MF (1996) Cytoplasmic ribonucleoprotein complexes containing human LINE-1 protein and RNA. EMBO J 15(3):630–639

29. Goodier JL et al (2007) LINE-1 ORF1 protein localizes in stress granules with other RNA-binding proteins, including components of RNA interference RNA-induced silencing complex. Mol Cell Biol 27(18):6469–6483

30. Luan DD et al (1993) Reverse transcription of R2Bm RNA is primed by a nick at the chromosomal target site: a mechanism for non-LTR retrotransposition. Cell 72(4):595–605

31. Cost GJ et al (2002) Human L1 element target-primed reverse transcription in vitro. EMBO J 21(21):5899–5910

32. Jurka J (1997) Sequence patterns indicate an enzymatic involvement in integration of mammalian retroposons. Proc Natl Acad Sci U S A 94(5):1872–1877

33. Monot C et al (2013) The specificity and flexibility of l1 reverse transcription priming at imperfect T-tracts. PLoS Genet 9(5):e1003499

34. Goodier JL (2016) Restricting retrotransposons: a review. Mob DNA 7:16

35. Boissinot S, Chevret P, Furano AV (2000) L1 (LINE-1) retrotransposon evolution and amplification in recent human history. Mol Biol Evol 17(6):915–928

36. Boissinot S, Furano AV (2001) Adaptive evolution in LINE-1 retrotransposons. Mol Biol Evol 18(12):2186–2194

37. Perepelitsa-Belancio V, Deininger P (2003) RNA truncation by premature polyadenylation attenuates human mobile element activity. Nat Genet 35(4):363–366

38. Han JS, Szak ST, Boeke JD (2004) Transcriptional disruption by the L1 retrotransposon and implications for mammalian transcriptomes. Nature 429(6989):268–274

39. Garcia-Perez JL et al (2010) Epigenetic silencing of engineered L1 retrotransposition events in human embryonic carcinoma cells. Nature 466(7307):769–773

40. Ostertag EM, Kazazian HH Jr (2001) Biology of mammalian L1 retrotransposons. Annu Rev Genet 35:501–538

41. Wheelan SJ et al (2005) Gene-breaking: a new paradigm for human retrotransposon-mediated gene evolution. Genome Res 15(8):1073–1078

42. Heras SR et al (2014) Control of mammalian retrotransposons by cellular RNA processing activities. Mobile Genetic Elements 4:e28439

43. Macia A, Blanco-Jimenez E, Garcia-Perez JL (2015) Retrotransposons in pluripotent cells: impact and new roles in cellular plasticity. Biochim Biophys Acta 1849(4):417–426

44. Ariumi Y (2016) Guardian of the human genome: host defense mechanisms against LINE-1 Retrotransposition. Front Chem 4:28

45. Harris RS, Liddament MT (2004) Retroviral restriction by APOBEC proteins. Nat Rev Immunol 4(11):868–877

46. Kinomoto M et al (2007) All APOBEC3 family proteins differentially inhibit LINE-1 retrotransposition. Nucleic Acids Res 35(9):2955–2964

47. Chiu YL, Greene WC (2008) The APOBEC3 cytidine deaminases: an innate defensive network opposing exogenous retroviruses and endogenous retroelements. Annu Rev Immunol 26:317–353

48. Levin HL, Moran JV (2011) Dynamic interactions between transposable elements and their hosts. Nat Rev Genet 12(9):615–627

49. van den Hurk JA et al (2007) L1 retrotransposition can occur early in human embryonic development. Hum Mol Genet 16(13):1587–1592

50. Kano H et al (2009) L1 retrotransposition occurs mainly in embryogenesis and creates somatic mosaicism. Genes Dev 23(11):1303–1312

51. Garcia-Perez JL et al (2007) LINE-1 retrotransposition in human embryonic stem cells. Hum Mol Genet 16(13):1569–1577

52. Malki S et al (2014) A role for retrotransposon LINE-1 in fetal oocyte attrition in mice. Dev Cell 29(5):521–533

53. Kitsou C et al (2016) Exogenous retroelement integration in sperm and embryos affects pre-implantation development. Reproduction 152(3):185–193

54. Wissing S et al (2012) Reprogramming somatic cells into iPS cells activates LINE-1 retroelement mobility. Hum Mol Genet 21(1):208–218

55. Klawitter S et al (2016) Reprogramming triggers endogenous L1 and Alu retrotransposition in human induced pluripotent stem cells. Nat Commun 7:10286

56. Moran JV et al (1996) High frequency retrotransposition in cultured mammalian cells. Cell 87(5):917–927

57. Muotri AR et al (2005) Somatic mosaicism in neuronal precursor cells mediated by L1 retrotransposition. Nature 435(7044):903–910

58. Hu WF, Chahrour MH, Walsh CA (2014) The diverse genetic landscape of neurodevelopmental disorders. Annu Rev Genomics Hum Genet 15:195–213

59. Jamuar SS, Walsh CA (2015) Genomic variants and variations in malformations of cortical development. Pediatr Clin N Am 62(3):571–585

60. Coufal NG et al (2009) L1 retrotransposition in human neural progenitor cells. Nature 460(7259):1127–1131

61. Muotri AR, Zhao C, Marchetto MC, Gage FH (2009) Environmental influence on L1 retrotransposons in the adult hippocampus. Hippocampus 19(10):1002–1007. doi:10.1002/hipo.20564

62. Abrusan G (2012) Somatic transposition in the brain has the potential to influence the biosynthesis of metabolites involved in Parkinson's disease and schizophrenia. Biol Direct 7:41

63. Maze I et al (2010) Essential role of the histone methyltransferase G9a in cocaine-induced plasticity. Science 327(5962):213–216

64. Hunter RG et al (2012) Acute stress and hippocampal histone H3 lysine 9 trimethylation, a retrotransposon silencing response. Proc Natl Acad Sci U S A 109(43):17657–17662

65. Liu S et al (2016) Inverse changes in L1 retrotransposons between blood and brain in major depressive disorder. Sci Rep 6:37530

66. Belancio VP et al (2010) Somatic expression of LINE-1 elements in human tissues. Nucleic Acids Res 38(12):3909–3922

67. Macia A et al (2017) Engineered LINE-1 retrotransposition in non-dividing human neurons. Genome Res 27(3):335–348

68. Sanchez-Luque FJ, Richardson SR, Faulkner GJ (2016) Retrotransposon capture sequencing (RC-Seq): a targeted, high-throughput approach to resolve somatic L1 Retrotransposition in humans. Methods Mol Biol 1400:47–77

69. Baillie JK et al (2011) Somatic retrotransposition alters the genetic landscape of the human brain. Nature 479(7374):534–537

70. Evrony GD et al (2012) Single-neuron sequencing analysis of L1 retrotransposition and somatic mutation in the human brain. Cell 151(3):483–496

71. Upton KR et al (2015) Ubiquitous L1 mosaicism in hippocampal neurons. Cell 161(2):228–239

72. Erwin JA et al (2016) L1-associated genomic regions are deleted in somatic cells of the healthy human brain. Nat Neurosci 19(12):1583–1591

73. Thomson JA et al (1998) Embryonic stem cell lines derived from human blastocysts. Science 282(5391):1145–1147

74. Yoder JA, Walsh CP, Bestor TH (1997) Cytosine methylation and the ecology of intragenomic parasites. Trends Genet 13(8):335–340

75. Kuwabara T et al (2009) Wnt-mediated activation of NeuroD1 and retro-elements during adult neurogenesis. Nat Neurosci 12(9):1097–1105

76. Muotri AR et al (2010) L1 retrotransposition in neurons is modulated by MeCP2. Nature 468(7322):443–446

77. Shpyleva S et al (2017) Overexpression of LINE-1 retrotransposons in autism brain. Mol Neurobiol. doi:10.1007/s12035-017-0421-x

78. Coufal NG et al (2011) Ataxia telangiectasia mutated (ATM) modulates long interspersed element-1 (L1) retrotransposition in human neural stem cells. Proc Natl Acad Sci U S A 108(51):20382–20387

79. Bundo M et al (2014) Increased l1 retrotransposition in the neuronal genome in schizophrenia. Neuron 81(2):306–313

80. Zhao K et al (2013) Modulation of LINE-1 and Alu/SVA retrotransposition by Aicardi-Goutieres syndrome-related SAMHD1. Cell Rep 4(6):1108–1115

81. Stetson DB et al (2008) Trex1 prevents cell-intrinsic initiation of autoimmunity. Cell 134(4):587–598

82. Burdette DL, Vance RE (2013) STING and the innate immune response to nucleic acids in the cytosol. Nat Immunol 14(1):19–26

83. Woo SR et al (2014) STING-dependent cytosolic DNA sensing mediates innate immune recognition of immunogenic tumors. Immunity 41(5):830–842

84. Stetson DB, Medzhitov R (2006) Recognition of cytosolic DNA activates an IRF3-dependent innate immune response. Immunity 24(1):93–103

85. Volkman HE, Stetson DB (2014) The enemy within: endogenous retroelements and autoimmune disease. Nat Immunol 15(5):415–422

86. Fowler BJ et al (2014) Nucleoside reverse transcriptase inhibitors possess intrinsic anti-inflammatory activity. Science 346(6212):1000–1003

87. Wei W et al (2001) Human L1 retrotransposition: cis preference versus trans complementation. Mol Cell Biol 21(4):1429–1439

88. Gilbert N, Lutz-Prigge S, Moran JV (2002) Genomic deletions created upon LINE-1 retrotransposition. Cell 110(3):315–325

89. Symer DE et al (2002) Human l1 retrotransposition is associated with genetic instability in vivo. Cell 110(3):327–338

90. Wagstaff BJ et al (2012) Rescuing Alu: recovery of new inserts shows LINE-1 preserves Alu activity through A-tail expansion. PLoS Genet 8(8):e1002842

91. Cano D et al (2016) Characterization of engineered L1 Retrotransposition events: the recovery method. Methods Mol Biol 1400:165–182

92. Ostertag EM et al (2000) Determination of L1 retrotransposition kinetics in cultured cells. Nucleic Acids Res 28(6):1418–1423

93. Brouha B et al (2002) Evidence consistent with human L1 retrotransposition in maternal meiosis I. Am J Hum Genet 71(2):327–336

94. Kimberland ML et al (1999) Full-length human L1 insertions retain the capacity for high frequency retrotransposition in cultured cells. Hum Mol Genet 8(8):1557–1560

Chapter 11

Estimation of LINE-1 Copy Number in the Brain Tissue and Isolated Neuronal Nuclei

Miki Bundo, Tadafumi Kato, and Kazuya Iwamoto

Abstract

The mammalian brain consists of heterogeneous cell types, including neurons and various glial cells. Because somatic mutations that include long interspersed element-1 (LINE-1) retrotranspositions are usually rare, the targeting of specific brain cell types in genomic analyses of these mutations is critical. We previously reported that isolated neuronal nuclei from the prefrontal cortex of patients with schizophrenia exhibit increased numbers of LINE-1 copies. In this chapter, we describe practical methods for isolating neuronal nuclei from frozen brain tissue and cultured cells, extracting genomic DNA, and estimating LINE-1 copy numbers with quantitative reverse transcription-polymerase chain reactions.

Key words Psychiatric disorder, Postmortem brain, Retrotransposon, Retrotransposition, Neuronal nuclei (NeuN)

1 Introduction

Schizophrenia and bipolar disorder are severe psychiatric disorders that each affect 1% of the worldwide population. The socioeconomic burden on these patients is tremendous, and the negative effects of these diseases are evident by the high rate of suicide of those affected. Twin, adoption, and family studies have suggested the involvement of strong genetic components in the etiology of psychiatric disorders [1]. Recent genome-wide association studies have revealed that these disorders are polygenic and involve many genetic factors with small effect sizes [2]. On the other hand, analyses of copy number variations and rare variations have successfully identified genetic components and specific pathways with large effect sizes [3–5]. However, in such cases, the occurrence of the mutations is usually very rare, and, despite exhaustive efforts, current genetic studies cannot fully explain the pathogenesis of these disorders.

Accumulating evidence indicates that genomic DNA in the brain contains various somatic mutations [6]. Recently, somatic

José María Frade and Fred H. Gage (eds.), *Genomic Mosaicism in Neurons and Other Cell Types*, Neuromethods, vol. 131, DOI 10.1007/978-1-4939-7280-7_11, © Springer Science+Business Media LLC 2017

mutations in known causative genes have been detected in the affected brain areas of patients with neurological disorders [7–10], which suggests that these somatic mutations may have an important role in the unexplained pathogenesis of psychiatric disorders. Among the reported somatic mutations in the brain, long interspersed element-1 (LINE-1) retrotransposition has been shown to have a distinctive mechanism. In somatic cells or tissues, retrotransposition activity is repressed by several mechanisms, including the DNA methylation of the 5′-untranslated region of LINE-1. However, neural progenitor cells, but not postmitotic neurons, show LINE-1 retrotransposition activity [11–13]. Although the frequencies and locations of neuronal retrotransposition are unclear, the number of copies of LINE-1 can be estimated with quantitative reverse transcriptase (RT)-polymerase chain reaction (PCR). Increased numbers of LINE-1 copies have been reported in the brain genomes of patients with Rett syndrome, ataxia-telangiectasia, and schizophrenia [13–15].

In our previous study [15], an initial assessment of the number of LINE-1 copies was performed on tissue from the prefrontal cortex and liver of patients with psychiatric disorders, including schizophrenia, bipolar disorder, and major depression, as well as controls. The copy numbers measured in the brain were normalized by those measured in the liver in the same subjects. The results showed an up to 1.2-fold statistical increase in all of the disease groups compared to the controls. However, due to the intrinsic instability of the quantitative PCR of genomic DNA that targets repetitive elements, which consist of thousands of perfect match and mismatch sequences, the resultant data tend to be highly variable with relatively low reproducibility. We then focused on schizophrenia because, of the patients tested with the three psychiatric disorders, patients with schizophrenia showed significant differences and the highest increase in LINE-1 copy numbers in two different internal controls. We performed a similar experiment on genomic DNA that was extracted from neuronal nuclei that had been isolated from the postmortem brains of an independent sample. In that analysis, we normalized the neuronal LINE-1 copy number with the non-neuronal LINE-1 copy number and assumed that the increase in the retrotranspositions primarily occurred within neuronal cells [12]. The results revealed clearer and more robust statistical increases in LINE-1 copy numbers in the neuronal nuclei from the postmortem brains of patients with schizophrenia, which highlighted the importance of genomic analyses that target specific brain cell types.

In this chapter, we describe the methods that are required to estimate LINE-1 copy numbers in isolated neuronal nuclei. We present the reagents and buffers that are required and the details of the following protocols: preparation of the nuclear fraction (Sect. 3.1), staining with the anti-neuronal nuclei (NeuN) antibody (Sect. 3.2), nuclear sorting with flow cytometry (Sect. 3.3), DNA extraction (Sect. 3.4), and LINE-1 copy number estimation (Sect. 3.5).

2   Materials

2.1   Required Reagents and Buffers

- Alexa Fluor 488-conjugated anti-NeuN antibody (P/N MAB377X, EMD Millipore Corporation, Billerica, MA, USA)
- cOmplete™, Mini, EDTA-free protease inhibitor cocktail (P/N 11836170001, Sigma-Aldrich Corporation, St. Louis, MO, USA)
- Potter-Elvehjem homogenizer (2 mL) with loose-fitting Teflon pestle (0.1 mm clearance)
- Optima TLX ultracentrifuge with a TLA110 rotor (Beckman Coulter Life Sciences, Brea, CA, USA)
- 4.7 mL OptiSeal™ ultracentrifuge tubes (P/N 36162, Beckman Coulter Life Sciences)
- 10% Nonidet P-40
- (For nuclei preparation from brain tissues) Homogenization buffer: 50 mM Tris-HCl (pH 7.4), 25 mM KCl, 5 mM $MgCl_2$, and 250 mM sucrose
- (For nuclei preparation from cells) Hypotonic buffer: 10 mM Tris-HCl (pH 7.4), 1.5 mM $MgCl_2$, and 10 mM KCl with cOmplete, Mini, EDTA-free protease inhibitor cocktail
- (For nuclei preparation from cells) 5X isotonic buffer: 210 mM Tris-HCl (pH 7.4), 19 mM $MgCl_2$, 85 mM KCl, and 250 mM sucrose
- 12%, 19%, 26%, 35%, and 57% Percoll solutions (GE Healthcare Bio-Sciences, Pittsburgh, PA, USA) in homogenization buffer

3   Methods

3.1   Preparation of the Nuclear Fraction

Anything that will touch the homogenate and nuclear fractions, such as tubes, pipettes, tips, and strainers, should be precoated with 1% bovine serum albumin (BSA)/phosphate-buffered saline (PBS) solution to prevent nuclear sticking. The protease inhibitor cocktail is dissolved in 10 mL of homogenization buffer. All steps should be performed on ice or in a cold room as much as possible.

The frozen brain sample (0.1-0.2 g) is placed in a Petri dish, which is covered with parafilm. The sample is wetted with a few drops of ice-cold homogenization buffer and then minced with a razor. The minced tissue is placed into a Potter-Elvehjem glass homogenizer. After the homogenization buffer is added to a final volume of about 1 mL, the tissue is homogenized at 1000 rpm with five strokes in an ice-water bath. The homogenate is filtered with a 40-μm cell strainer (P/N 352340, Corning, Corning, NY). The homogenization buffer is added to the cell strainer to a final volume of 2 mL. Twenty microliter of 10% Nonidet P-40 is added to the filtered homogenate. The homogenate is then gently mixed by rolling the tube, which is then incubated for 15 min in an ice-water bath. A total of 1 mL of 57% Percoll/homogenization buffer is added to the homogenate so that the final concentration of Percoll is 19%. The reaction tube is gently mixed by rolling the tube again.

For centrifugation of the Percoll density gradient, the Percoll solutions are carefully layered in an ultracentrifuge tube with syringes with long (70 mm) 20-gauge needles in the following order: 0.3 mL of 12% Percoll, 3 mL of homogenate (19%), 0.8 mL of 26% Percoll, and 0.3 mL of 35% Percoll. The tube is centrifuged at $30,700 \times g$ for 10 min at 4 °C. If the Optima TLX ultracentrifuge is used, the acceleration and deceleration speeds should be set to 6 and 3, respectively, to protect the gradient and sample-to-gradient interface. After centrifugation, the bottom of the tube is pierced with a 21-gauge needle to collect the nuclear fraction (350 μL) in a 1.5-mL microtube with gravity. The tube is then gently tapped to loosen the nuclear clump. The nuclear yield can be calculated by counting with a hemocytometer.

For cells, such as neurons that are induced from induced pluripotent stem cells, the nuclear fraction can be prepared according to a previously reported method [16] with some modifications. The cells ($\sim 10^7$) are triturated 10 times with a 21-gauge needle and syringe in 200 μL of hypotonic buffer. The sample tubes are filled with hypotonic buffer to a volume of 800 μL and mixed by inverting the tube. For cell swelling, the samples are chilled on ice for 10 min and then triturated 10 more times with a 23-gauge needle and syringe. To return the samples to the isotonic condition, 200 μL of 5× isotonic buffer and 45 μL of 30% BSA/PBS are added to the cell suspension. For centrifugation of the Percoll density gradient, 1 mL of homogenization buffer and 1 mL of 57% Percoll/homogenization buffer are added to the samples. The Percoll solutions are carefully layered in an ultracentrifuge tube with syringes with long (70 mm) 20-gauge needles in the following order: 0.3 mL of 12% Percoll, 3 mL of cell sample (19%), and 1 mL of 26% Percoll. The tubes are subjected to centrifugation as described above. After centrifugation, the nuclear clump is collected directly from the bottom of the tube with a long 20-gauge needle, and the tube is then filled with homogenization buffer to a volume of 350 μL.

3.2 Staining with the Anti-NeuN Antibody

First, 150 μL of homogenization buffer and 45 μL of the 30% BSA/PBS solution are added to the collected solution. The solution is incubated at 4 °C for 2 h with continuous gentle rolling to prevent background staining. The unstained control sample is prepared for flow cytometry by mixing 20 μL of the sample and 180 μL of 1% BSA/PBS in another tube. The Alexa Fluor 488-conjugated anti-NeuN antibody (2.6 μL) is added to the rest of the sample. The sample is then incubated at 4 °C overnight with gentle rolling. The presorted sample, which is prepared with 10 μL of the sample, is used to confirm the staining with fluorescence microscopy. Both the unstained control and stained samples are filtered with a 35-μm cell strainer (P/N 352235, Corning) with a polystyrene tube. Then, 1% BSA/PBS at 2–3 volumes of the sample is added through the strainer.

3.3 Nuclear Sorting by Flow Cytometry

If a BD FACSAria™ III cell sorter (BD Biosciences, San Jose, CA, USA) is used, nuclear sorting is performed in the purity mode at a rate of 500–1000 events/s with an 85-μm nozzle. The forward scatter (FSC) and side scatter (SSC) voltages are optimized with an unstained control sample. The voltage of the fluorescence photomultiplier tube for the signal from the blue laser is optimized with the stained sample. For more details, refer to the cytometer manual. The first (P1) gate is set in a FSC-area/SSC-area dot plot to discriminate the nuclei and cell debris. The P2 and P3 gates are set in FSC-height/FSC-width and SSC-height/SSC-width dot plots, respectively, to remove doublets and clumps of nuclei. The P4 and P5 gates are set in an Alexa Fluor 488-area/FSC-area dot plot to separate the stained and unstained samples. The nuclear fraction of the P4 (NeuN$^-$) or P5 (NeuN$^+$) gate is collected in the tube. Part of the sorted samples (about 5000 events) is reanalyzed with flow cytometry to verify the purity.

3.4 DNA Extraction

To collect the nuclear pellet, the sample tubes are centrifuged with a swing rotor at $1000 \times g$ at 4 °C for 20 min. Genomic DNA is extracted with the standard phenol-chloroform method. Extracted DNA is then subjected to RNase treatment and purification with the standard method. All tissues or cells in the same project must be processed with the same DNA extraction protocol. The DNA concentration is strictly adjusted with a PicoGreen® dsDNA Quantitation Kit (Thermo Fisher Scientific Inc., Waltham, MA, USA) or equivalent.

3.5 LINE-1 Copy Number Estimation

Estimation of the LINE-1 copy numbers is performed with the quantitative RT-PCR method. Both TaqMan-based and SYBR Green-based methods are applicable. In both of these methods, single amplicon analyses are performed in triplicate.

The TaqMan-based assay is performed based on the method reported by Coufal et al. [17] with minor modifications. Twenty microliter of the PCR mixture containing the 1× TaqMan Gene Expression Master Mix (Cat# 4369016, Thermo Fisher Scientific Inc.), 0.4 µM of each primer, 0.25 µM of the TaqMan probe, and 100 or 500 pg of DNA are prepared in a 384-well plate. The PCR is performed with an ABI Prism 7900HT Real-Time PCR System (Thermo Fisher Scientific Inc.) and the following conditions: 10 min at 95 °C for initial activation of the AmpliTaq Gold, 40 cycles of 15 s at 95 °C, and 1 min at 60 °C.

The SYBR Green-based assay is performed according to Muotri et al. [13]. Ten microliter of PCR mixtures containing 1× Power SYBR Green Master Mix (Cat# 4368577, Thermo Fisher Scientific Inc.), 0.5 µM of each primer, and 500 pg of DNA are prepared in a 384-well plate. The PCR is performed with the same conditions described above, with the addition of a cycle for the dissociation curve.

In addition to LINE-1 copy number, the copy numbers of internal control genomic regions (HERV and SATA in human and 5 sr RNA in mouse and crab-eating monkey) should be quantified to correct for differences in the DNA contents in the PCR reactions. In addition, to detect brain- or neuron-specific LINE-1 copy number differences, DNA from non-brain tissues (liver) or non-neuronal cells from the same individuals should also be quantified. The relative LINE-1 copy numbers in the brain and neurons are calculated with the comparative Ct method and the following respective formulas:

Relative LINE-1 copy number in brain: $2^{-[\Delta Ct\ of\ brain\ (Ct\ of\ LINE1 - Ct\ of\ the\ internal\ control\ of\ brain\ DNA) - \Delta Ct\ of\ non-brain\ (Ct\ of\ LINE1 - Ct\ of\ the\ internal\ control\ of\ non-brain\ DNA)]}$ and Relative LINE-1 copy number in neurons: $2^{-[\Delta Ct\ of\ neuron\ (Ct\ of\ LINE1 - Ct\ of\ the\ internal\ control\ of\ neuronal\ DNA) - \Delta Ct\ of\ non-neuron\ (Ct\ of\ LINE1 - Ct\ of\ the\ internal\ control\ of\ non-neuronal\ DNA)]}$.

The primers are listed in Table 1. In addition to the primers previously reported, several primers that are applicable to mice and crab-eating monkey are included.

4 Notes

1. *Neuronal nuclei isolation*

   The method for nuclear sorting of the neuronal and non-neuronal nuclei of human postmortem brain has been described previously [15, 18]. The method described here was based on our recently updated protocol, and more detailed information can be found elsewhere [19]. The current protocol has been applied to various species, including human, chimpanzee, macaque, marmoset, pig, and mice brains, in our laboratory (Bundo et al., unpublished).

Table 1
List of primers

Species	Method	Primer name	Location		Sequence	Reported name and Ref.
Human	Taqman	h5UTR#2	L1 5′ UTR	PCR primer_F PCR primer_R Taqman probe	ACAGCTTTGAAGAGAGCAGTGGTT AGTCTGCCCGTTCTCAGATCT TCCCAGCACGCAGC	Coufal et al. (2009) (L1 5′ UTR#2)
Human	Taqman	hORF1#5	L1 ORF1	PCR primer_F PCR primer_R Taqman probe	GAATGATTTTGACGAGCTGAGAGAA GTCCTCCGTAGCTCAGAGTAATT AAGGCTTCAGACGATC	Coufal et al. (2009) (L1 5′ UTR#1)
Human	Taqman	hORF1#1	L1 ORF1	PCR primer_F PCR primer_R Taqman probe	ATAACCAATACAGAGAAGTGCTTAAAAGGA GCTTCTGCATTCTTCACGTAGTTC CTGATGGAGCTGAAAACCAAGGCTCG	Bundo et al. (2014)
Human	Taqman	hORF1#3	L1 ORF1	PCR primer_F PCR primer_R Taqman probe	AAACCAAGGCTCGAGAACTACGT CATTGCTGATACCCTTTCTTCCA AGCCTCGGGAGCCGATGCG	Bundo et al. (2014)
Human	Taqman	hORF2#1	L1 ORF2	PCR primer_F PCR primer_R Taqman probe	TGCGGAGAAATAGGAACACTTTT TGAGGAATGCCACACTGACT CTGTAAACTAGTTCAACCATT	Coufal et al. (2009) (L1 ORF2#1)
Human	Taqman	hORF2#2	L1 ORF2	PCR primer_F PCR primer_R Taqman probe	CAAACACCGCATATTCTCACTCA CTTCCTGTGTCCATGTGATCTCA AGGTGGGAATTGAAC	Coufal et al. (2009) (L1 ORF2#2)
Mouse	SYBR Green	m5UTR	L1 5′ UTR	PCR primer_F PCR primer_R	TAAGAGAGCTTGCCAGCAGAGA GCAGACCTGGGAGACAGATTCT	Muotri et al. (2010) (primer set A:5′ UTR)
Mouse	SYBR Green	mORF1	L1 ORF1	PCR primer_F PCR primer_R	TGGAAGAGAGAATCTCAGGTGC TTGTGCCGATGTTCTCTATGG	Bundo et al. (2014)

(continued)

Table 1
(continued)

Species	Method	Primer name	Location		Sequence	Reported name and Ref.
Mouse	SYBR Green	mORF2	L1 ORF2	PCR primer_F PCR primer_R	CTGGCGAGGATGTGGAGAA CCTGCAATCCCACCAACAAT	Muotri et al. (2010) (ORF2)
Crab-eating monkey	SYBR Green	cORF2#4	L1 ORF2	PCR primer_F PCR primer_R	CAGGGCTCTGAAATTGAGGCAA CCAGGCTTTGGTATCAGGATG	Bundo et al. (2014)
Crab-eating monkey	SYBR Green	cORF2#3	L1 ORF2	PCR primer_F PCR primer_R	GCCCTTCATGCTAAAAACGCT TTGCCCTAGCCAGAACTTCCA	Bundo et al. (2014)
Human	Taqman	hSATA	internal control	PCR primer_F PCR primer_R Taqman probe	GGTCAATGGCAGAAAAGGAAAT CGCAGTTTGTGGGAATGATTC TCTTCGTTTCAAAACTAG	Coufal et al. (2009) (SATA)
Human	Taqman	hHERVH	internal control	PCR primer_F PCR primer_R Taqman probe	AATGGCCCCACCCCTATCT GCGGGCTGAGTCCGAAA CCCTTCGCTGACTCTC	Coufal et al. (2009) (HERV-H)
Mouse	SYBR Green	m5srRNA	internal control	PCR primer_F PCR primer_R	ACGGCCATACCACCCTGAA GGTCTCCCATCCAAGTACTAACCA	Muotri et al. (2010) (5S rRNA)
Crab-eating monkey	SYBR Green	c5srRNA	internal control	PCR primer_F PCR primer_R	TCTACGGCCATACCACCCTGAA AGGCGGTCTCCCATCCAAGT	Bundo et al. (2014)

2. *Practical example of the estimation of LINE-1 copy numbers in postmortem brains: limitations and interpretation*

Several items need to be noted in order to overcome some of the potential pitfalls with the use of quantitative RT-PCR and genomic DNA. These include the use of highly purified genomic DNA, strict adjustments of the DNA concentration (with a kit, such as the PicoGreen® dsDNA quantitation kit), and optimization of the DNA amount per reaction. We prefer a single amplicon analysis rather than multiplex measurements in one reaction. Other efforts include using samples in the PCR plate that are equally chosen from controls and patients, as well as negative controls, to minimize possible batch effects. Likewise, several lots of chemicals, such as the PCR master mix, are first combined, and then aliquots from these batches are used in the same project.

References

1. Cardno AG, Marshall EJ, Coid B et al (1999) Heritability estimates for psychotic disorders: the Maudsley twin psychosis series. Arch Gen Psychiatry 56:162–168

2. Schizophrenia Working Group of the Psychiatric Genomics Consortium (2014) Biological insights from 108 schizophrenia-associated genetic loci. Nature 511:421–427

3. Cnv, Schizophrenia Working Groups of the Psychiatric Genomics Consortium, Psychosis Endophenotypes International Consortium (2017) Contribution of copy number variants to schizophrenia from a genome-wide study of 41,321 subjects. Nat Genet 49:27–35

4. Purcell SM, Moran JL, Fromer M et al (2014) A polygenic burden of rare disruptive mutations in schizophrenia. Nature 506:185–190

5. Fromer M, Pocklington AJ, Kavanagh DH et al (2014) De novo mutations in schizophrenia implicate synaptic networks. Nature 506:179–184

6. Muotri AR, Gage FH (2006) Generation of neuronal variability and complexity. Nature 441:1087–1093

7. Lee JH, Huynh M, Silhavy JL et al (2012) De novo somatic mutations in components of the PI3K-AKT3-mTOR pathway cause hemimegalencephaly. Nat Genet 44:941–945

8. Poduri A, Evrony GD, Cai X et al (2012) Somatic activation of AKT3 causes hemispheric developmental brain malformations. Neuron 74:41–48

9. Lim JS, Kim WI, Kang HC et al (2015) Brain somatic mutations in MTOR cause focal cortical dysplasia type II leading to intractable epilepsy. Nat Med 21:395–400

10. Nakashima M, Saitsu H, Takei N et al (2015) Somatic mutations in the MTOR gene cause focal cortical dysplasia type IIb. Ann Neurol 78:375–386

11. Muotri AR, Chu VT, Marchetto MC et al (2005) Somatic mosaicism in neuronal precursor cells mediated by L1 retrotransposition. Nature 435:903–910

12. Kuwabara T, Hsieh J, Muotri A et al (2009) Wnt-mediated activation of NeuroD1 and retro-elements during adult neurogenesis. Nat Neurosci 12:1097–1105

13. Muotri AR, Marchetto MC, Coufal NG et al (2010) L1 retrotransposition in neurons is modulated by MeCP2. Nature 468:443–446

14. Coufal NG, Garcia-Perez JL, Peng GE et al (2011) Ataxia telangiectasia mutated (ATM) modulates long interspersed element-1 (L1) retrotransposition in human neural stem cells. Proc Natl Acad Sci U S A 108:20382–20387

15. Bundo M, Toyoshima M, Okada Y et al (2014) Increased l1 retrotransposition in the neuronal genome in schizophrenia. Neuron 81:306–313

16. Dignam JD, Lebovitz RM, Roeder RG (1983) Accurate transcription initiation by RNA polymerase II in a soluble extract from isolated mammalian nuclei. Nucleic Acids Res 11:1475–1489

17. Coufal NG, Garcia-Perez JL, Peng GE et al (2009) L1 retrotransposition in human neural progenitor cells. Nature 460:1127–1131

18. Iwamoto K, Bundo M, Ueda J et al (2011) Neurons show distinctive DNA methylation profile and higher interindividual variations compared with non-neurons. Genome Res 21:688–696

19. Bundo M, Kato T, Iwamoto K (2016) Cell type-specific DNA methylation analysis in neurons and glia. Springer, New York, pp 115–123

Chapter 12

Analysis of Somatic LINE-1 Insertions in Neurons

Francisco J. Sanchez-Luque, Sandra R. Richardson,
and Geoffrey J. Faulkner

Abstract

The method described here is designed to detect and localize somatic genome variation caused by the human retrotransposon LINE-1 (L1) in the genome of neuronal cells. This method combines single-cell manipulation and whole genome amplification technology with a hybridization-based, high-throughput sequencing method called Retrotransposon Capture sequencing (RC-seq) for the precise analysis of the L1 insertion content of single cell genomes. The method is divided into four major sections: extraction of neuronal nuclei and single nuclei isolation; whole genome amplification; RC-seq; and experimental validation of putative insertions.

Key words Retrotransposition, Somatic mosaicism, Single-cell, Whole genome sequencing (WGS), LINE-1, Mobile genetic element, Neurogenesis

1 Introduction

The genomes of all living forms studied to date are replete with repeated sequences [1–3]. The accumulation of repetitive sequences is largely due to the activity of mobile genetic elements, including transposons and retrotransposons (detailed in [4]). These elements are able to distribute copies of themselves throughout the genome. Retrotransposons in particular propagate efficiently due to their replicative mobilization mechanism [5, 6]. Although individual retrotransposon copies accumulate internal mutations over time, sequence analysis has revealed families and lineages of mobile genetic elements of polyphyletic origin [7–10]. Around one-third of the human genome comprises identifiable mobile element sequence [1], and an additional one-third may comprise ancient mobile elements too highly diverged for identification [11].

Due to the accumulation of mutations during and after mobilization, the vast majority of retrotransposon copies are inactive for further mobilization [1, 12]. In modern humans, there are three

José María Frade and Fred H. Gage (eds.), *Genomic Mosaicism in Neurons and Other Cell Types*, Neuromethods, vol. 131, DOI 10.1007/978-1-4939-7280-7_12, © Springer Science+Business Media LLC 2017

types of active retrotransposons: Long Interspersed Element 1 (LINE1 or L1), *Alu*, and the SINE-VNTR-*Alu* element (SVA). Among them, only L1 is autonomous [13], meaning that it encodes its own retrotransposition machinery and contains its own internal promoter [14]. To facilitate its own mobilization, an intact L1 element in the genome generates an RNA transcript that serves both as messenger RNA for the synthesis of the L1 mobilization machinery and as a template for reverse transcription of a new L1 copy that is ultimately integrated at a new genomic location. The same machinery is able to reverse transcribe and mobilize transcripts of the other two elements, *Alu* and SVA, as well as cellular transcripts (generating processed pseudogenes [15–18]). *Alu* and SVA are therefore known as non-autonomous elements [19, 20]. L1-mediated reverse transcription is primed by a 3' hydroxyl DNA end generated at the genomic insertion site by an endonucleolytic cleavage, hence the term Target-Primed Reverse Transcription (TPRT, Fig. 1) [6, 21]. Characteristically, the new copy ends with a duplication of target site sequence, typically 10–20 nt in length (Target Site Duplication, TSD) [22].

The existence of polymorphic retrotransposon insertions indicates recent activity of these mobile elements in the human germline [1, 23]. Beginning with the work of Kazazian and colleagues in 1988 [24], de novo L1 mobilization events have been characterized in humans, and donor L1 elements identified in the genome, particularly those related to disease (around 100 of cases of disease has been linked to recent mobilization of L1, Alu, and SVA) (reviewed in [25, 26]). The extent to which individual L1 elements are active within the human genome is the subject of ongoing study, but the existence of a subset of particularly active elements called "hot" L1s has been established [27–30]. Active L1 reported elements belong to the L1 families Ta and pre-Ta (from Transcriptionally Active [13]; reviewed in [26, 31]).

The development of assays able to report L1 mobilization in vitro [31] together with the development of transgenic animal models pointed, for first time, to certain somatic niches where mobile elements are especially active and presumably non-deleterious, including neurogenic areas of the brain [32–35] (reviewed in [36]). Modulation of gene expression levels via transcriptional regulation and epigenetic marks is necessary for the function of complex organisms, but actual editing of genomic DNA sequence is highly regulated and restricted to specific scenarios (e.g., V(D)J recombination and the somatic hypermutation related to immune system receptors) [37, 38]. Somatic mosaicism in the brain due to insertional mutagenesis appears to be associated to a specific and regulated retrotransposition machinery [33, 35], and it is known that the new retrotransposon insertions can affect gene expression in a variety of ways (reviewed in [39, 40]). Thus, it is intriguing to speculate that the activity of mobile elements in

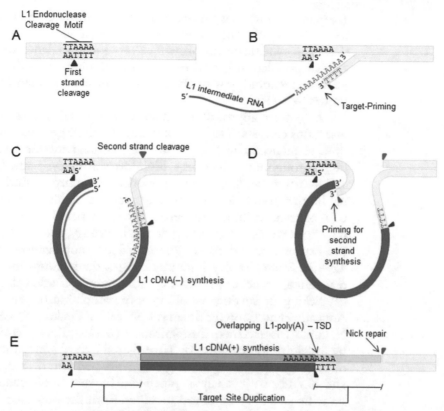

Fig. 1 Conventional retrotransposition mechanism by Target-Primed Reverse Transcription (TPRT). (**a**) An endonucleolytic cleavage (*red arrowhead*) is catalyzed by the L1 Endonuclease and exposes a single strand 3' hydroxyl end of the target DNA. (**b**) This 3' end will anneal over the 3' polyA of the L1 intermediate RNA, and (**c**) prime the reverse transcription of the cDNA(−) strand of the new copy. (**d**) A second endonucleolytic cleavage in the opposite strand of the target region will generate the 3' hydroxyl end that will be used as primer for the synthesis of the second strand of the insertion. (**e**) The resolution of the new insertion will show the typical TPRT hallmarks: the duplication of a short stretch of the original target site DNA (10–20 nt); the inclusion of a long polyA tract in the 3' end of the insertion as a consequence of the untemplated polyadenylation of the L1 intermediate RNA; and the presence of an L1 Endonuclease motif in the 5' boundary of the target site duplication (TSD)

the neurogenic areas of the brain could potentially constitute a form of genome editing that contributes to brain function.

High-throughput DNA sequencing and computational analysis provide the technology to study the distribution pattern of endogenous retrotransposition events in the brain and other tissues [41, 42]. However, the large amount of germline insertions, present in every cell of the body, compared to the low prevalence of somatic variants, hinders efforts to detect and map the genomic coordinates of these variants. A number of enrichment strategies have succeeded in biasing the content of sequencing libraries towards fragments of L1 or other elements, specifically targeting

L1-genome junctions, in order to obtain information about their location [41, 43–46]. However, the current state of research on detecting somatic events present at very low prevalence within the cellular population has arisen from the combination of single-cell genome technology with the above mentioned sequencing techniques [47–49].

A general approach to identify low-prevalence somatic L1 insertions consists of single-cell isolation, whole genome amplification, sequencing library preparation, enrichment for L1 copies, high-throughput sequencing, bioinformatic analysis and filtering, and insertion validation. Recent works have aimed to study neuronal human populations in this way, yielding highly divergent rates of retrotransposition between 1 insertion per 300 cells to ~10 insertions per cell, depending on methodology, and likely marking the extreme boundaries of the actual L1 mobilization rate [44, 47–49]. Strategies developed by independent groups incorporate different approaches to each step, for example, single-cell isolation by picking or sorting; whole genome amplification by Multiple Annealing and Looping Based Amplification Cycles (MALBAC) or Multiple Displacement Amplification (MDA); library enrichment by hybridization capture, by L1-targeted library generation, or by direct, high depth sequencing of unenriched libraries; and different bioinformatic analysis pipelines. In this book chapter, we describe an RC-seq method modified from the one used in reference [49] to incorporate MDA whole genome amplification instead of MALBAC due to more uniform coverage and reliability in subsequent PCR validation of L1 insertions, which has further enabled detection of somatic L1 insertions found in neurons.

2 Materials

Solutions should be prepared with molecular grade water, i.e., water purified by filtration and deionization to achieve a resistivity of 18.2 MΩ cm at 25 °C (such as Mili-Q water produced by Millipore Corporation water filtration stations) or water distilled and filtered by 0.1 μm membrane filters (such as Ultrapure Water provided by Invitrogen—Life Technologies). All the other reagents used in solutions must be Molecular Biology Grade certified by the manufacturers. Solutes used for solutions should also be high purity level. It is recommended to reserve a set of general molecular biology materials (bench, racks, pipettes, tip boxes, thermocyclers, pre-PCR hood, etc.) for nuclei picking and whole genome amplification.

2.1 Single Nuclei Isolation from Neurons

1. Sucrose (Sigma-Aldrich).

2. Tris–HCl pH 8.0.

3. Calcium Chloride (Sigma-Aldrich).

4. Magnesium Acetate Tetrahydrate (Sigma-Aldrich).

5. 0.5 M ethylenediaminetetraacetic acid disodium salt solution (EDTA, Sigma-Aldrich).

6. DL-Dithiothreitol solution, BioUltra (DTT, Sigma-Aldrich).

7. Triton™ X-100 (Sigma-Aldrich).

8. cOmplete™ ULTRA Tablets, Mini, *EASYpack* Protease Inhibitor Cocktail (Roche).

9. Refrigerated centrifuge for 1.5 mL tubes.

10. Dry ice.

11. Surgical material (scalpel and forceps). Scalpel blades size 20–24.

12. 6 mm Vented Petri Dish P-560 (Greiner BIO-One).

13. Douncer.

14. Corning™ Falcon™ Test Tube with Cell Strainer Snap Cap 40 nm (Beckton Dickinson).

15. 2.0, 1.5, and 0.2 mL tubes.

16. Dulbecco's Phosphate-Buffered Saline with Calcium and Magnesium (PBS+) (ThermoFisher Scientific/Life Technologies).

17. Dulbecco's Phosphate-Buffered Saline No Calcium and Magnesium (PBS) (ThermoFisher Scientific/Life Technologies).

18. Goat Serum Standard, Sterile-Filtered (ThermoFisher Scientific/Life Technologies).

19. Bovine Serum Albumin (BSA), lyophilized powder (Sigma-Aldrich).

20. Sodium Azide, *ReagentPlus*® (Sigma-Aldrich).

21. Millex-GP Syringe Filter Unit, 0.22 μm, polyethersulfone, 33 mm, gamma sterilized (Millipore).

22. Mouse anti-NeuN antibody, clone A60 (Millipore).

23. Goat Anti-Mouse IgG1 APC (Abcam). Other antibodies like Goat anti-mouse Alexa488 (Invitrogen) or Goat anti-Mouse IgG (H + L) Secondary Antibody, DyLight® 488 conjugate (ThermoFisher) have been also tested successfully.

24. SYTOX® Blue Dead Cell Stain, for flow cytometry (Life Technologies).

25. MOFLO Astrios High Speed Cell Sorter (Beckman Coulter).

26. Axiovert 200 M Fluorescence/Live Cell Imaging Microscope (Zeiss) coupled to a Transferman® NK 2 (Eppendorf) micromanipulator and a CellTram® Vario (Eppendorf). TransferMan and CellTram can be adapted to other microscopes, such as IX71 Inverted Microscope (Olympus).

27. Non-bevelled embryo biopsy pipette (biopsy needle), OD 35 μm and 30° bend (The Pipette Company).

28. DNA AWAY™ Surface Decontaminant (ThermoFisher Scientific).

29. Precleaned Forsted end Microscope Slide—Hybridization Slides (ProSciTech).

30. Ethanol, not molecular grade (for cleaning).

31. Low lint paper wipes (such as Kimtech Science Kimwipes).

2.2 Whole Genome Amplification by "Multiple Displacement Amplification" (MDA)

1. Thermocycler.

2. Pre-PCR hood with ultraviolet lamp.

3. DynaMag™-96 Side Magnet plate (Life Technologies).

4. Safe Blue Light Imager or UV trans-illuminator (Safe Blue Light source is preferable).

5. Benchtop centrifuge for 0.2 mL tubes.

6. Gel tray, gel combs, electrophoresis tank, and power pad.

7. 0.2 and 1.5 mL tubes, 0.2 mL tubes, PCR-grade.

8. Agarose, Molecular Grade (Bioline).

9. Tris Base ULTROL Grade (Calbiochem).

10. Agencourt® AMPure® XP beads (Beckman Coulter).

11. 0.5 M ethylenediaminetetraacetic acid disodium salt solution (EDTA, Sigma-Aldrich).

12. SYBR® Safe Nucleic Acid Gel Stain (Life Technologies).

13. DNA ladder. Recommended: a ladder with several bands in the 200–700 bp rage, and bands up to 20 kb, e.g., GenRuler 1 Kb Plus DNA ladder 0.5 μg/μL (Thermo Scientific).

14. RepliPHI™ Phi29 Reagent Set (Epicentre).

15. MyTaq™ MS DNA Polymerase (BioLine).

16. Mung-Bean Nuclease (New England Biolabs).

17. Potassium hydroxide (KOH, Sigma-Aldrich).

18. Hydrochloric acid (HCl, Ajax Finechem Pty Ltd. B).

19. Ethyl-alcohol, Pure (Sigma).

20. Resuspension buffer from Illumina® Library Prep Kits.

21. TAE Buffer (40 mM Tris, 20 mM Acetic acid, 1 mM EDTA).

22. Gel Loading Buffer (20% glycerol, 0.04% Orange G stain).

23. Thio-phosphate modified random hexamers (1 mM). Sequence: 5'NNNN*N*N3', where * indicates a thio-phosphate linkage modification (see **Note 1**).

24. Primers for amplification quality control PCR [47]: chr5_fwd 5'GGAGTCATCCTCCAGGTTATTGTTACCATC3'; chr5_rev

5'CCTTGGAAGAGGGAGAAATTCCTTGGTTA3'; chr10_ fwd 5'CTTTCCGCCTAACTAGAATGCAGACCA3'; chr10_ rev 5'CGCTCGTGTTGGGAAGAAGACTCC3'; chr15_fwd 5'TGCTGGAGCAATACTCAGAACTGTTGC3'; chr15_rev 5'GCTAATCCCTGCAGTAATTTCAAATGGCT3'; chr20_ fwd 5'CTGGACCAAGTGGCTTCTTCGACTAG3'; chr20_ rev 5'GCGTGCCGAAGTCTAGGTCTTTATATCTAG3'. All primer stocks should be prepared at 100 μM.

2.3 Retrotransposon Capture Sequencing (RC-Seq)	Materials for this section are described in reference [50], also published by Springer Protocols.
2.4 Validation	The PCR validation method reported here must be considered as a general guideline, since each retrotransposon insertion is unique and will require individual sequence analysis and modulation of PCR conditions to achieve a successful validation. Software and reagents described here are suggestions, but a wide range of alternatives are available to achieve the same goals.

For Insertion Characterization and Primer Design

1. Genome Browser (https://genome.ucsc.edu/, from the University of California, Santa Cruz—UCSC).

2. BLAST® (https://blast.ncbi.nlm.nih.gov/Blast.cgi).

3. Reverse Complement tool (http://www.bioinformatics.org/sms/rev_comp.html).

4. Primer3 software (PrimerBasics).

For Amplification

5. MyTaq™ MS DNA Polymerase (BioLine).

6. Roche Expand Long Range dNTPack.

7. Platinum® Taq DNA Polymerase High Fidelity.

For DNA Imaging and Purification

8. Agarose electrophoresis material and Safe Blue Light Imager or UV trans-illuminator (Safe Blue Light source is preferable).

9. Material for agarose gel-purification of DNA.

For Cloning

10. Promega pGEM®-T vector system.

11. Life Technologies TOPO® PCR cloning system.

12. Material for molecular cloning, bacteria transformation, culture, and DNA extraction.

3   Methods

The aim of this protocol is to identify somatic L1 insertions present in specific neurons of a given human donor. As a general overview, the protocol comprises the isolation of single neuronal nuclei followed by amplification of the single nuclei genomic DNA (whole genome amplification) to generate sufficient material to complete RC-seq and future validation experiments. RC-seq is a well-established method that has been previously described in Volume 1400, "Transposons and Retrotransposons—Methods and Protocols"; from the series "Methods in Molecular Biology" also from Springer Protocols, with the title "Retrotransposon Capture Sequencing (RC-seq): A Targeted, High-Throughput Approach to Resolve Somatic L1 Retrotransposition in Humans," by Sanchez-Luque et al. 2016 [50].

The approach described here to obtain and isolate nuclei is specifically centered in neuronal cells, although the same technique has been used for the detection of somatic L1 insertions in neurons as well as in non-neuronal brain cells [49] (using a protocol adapted from [51, 52]). However, the Multiple Annealing and Looping Based Amplification Cycles (MALBAC) whole genome amplification method used in combination with RC-seq requires the addition of several modifications to the standard RC-seq protocol [41, 42, 49]. The alternative Multiple Displacement Amplification (MDA) approach reported here for whole genome amplification generates very high molecular weight amplicons and does not require any fixed sequence introduced by the primer oligonucleotides [47, 53]. Thus, this MDA-generated DNA is perfectly compatible with the RC-seq protocol mentioned above [50] and the indications necessary to link it with the current protocol are given in Sect. 3.3.

MDA is a PCR-free method used for whole genome amplification (WGA) [53]. The amplification is performed in an isothermal reaction using the highly processive Phi29 DNA polymerase. This enzyme is able to $5' \rightarrow 3'$ displace annealed DNA strands while catalyzing an elongation reaction. The displaced DNA strands can therefore bind new hexamers and serve as templates, rendering denaturing steps of cycling-based methods unnecessary. Cycling-based MALBAC, previously used in combination with RC-seq [49], generates 500–1500 bp amplicons [54] while the MDA method reported here (adapted from [47]) allows the generation of amplicons several kilobases in length. Although validation of $5'$ L1-genome junctions is possible using MALBAC amplified material [49], PCR amplification of entire insertions (empty/filled PCR) is compromised for insertions longer than the MALBAC amplicon size. MDA libraries comprising longer DNA fragments enable PCR amplification across full-length L1 insertions, as well as identification of possible alterations of the 5' L1-genome junction (i.e., inversions or transductions) [47, 48]. Details about these features are described in Sect. 3.4.

RC-seq is a hybridization-based capture method for enriching sequencing libraries in L1 junctions that has been repeatedly used for the detection of L1 insertions in human tissue and single cells, as well as human-induced pluripotent stem cell lines [41, 42, 49, 55, 56]. However, similar enrichment can be achieved by different approaches of L1-targeted library generation as L1 profiling [45], SIMPLE [57], ME-Scan [46], or the most recent SLAV-seq presented in Chap. 13 of this book [44].

As explained in the Introduction, somatic L1 insertions will be identified by comparison to the germline genome of the individual. Each individual's genome contains a cohort of polymorphic L1 insertions, of which a significant number may be absent from the reference genome or any other database. Further validation experiments will consist of PCR for detection of the appropriate L1 insertion in independent neurons, but to rule out the possibility of an unreported polymorphic insertion, control reactions with input DNA from bulk brain and other unrelated tissue from the same donor need to be included. The L1 insertion will be considered somatic only if the PCR product is not detected in unrelated control tissue. Thus, genotyping the polymorphic insertions of the individual is essential to properly filter the sequencing calls that are more likely to be true somatic insertions. Any new insertion detected by bulk RC-seq in both brain and liver, for example, will be automatically discarded from the potential somatic L1 variants called in neurons. Therefore, RC-seq experiments with bulk DNA from brain and other control tissue from the same individuals need to be performed as described in the abovementioned protocol [50].

3.1 Single Nuclei Isolation from Neurons

Preparation

1. Set up a refrigerated 1.5 mL tube centrifuge to 4 °C.

2. Prepare Nuclear Extraction Buffer (NEB): 0.32 M Sucrose, 10 mM Tris–HCl ph 8.0, 5 mM Calcium Chloride, 5 mM Magnesium Acetate, 0.1 mM EDTA, 1 mM DTT, 0.1% Triton X-100, and 1× Protease Inhibitor Cocktail (1 tablet/10 mL). Allow 2 mL of NEB for every 5 mg of sample. A stock of NEB buffer can be prepared with all the components but DTT, Triton X-100, and Protease Inhibitor Cocktail. The last three components are to be added to small aliquots immediately before use. Here are the calculations for a 22 mL solution of NEB stock (without DTT, Triton X-100, and protease inhibitors) and for a 2 mL fresh, ready-to-use NEB final solution (see **Note 2**).

NEB stock, 22 mL		NEB final solution, 2 mL (µL)	
Sucrose 1.82 M	4.4 mL	NEB stock solution	1760
Tris–HCl 1 M pH 8.0	250 µL	DTT 0.1 M	20
Calcium chloride 1 M	125 µL	Protease inhibitor	200
Magnesium acetate 1 M	125 µL	cocktail—Roche	
EDTA 500 mM	5 µL	complete ultra mini	
		(1 tablet/mL	
		solution)	
Molecular grade H$_2$O	17.1 mL	Triton X-100 10%	20

3. Prepare blocking buffer: 1× PBS+, 10% Goat Serum Standard, 0.5% BSA, and 0.1% Sodium Azide. Here are the calculations for a 100 mL solution of Blocking Buffer. Once the buffer is prepared, it must be filtered through a 0.2 µm low protein binding filter. Keep the buffer on ice until its use (see **Note 3** about the PBS to use).

Blocking buffer, 100 mL	
PBS+	90 mL
Goat serum standard	10 mL
BSA	0.5 g
Sodium Azide	0.1 g

4. Prepare the following pre-mix of antibodies and incubate them at 4 °C protected from light for 20 min (see **Note 4** about the considerations on the antibody mix preparation).

Component	Positive mix (µL)	Negative mix (µL)
Blocking buffer	150	150
Anti-NeuN (1:300)	0.4–1	–
Anti-mouse IgG1 APC (1:300)	0.33–1	0.2–0.5

5. Prepare a generous dry ice bed and place Petri dish, scalpel, and forceps on it for cooling.

6. In a pre-PCR hood free of DNA contamination, UV irradiate an aliquot of PBS solution and a number of 0.2 mL tubes into which nuclei will be picked for 40 min (see **Notes 5** and **6**). Allow 1 mL of PBS solution for a picking of up to 96 nuclei.

7. Close the 0.2 mL tubes inside the hood and pay attention to not touch the rim or the interior of the cap of the tubes after they are irradiated.

8. Keep the 0.2 mL tubes, a Blocking Buffer aliquot (allow 1 mL), and the irradiated PBS aliquot on ice.

9. Prepare a container with a 0.2 mL tube-rack in a dry ice bed (for snap-freezing the picked nuclei).

10. Set up the Microscope. Usually the main switch must be turned on for a certain amount of time in advance in order to the light bulb to warm up.

11. Clean door handles, the tip of the arm, the stage, the tube holder, and the benchtop centrifuge with DNase away.

12. Load the Biopsy Needle. First remove bubbles from the oil in the arm by purging them out; place the needle with the tip up and push it all the way into the arm (see **Note 7**); then tighten it by holding the plastic tip of the arm while turning the metal screw (Fig. 2a); and finally move the oil surface to the front of the needle but stop ~5 mm before the end of the tip (see **Note 8**).

13. Load the slide. First clean the slide with 70–80% ethanol and a low lint paper wipe; then move the stage to fit the tube holder on the opposite side to the arm and clamp the slide by moving the right side of the stage only (Fig. 2a); finally, place a 40 μL drop of PBS on the left side of the slide (for washing the nuclei, see **Note 9**) and a 20 μL drop of Blocking Buffer on the right side (that will be combined with the sample at the moment of the picking, see **Note 10**).

14. Prepare the Biopsy Needle. Wash the needle by aspirating ~10 μL of Blocking Buffer (leaving the air drop gap between the oil and the buffer) and pumping them out, and then draw ~5 μL of PBS into the needle (Fig. 2b, see **Note 11**).

15. Focus the microscope on the edge of the drop to more easily find the focus plane for the nuclei.

16. Set up needle positions in the Transferman software. Center the needle within focus window and set this position as "Center" (see **Note 12**), usually by pushing the button in the center of the joystick. Set up "Position 1" with the needle right above the liquid surface. Set up "Position 2" with the needle at the height of the tube (Fig. 2c).

Procedure

1. Working on a bed of dry ice, place the tissue sample in a pre-cooled Petri dish using the forceps and shave ~5 mg of sample using the scalpel.

2. Transfer the sample into the douncer and immediately proceed to homogenization.

3. Homogenize with slow up and down moves of the douncer avoiding the generation of bubbles, for 90 s. Time is critical for this step.

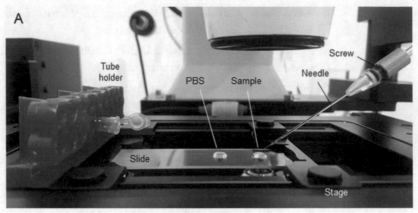

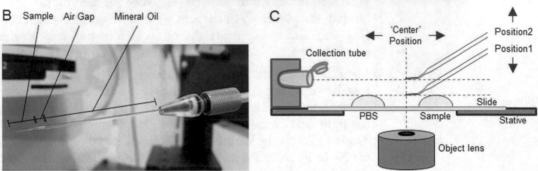

Fig. 2 Typical settings for single nuclei picking at the Zeiss Axiovert 200 M Microscope. (**a**) The standard disposition of the different elements involved in the sample picking within the working frame of the microscope. Note that the Tube Holder is set for the horizontal disposition of the collection 0.2 mL tube. The closer drop to the tube is the PBS, and the further one is the sample. In this way, the workflow of the needle is from right to left in the picture: a nucleus will be picked from the sample; move leftwards to the PBS drop for wash; and move further leftwards to the collection tube. (**b**) Detail of the needle (Biopsy pipette) with a representative distribution of the fluids inside: sample in the tip and mineral oil in the body, both separated by an air gap. Notice the angled tip of the needle. (**c**) Schematic representation of the position settings of the needle. Notice that position "Center" is established within the horizontal plane in reference to the objective lens; "Position 1" and "Positon 2" are established in the vertical axis in reference to the surface of the drops and the height of the collection tube, respectively

4. Transfer the homogenized solution to the filter cap of the 40 nm cell strainer tube and allow it to filter through.

5. Transfer the filtrate to a fresh 2 mL Eppendorf tube.

6. Centrifuge at $900 \times g$, 4 °C for 7 min.

7. Carefully discard the supernatant by gently inverting the tube. Quickly aspirate the remaining buffer with a pipette without disturbing the pellet.

8. Add 175 or 500 μL of Blocking Buffer to the sample and incubate for 15 min on ice (see **Note 13**).

9. Vortex briefly to resuspend.

10. Combine sample with antibodies premixes (see **Note 14**):

Positive tube (µL)		Negative tube (µL)	
Sample	150	Sample	25
Positive mix of antibodies	150	1× Block buffer	125
		Negative mix of antibodies	150

11. Incubate the staining mixes for 45 min to 1 h at 4 °C, protected from light.

12. Bring each sample to 1 mL final volume with PBS+.

13. Centrifuge the samples at $900 \times g$, 4 °C for 10 min.

14. Carefully aspirate the supernatant by pipetting and discard it.

15. Add 400–500 µL of chilled PBS+. Gently dislodge the pellet from the side of the tube by pipetting or by brief vortexing.

16. Add 1 µL of SYTOX blue stain to each tube.

17. Incubate the samples for 20 min at 4 °C and gently resuspend the pellet by pipetting.

18. Transfer into cell sorting tubes, and prepare sorting collection tubes with 150 µL of Blocking Buffer per every sample.

19. Perform FACS sorting using MOFLO Astrios High Speed Cell Sorter at a slow flow rate of 100–400 event/s with a 70 µm nozzle at 60 psi. Set the collecting chamber to room temperature to maintain the nuclei integrity. Events must be first gated using forward and side scatter of cells, and then by positive staining for NeuN and SYTOX. Use the control sample stained with SYTOX plus the secondary antibody only, to rule out unspecific staining. An example of the gating parameters can be found in reference [49].

20. After collecting the NeuN+ nuclei population in a sorting tube, place the tube on ice. Add 20 µL of the sorted nuclei sample to the Blocking Buffer drop within the slide. Let the combined sample sit for 10 min to allow nuclei to settle on the slide (see **Note 15**).

21. Proceed to picking nuclei. Once the positions of the needle have been set in Step 16, *Preparation*, horizontal movements should be done only with the stage, while the needle should be moved only vertically between "Position 1" and "Position 2" and in the approaching to the nuclei. Use 10× magnification.

22. Aim the needle towards a nucleus, rotating the handle of the joystick of the Transferman to move it down to surface of the slide within the drop containing the nuclei. Start with nuclei that are fairly isolated from other nuclei.

23. Gently aspirate a single nucleus into the needle using the precision rotary knob of the CellTram and bring needle to "Position 1."

24. Move the stage to place the needle above the PBS drop and move the needle down into the drop rotating the handle of the joystick of the Transferman (see **Note 16**).

25. Pump the nucleus out into the PBS and re-aspirate it into the needle using the rotary knob of the CellTram. Do this procedure as fast as possible.

26. Move the needle to "Position 2." Open the 0.2 mL collection tube within the Microscope Chamber and pay special care not to touch the rim. Bring the tube in the tube holder to the needle by operating the stage.

27. After depositing the nucleus in the tube, bring the tube away from the needle by operating the stage. Close the tube and spin it down in the benchtop centrifuge for 10 s.

28. Place the tube on the rack on dry ice.

29. Repeat Steps 21–28 until you have picked the desired number of nuclei. The PBS washing drop should be replaced after every 12–24 nuclei picked. Also, before replacing the PBS, take a sample of the drop into a 0.2 mL tube as a negative control (see Step 3 of the *Procedure* for Whole Genome Amplification below).

SAFE STOPPING POINT. You can store the picked nuclei at −80 °C for few days, although it is recommendable to proceed immediately to whole genome amplification.

3.2 Whole Genome Amplification by "Multiple Displacement Amplification" (MDA)

Preparation Day 1

1. Prepare Lysis Buffer 1×: 200 mM Potassium Hydroxide (KOH), 5 mM EDTA, and 40 mM DTT. A 2× solution without DTT (400 mM KOH; 10 mM EDTA) can be prepared and stored at room temperature away from light. KOH is typically supplied in pearl format, so it has to be weighed first and the volumes of the other components calculated accordingly. As a guideline, here are the calculations for 15 mL of 2× KOH-EDTA solution, and for a 15 reactions (42 μL) of 1× lysis buffer (see **Notes 17** and **18**).

KOH-EDTA solution 2× (15 mL)		1× Lysis buffer for 15 rxn (42 μL)	
KOH	320 mg	2× KOH-EDTA	21 μL
EDTA 0.5 M	285 μL	DTT 0.1 M	16.8 μL
H_2O	13.96 mL	H_2O	4.2 μL

2. Prepare neutralization buffer 1×: 400 mM HCl; 600 mM Tris–pH 7.5. A stock solution can be prepared and stored at room

temperature away from light (NOTE for clean environment before preparation). Here are the volumes for an 8 mL stock solution (see **Notes 18** and **19**).

1× Neutralization buffer (8 mL)	
HCl (1.16 g/mL, 32% purity)	314 μL
Tris	580 mg
H_2O	7.68 mL

3. UV irradiate pipettes, racks, tips (open the boxes for irradiation), and tubes (open and in the rack) in a pre-PCR hood for 10–20 min. UV irradiate the Lysis and Neutralization Buffer stocks.

4. Set up a thermocycler with a 30 °C block, 40 °C lid.

Procedure Day 1

1. Prepare Lysis Buffer 1× from the stock.

2. After irradiation, thaw the components for the reaction mix in the pre-PCR hood and prepare the master mix (see **Note 20**). Keep the mix on ice until use.

Component	1 rxn (μL)	14 rxn (μL)
10× Reaction Buffer	2	28
25 mM each dNTP solution	1.6	22.4
1 mM Random thio-phosphate Hexamers	1	14
H_2O	10.8	151.2
Phi29 enzyme (100 U/μL; 0.1 μg/μL)	0.4	5.6

3. Place the nuclei tubes and PBS control at room temperature. Include an additional empty tube labelled as Blank control (see **Note 21**).

4. Add 2.8 μL of Lysis Buffer 1× to each tube (see **Note 22**).

5. Add 1.4 μL of Neutralization Buffer to each tube (see **Note 22**).

6. Add the enzyme to the master mix as detailed in Step 2. Pipette the entire volume of the master mix up and down for 2–3 times to ensure a homogeneous solution.

7. Add 15.8 μL of master mix to each tube (see **Note 22**).

8. Place the tubes in the thermocycler at 30 °C for 16 h (see **Note 23** for an appropriate timing of the reaction).

Preparation Day 2

5. Prepare two 0.8% agarose gels in 1× TAE with sufficient wells for samples and controls, plus flanking wells for molecular weight markers.

6. Prepare a 2% agarose gel in 1× TAE with sufficient wells for samples and controls, two additional wells, and wells for molecular weight markers.

7. Place an aliquot of AMPure® XP beads at room temperature. Allow 60 μL of beads per reaction.

8. Prepare 80% of ethanol solution. Allow 900 μL per reaction.

9. Prepare sufficient 0.2 mL tubes for samples and controls, and add 49 μL of molecular grade H_2O to each of them. They will be used to prepare a dilution 1:50 of each raw amplified material.

10. Set up a thermocycler with the protocol: 30 °C, 30 min; 4 °C forever; for a reaction volume of 50 μL. Lid must be at 40 °C.

11. Set up a thermocycler with the following protocol: (94 °C, 3 min) × 1; (94 °C, 1 min; 68 °C*, 1 min; 72 °C, 1 min) × 13; (92 °C, 1 min; 55 °C, 1 min; 72 °C, 1 min) × 27; (72 °C, 10 min; 4 °C, hold) × 1. In the step marked with an asterisk, include a modification of −1 °C/cycle.

Procedure Day 2

9. Take the amplification reactions from the thermocycler immediately after the 16 h incubation (see **Note 24**).

10. Transfer 1 μL of each tube to the corresponding prepared 1:50 dilution tubes with 49 μL of molecular grade H_2O. Vortex them, spin them in a benchtop centrifuge, and store them at −20 °C.

11. Add 20 μL of molecular grade water to each sample.

12. Add 1.3 volumes of AMPure® XP beads to each tube (53 μL), mix by pipetting ten times, and incubate at room temperature for 15 min (see **Note 25**).

13. Place the tubes on a magnetic rack for 2 min. Aspirate and discard the supernatant.

14. Add 200 μL of 80% ethanol to each tube without disturbing the beads and incubate at room temperature for 30 s.

15. Remove the ethanol and repeat Step 14 once.

16. Remove the ethanol (use a 10 μL pipette to remove the residual liquid) and allow the tubes to air-dry on the magnetic rack for 15 min.

17. Add 26 μL of resuspension buffer to each tube. Remove the tubes from the magnetic rack and flick until beads are completely

resuspended (see **Note 25**). Spin the tubes briefly in a benchtop centrifuge to collect all the volume in the bottom of the tube, but avoid pelleting the beads during the spin.

18. Incubate at room temperature for 2 min and place the tubes back in the magnetic rack. Incubate at room temperature for 5 min (the liquid must appear clear).

19. Transfer the supernatant to new fresh tubes.

SAFE STOPPING POINT. You can store the samples at −20 °C or proceed immediately for the next step.

20. Quantify DNA concentration of the 1:50 dilution prepared at Step 10 using Qubit® Fluorometric Technology. As a first attempt, use 10 μL of the 1:50 dilution for a standard Qubit quantification following manufacturer's instructions. If the concentration is too high, proceed to measure smaller volumes (Qubit quantification preparations can tolerate volumes from 1 to 10 μL). If the concentration is too low to be measured, proceed to Step 21 and check **Note 21** assessment of WGA performance. Calculate the concentration and the total amount of DNA of the original undiluted samples (see **Notes 26** and **27**).

21. Run 10 μL of the 1:50 dilution in a 0.8% agarose gel electrophoresis in 1× TAE. Amplification should be detected as a thick ~20 kb band (Fig. 3a, see **Note 21**).

22. Proceed to the debranching reaction with the successfully amplified nuclei. Thaw the cleaned-up 25 μL volume WGA samples on ice if they were stored at −20 °C.

23. Prepare a debranching master mix for all the amplified nuclei plus the controls (including the Blank and PBS amplified controls). Here are the calculations for a 14 reaction master mix:

Component	1 rxn (μL)	14 rxn (μL)
Mung-Bean Nuclease 10× reaction buffer	5	70
Molecular grade H$_2$O	19	266
Mung-Bean DNase 10 U/μL	1	14

24. Add 25 μL of master mix to each tube. Incubate the tubes at 30 °C for 30 min in a thermocycler. Set a 4 °C hold after the reaction (see **Note 28**).

25. Place an aliquot of AMPure® XP beads at room temperature for 30 min (do that while the reaction at Step 24 takes place). Allow 70 μL of beads per reaction.

26. Add 1.3 volumes of beads to each tube (65 μL), mix by pipetting ten times, and incubate at room temperature for 15 min.

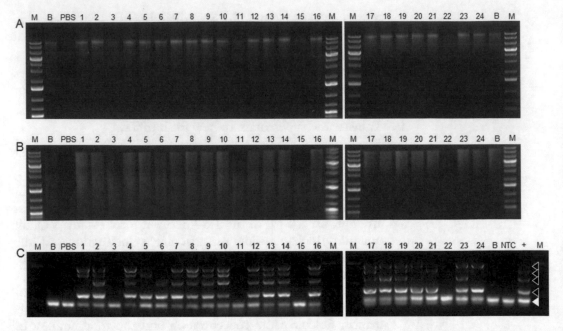

Fig. 3 Representative control analysis of a MDA experiment. (**a**) Imaging of a gel electrophoresis run using 10 μL of the 1:50 dilution of the original whole genome amplification reaction described in Step 10, *Procedure* in Sect. 3.2. Notice how the bulk of the amplification appears as a band of ~20 kb. (**b**) Imaging of a gel electrophoresis run using 10 μL of the 1:50 dilution of the debranched whole genome amplification reaction as described in Step 43 in the same section. Notice how, after debranching, the original single band in (**a**) becomes a smear that extends from ~20 kb to fragments below 500 bp. (**c**) Imaging of a gel electrophoresis of the Quality Control PCR for assessing the MDA coverage. The *black arrow* heads on the *right point* the four different fragments expected from the multiplex PCR. The *white arrowhead points* the primer "leftover" of the reaction. For all the gels: *M* marker, *B* blank control (duplicated), *PBS* PBS control, numbers represent different nuclei. In (**c**), NTC is the No-Template Control of the PCR, and + is the positive control of the PCR, performed with 10 ng of genomic DNA input

27. Place the tubes on a magnetic rack for 5 min. Aspirate and discard the supernatant.

28. Add 200 μL of 80% ethanol to each tube without disturbing the beads and incubate at room temperature for 30 s.

29. Remove the ethanol and repeat Step 28 once.

30. Remove the ethanol (use a 10 μL pipette to remove the residual solution) and let the tubes air-dry on the magnetic rack for 15 min.

31. Add 21.5 μL of resuspension buffer to each tube. Remove the tubes from the magnetic rack and flick until beads are completely resuspended. Spin the tubes in a benchtop centrifuge to collect all the volume in the bottom of the tube, but avoid pelleting the beads during the spin.

32. Incubate at room temperature for 2 min and place the tubes back in the magnetic rack. Incubate at room temperature for 5 min (the liquid must appear clear).

33. Transfer 21 µL of the supernatant to new fresh tubes.

34. Prepare a 1:4 or 1:8 dilution by transferring 1 µL of the eluted DNA to a fresh new tube and adding 3–7 µL of molecular grade water (see **Note 29**).

SAFE STOPPING POINT. You can store the samples at −20/−80 °C until you proceed with RC-seq experiments.

35. Quantify DNA concentration of the 1:4 or 1:8 dilution prepared at Step 10 using Qubit® Fluorometric Technology. As a first attempt, use 1 µL of the dilution for a standard Qubit quantification following manufacturer's instructions. If the concentration is too high when using the 1:4 dilution, proceed to a 1:8 dilution. Calculate the concentration and the total amount of DNA in the original undiluted samples. Check that the total amount of DNA in every sample is consistent with the previous values obtained directly after the WGA reaction (you should expect close but slightly lower values).

36. Prepare a final 1:50 dilution with the remaining 1:4 or 1:8 dilution by adding 37.5 or 43.72 µL, respectively, of molecular grade water to the remaining 1:4 or 1:8 dilutions.

37. Proceed to the WGA Quality Control PCR. Transfer 2 µL of each 1:50 debranched DNA dilution, as well as the WGA control reactions, to a fresh tube. Prepare an additional tube with 2 µL of molecular grade water for No-Template Control PCR, as well as a tube with 2 µL of a 5 ng/µL dilution of human genomic DNA as a positive control for the PCR (see **Note 30**).

38. Prepare the WGA Quality Control PCR master mix. Here are the volumes for a 15 reaction master mix:

Component	1 rxn (µL)	15 rxn (µL)
MyTaq 5× reaction buffer	4	60
Primer chr5_fwd 100 µM	1	15
Primer chr5_rev 100 µM	1	15
Primer chr10_fwd 100 µM	1	15
Primer chr10_rev 100 µM	1	15
Primer chr15_fwd 100 µM	1	15
Primer chr15_rev 100 µM	1	15
Primer chr20_fwd 100 µM	1	15
Primer chr20_rev 100 µM	1	15
Molecular grade H$_2$O	5.8	87
MyTaq Enzyme 5 U/µL	0.2	3

39. Add 18 µL of master mix to each PCR reaction tube. Place the tubes in the thermocycler indicated in *Prepare Day2*, Item 11 and proceed to the PCR reaction (see **Note 31**).

40. After the PCR reaction, add 4 µL of Gel Loading Buffer 6× to each tube and load the 24 µL samples in the 2% agarose gel. Include molecular weight marker lanes.

41. Run the samples at 120 mA until proper separation of the different multiplexed amplicons is achieved.

42. Image the gel and analyze the performance of each nuclei in the WGA reaction. The WGA Quality Control PCR consists of a multiplex amplification of four different amplicons within the human genome. The amplicons have sizes of 532, 423, 291, and 140 bp. Only nuclei with minimum three out of the four bands are considered adequate for downstream analysis. Figure 3c shows a representative example of a WGA Quality Control PCR (see **Note 32**).

43. Resolve 10 µL of the remaining 1:50 dilution from Step 37 in a 0.8% agarose gel in 1× TAE. Combine 10 µL of the 1:50 solution with 2 µL of 6× Loading Buffer. Run at 100 mA until proper separation of the DNA fragments is achieved.

44. Image the gel and analyze the efficiency of the debranching reaction. A properly debranched DNA should appear as a smear expanding downwards from ~20 kb (Fig. 3b shows an example of typically debranched DNA).

45. Satisfactorily debranched DNA samples with ≥3 bands in the WGA Quality Control PCR can be used for subsequent steps of the protocol.

SAFE STOPPING POINT. You can store the samples at −20 °C until further use.

3.3 Retrotransposon Capture Sequencing (RC-Seq)

As indicated in Sect. 3, it is necessary to perform parallel RC-seq experiments using genomic DNA extracted from the same brain area from which the neurons were obtained and another tissue from the same donor. This experiment must be performed as described in previously published RC-seq method [50]. Nonetheless, a few modifications are required to adapt the RC-seq protocol from reference [50] to single neurons: disregard the sections describing **DNA extraction** and proceed directly to the sections for **DNA shearing** using a total of 1 µg of debranched WGA DNA from the positively amplified neurons instead of genomic DNA.

RC-seq will, in principle, detect any L1 3' breakpoint, and will detect 5' breakpoints of full-length L1 insertions, and heavily 5' truncated L1 insertions. Sequencing outcomes can be computationally analyzed following the general guidelines specified in the referred RC-seq protocol [50], with modifications as described in reference [49] to exclude molecular artifacts generated during WGA and Illumina library preparation. In addition to the RC-seq computational pipeline, there are additional software tools capable of analyzing data generated by the RC-seq bench protocol. TEBreak (https://github.com/adamewing/tebreak), which uses a breakpoint consensus and local sequence assembly approach to detecting insertions, is a suitable recommendation. Paired-end sequence data generated by RC-seq should be aligned to a suitable reference genome using a short-read aligner capable of soft-clipping reads such as BWA [58] or Bowtie2 [59]. For BWA we recommend using the -Y parameter to soft-clip supplementary alignments, and we recommend marking PCR duplicates with a suitable tool such as Picard MarkDuplicates (https://broadinstitute.github.io/picard). TEBreak can be run with default parameters on most RC-seq datasets, following the instructions in the documentation will yield predicted transposable element insertions in a tab-delimited format.

3.4 Validation

RC-seq and other high-throughput sequencing-based methods for insertion calling are useful for capturing junctions of unreported insertions [41, 45–47, 60]. The RC-seq analysis pipeline will provide a stack of different reads crossing the putative L1-genome junction, while TEBreak will generate a consensus of the different unique reads that cross a potential L1-genome junction. The validation of a somatic insertion is the experimental confirmation that a computationally derived prediction actually corresponds to a real insertion using an orthogonal sample preparation method on the original DNA sample. Validation is necessary because there is a high risk of chimera formation during the preparation of sequencing samples. A chimera is a DNA fragment comprising sequences from two or more distal locations in the genome that has been artifactually generated during sample processing. DNA polymerases and ligases used during library preparation and PCR-based enrichment methods can generate chimeras in a variety of ways. Thus, the null hypothesis must always be that a putative insertion is actually a chimera.

An ideal PCR validation entails amplification of the entire insertion, or filled site, comprising both 5' and 3' L1-genome junctions and the body of the integrated retrotransposon. PCR products should be Sanger sequenced in their entirety [41, 42, 47, 48, 61]. Validated insertions are expected to bear the structural hallmarks of TPRT (Fig. 1) [6, 16, 21, 31, 62, 63]. In WGA-based experiments, the performance of the WGA amplification and average amplicon size may hinder the amplification of an entire filled site. In this case, independent amplification of both 5' and 3' L1-genome junctions is sufficient for TSD annotation as well as junction-specific features [49].

A true somatic insertion will not be detected by PCR in the bulk genomic DNA from the control tissue of the donor (i.e., liver for insertions detected in single neurons) [41, 42, 49]. Depending on how prevalent a somatic insertion is in the tissue of interest (i.e., brain), it may be detected by PCR in bulk genomic DNA from that tissue [41, 42], and it may also be detected by PCR in WGA material from additional single neurons where it was missed by high-throughput sequencing [47–49].

Here, we present some general guidelines for optimal primer design and PCR cycling conditions, beginning with characterization of an L1-genome junction spanning sequencing read or consensus sequence obtained from a high-throughput sequencing approach.

1. Identify the fragment of the junction sequence that corresponds to the L1 element. Online tools such as BLAST® using the consensus sequence of L1 are very useful (L1 consensus can be found at RepBase [64]). For 3' L1-genome junctions, a long polyA tract should be readily identified.

2. Orient the junction sequence so that the L1 element is in the sense orientation.

3. If the insertion has been detected at both ends, identify the TSD as the overlapping sequence between the 5' and 3' genomic flanks. Reconstruct the predicted empty site by fusing the sequence of both genomic flanks with only one TSD (see **Note 33**).

4. If the insertion has been detected only at one end, select the genomic flank.

5. Mark additional information like the orientation of the element sequence at the 5' junction with respect to the consensus (see **Note 34**) or the presence of untemplated nucleotides.

6. Use BLAT tool of the Genome Browser for identifying the location of the flanking genomic DNA.

7. Insertions occurring into repeated regions may produce multiple BLAT hits, so it may be necessary to evaluate the most likely location (the BLAT Score, the calculated Identity and the Start and End points may help for this). Determine whether the insertion is on the sense or antisense strand relative to reference genome. Proceed to inspect the Browser for the best hit. Zoom out 1.5–3 times and inspect the genomic environment of the insertion (see **Note 35**).

8. Retrieve the DNA of the target site from the menu "View" and the option "DNA." Choose the appropriate strand and select "Mask Repeats to Lower Case."

9. Transfer the retrieved sequence to the text editor file. Annotate the break point, the TSD, and the potential Endonuclease motif if possible.

Characterization of insertions in this way should help to discriminate potential chimeras. Primer design and further validation is a tedious process, so it is helpful to shortlist insertions in this step to prioritize validation attempts (see **Note 36**). Given the prevalence of L1 3′ end sequences in the human genome [1], it is preferable to design the most specific primers possible for the 3′ L1-genome junction, and then design compatible 5′ L1-genome junction primers.

1. Reconstruct the 3′ junction comprising ~130 nt of the consensus 3′ end of the L1 element and a ~20 nt polyA tract, followed by the genomic 3′ flank (see **Note 37**).

2. Choose a standard primer in forward within the 3′ end of the L1 consensus. A primer ending in the specific L1_Ta trinucleotide [13, 65, 66] increases the specificity for the highly active L1-Ta family [43, 45] (see **Note 38**).

3. Use a primer design software to minimize mispriming and primer self-complementarity. Here we use Primer3, input version 0.4.0.

4. Introduce the 3′ junction. Flank with squared brackets "[]" from the first A in the polyA to the end of the TSD (estimate the TSD as 15–20 nt if the 5′ junction is not detected) to mark the region that the primers must flank. Force the use of the primer from Step 2 as "left primer." Set Product Size Range to 150–250 bp and set "Min. Primer. Tm." to 60 °C. Use the default values for the rest of the parameters.

5. Select a genomic 3′ flank primer. If there are repeated regions within the 3′ genomic flank, choose a primer within non-repetitive sequence, or at least place the 3′ end of the primer within non-repetitive sequence (see **Note 39**).

6. Re-set the Primer3 interface and introduce the sequence of the empty site. Mark the TSD with square brackets as in Step 4. Force the software to use the primer designed in Step 5 as "right primer." Set the rest of the parameters as in Step 4. Run the software and select a genomic 5′ flank primer, avoiding repeat sequences as described in Step 5.

A range of different polymerases can be used for the PCR validation of the junctions and the PCR validation of the full length insertion (empty/filled PCR). A robust *Taq* polymerase, frequently without proofreading activity (e.g., MyTaq from Bioline) is a good choice for the 3′ junction PCR.

As a first approach, use the recipe: 4 μL of 5× MyTaq Reaction Buffer; 20 pmol of L1_3′ primer; 20 pmol of genomic 3′ flank primer; 0.2 μL of MyTaq 5 U/μL; 10 ng of DNA; and up to 20 μL of molecular grade water for a single reaction; and the following cycling conditions: (95 °C, 2 min) × 1; (95 °C, 30 s; 60 °C, 30 s;

72 °C, 30 s) × 25; (72 °C, 5 min; 4 °C, hold) × 1 (see **Note 40**). Resolve the PCR products on a 2% agarose gel and be aware that amplicons may slightly differ with the estimated expected size, since the real length of the polyA tract in the new insertion cannot be estimated from the high-throughput sequencing read. Include a non-template PCR control, and control reactions with 10 ng input genomic DNA extracted from the same brain area from which the neurons were isolated and a control tissue.

In the case of empty/filled PCRs, the fact that the primers target flanking, non-repeated sequences will increase their specificity. Processive, proof reading enzymes like Expand Long Range dNTPack (Roche) or Platinum® Taq DNA Polymerase High Fidelity (ThermoFisher Scientific) are recommended. As a first approach, use the recipe: 5 µL of 5× Reaction Buffer with $MgCl_2$; 1.25 µL of 10 mM each dNTPs; 0.67 µL of 100% DMSO; 10 pmol of each franking primer; 0.35 µL of Expand Long Range enzyme blend; 4 ng of input DNA; up to 25 µL of molecular grade water for a single reaction; and the following cycling conditions: (92 °C, 2 min) × 1; (92 °C, 10 s; 58 °C, 15 s; 68 °C, 6 min) × 10; (92 °C, 10 s; 58 °C, 15 s; 68 °C, 6 min + Δ20 s/cycle) × 30; (68 °C, 10 min; 4 °C, hold) × 1 (see **Note 41**). Design the experiment and controls as for the 3' junction validation described above. The empty product must be present in all the single neurons and positive controls, but the filled product must be detected in the expected single neuron and absent from the control tissue.

If the empty/filled PCR is not successful, a 5' junction PCR can be attempted for insertions captured only at the 3' junction. Since the length of an insertion is a priori unknown, we recommend multiple PCR reactions using the fixed genomic 5' flank primer in combination with battery of L1 antisense primers evenly distributed across the element sequence (see **Note 42**). Additionally, it is worth mentioning that alternative TPRT events like 5' inversions or transductions may hinder this validation strategy. Sanger sequencing of the detected amplicons in the validations experiments can be performed using conventional molecular biology techniques.

It is worth noting that in some cases, insertions detected in a single neuron by next-generation sequencing techniques were later revealed by PCR validation to be present in additional analyzed neurons [47–49], indicating that high-throughput sequencing analysis had underestimated the true prevalence of the insertion. Thus, we recommend that once an insertion has been validated in the single neuron in which it was originally detected, the full panel of neurons should be genotyped for the presence of the insertion (see **Note 43**). Attempt 3' and 5' junctions, as well as empty/filled PCR, since amplification drop out may have affected either of the junctions in a given cell.

4 Notes

1. Phi29 DNA polymerase has $3' \rightarrow 5'$ exonuclease activity as part of its proofreading activity. In the initial stages of the reaction, when the hexamers are in abundance but input DNA has not been sufficiently amplified, the highly processive $3' \rightarrow 5'$ exonuclease activity of Phi29 can degrade the hexamers and impair the amplification reaction. To avoid this, it is extremely important to use thio-linkage modified hexamers in the two $3'$-most nucleotides [67].

2. Protease inhibitor cocktail is supplied in tablets in foil blister packs. Prepare the required cocktail by dissolving the tablets at a ratio of 1 tablet per 1 mL of molecular grade water.

3. The original recipes use PBS without magnesium and calcium [51, 52], but the presence of these two cations in the PBS+ as well as the other components of the Blocking Buffer detailed here helps to maintain the integrity of the nuclei.

4. The mixes can be prepared and left for incubation while performing Steps 1–10. Note also that the antibody dilutions indicated here should be considered as guidelines, since every batch of antibodies must be optimized independently. As an example, in other recipes the Alexa Fluor488 secondary antibody is used at a 1:500 dilution, although in our hands this condition results in unspecific staining. Prepare the antibody dilution in Blocking Buffer. The volumes indicated in the table for use of antibody dilutions are a starting point for adjusting the antibodies batches. The added Blocking Buffer can be diluted up to 1:4 with PBS+ if the level of staining is not satisfactory.

5. Use a disposable laboratory coat and long cuff gloves for picking and WGA.

6. Items 6–12 can be prepared during the cytometry sorting of the nuclei in Step 19 if the sorting is going to be performed by an operator in a central cytometry facility.

7. The needle angled tip needs to be close to horizontal in the "picking position" of the arm. The device is designed in such a way that loading the needle with the tip pointing up when the arm is in the "loading position" will result in the optimal needle angle when in the "picking position" (nearly parallel to the slide). Be aware of the potential presence of bubbles in the needle; purge and repeat the cleaning and preparation of the needle if they appear during the process.

8. The reason for not bringing the oil to the end of the needle is to prevent the oil from reaching the section of the needle used for picking the nuclei. However, if the gap between the PBS in

the tip of the needle and the oil is too big, it will be difficult to aspirate liquid into the needle since there will be a delay between the rotary knob turning and the actual aspiration.

9. Do not use PBS+ for washing the nuclei. The nuclei have to be washed in PBS without magnesium and calcium because these cations as well as goat serum and BSA in the Blocking Buffer may interfere with subsequent amplification.

10. Items 13–16 can be prepared once the sorted solution is ready. This way, the Blocking Buffer and the PBS will be cold when picking starts.

11. It is important to have PBS always present in the tip of the needle when aiming to pick a nucleus. Don't attempt picking with a dry needle tip.

12. The needle must be in the center of the field observed through the lens, but not inside or touching the drop, so it will be slightly out of focus as it will be above the focal plane.

13. The nuclei are very unstable at this moment, so dispense the Blocking Buffer carefully on the wall of the tube and don't disturb the pellet. This incubation eases the stress of the nuclei before adding the antibodies mix. The volume of Blocking Buffer to resuspend the nuclei must be optimized empirically, since it should be optimized for the amount of nuclei that the sample/extraction generates. Recommended volumes for setting up tries are 75, 125, 300, or 500 μL.

14. The distribution of the sample along the Positive and Negative Mixes of antibodies is asymmetrical. The Control Tube will be used to set up the gating parameters in the cytometer and does not need as many events as the Positive Tube, so the amount of cell suspension added can be reduced as indicated in the table.

15. The amount of sample used as source for picking must be adjusted according to the yield of the sorting. If the amount of nuclei is too high, they will be too close together for a successful individual picking. As a starting reference for adjusting, for a sorting with ~5000 events, combine 10 μL of sample with 30 μL drop of Blocking Buffer on the slide.

16. Notice that as soon as the needle enters the PBS drop, the volume inside can be partially pumped out due to pressure differences between inside and outside the needle. Since only the very surface of the slide is in focus, there is high risk of losing the nucleus if this is expelled before reaching the bottom of the drop. Therefore, it is recommended to bring the needle as fast as possible within the focus plane, close to the slide. Be aware that if the nucleus is expelled and not recovered, the drop of PBS is considered contaminated and, consequently, it should be replaced.

17. Potassium hydroxide from the lysis buffer tends to crystallize and precipitate after a time. Do not use the lysis buffer if it is more than 3 weeks old or if it has clear crystals in the solution. The concentration of potassium hydroxide in a crystallized solution is altered and, therefore, pH will not be neutralized by the Neutralization Buffer in Step 5. See **Note 21** for how to deduce a failure of the Lysis Buffer from the control reactions of the WGA.

18. The WGA reaction is very sensitive to contamination. In these early steps of the process, any contaminating DNA is susceptible to being amplified, which will severely confound downstream results. Prepare all these buffers in a DNA-free area and use a DNA removing product to clean pipettes and surfaces (hydrogen peroxide, SDS-based products, or sodium hypochlorite is recommended). Use only brand new manufacturer-sealed dispensable serological pipettes, fresh new molecular grade water, and clean bottles. Do not open the buffers outside a pre-PCR hood after they are prepared.

19. Be very careful when preparing HCl dilution. Calculations are made for a specific HCl stock solution with the parameters indicated in the table. Molar concentration of different HCl stock solutions can be calculated using the formula: Concentration (Molar) = density (g/mL) × percentage (%) × 10/36.46. For an example percentage of 32%, introduce value 32, not 32/100.

20. Prepare the reaction master mix without enzyme and keep it at 4 °C. Keep the manufacturer's enzyme tube in a −20 °C cooler and only add the enzyme to the mix after Steps 3–5 are completed.

21. The PBS control is necessary for the quality of the amplification. It consists of a WGA reaction using a drop of the PBS for "washing" the nuclei used during the picking. DNA from degraded nuclei in the sample is a potential source of contamination of the input DNA material. This PBS control WGA reaction will serve to evaluate the presence of such DNA in the nuclei solution. On the other hand, the Blank control is a reaction that will provide information about the quality of the Lysis and Neutralization Buffers used for the nuclei and DNA denaturation. Spoiled solutions will not buffer the pH properly and will result in a failure of the amplification. The Phi29 enzyme is able to produce amplified material using only the input hexamers, even without input DNA. A failure of the amplification detected in Steps 20 and 21 for nuclei and also Blank control likely indicates that lysis and neutralization solutions were not prepared properly. Check and replace your solutions in that case. A bulk sorted nuclei control and a 10 pg DNA input control are optional positive controls of amplification.

22. During and after the lysis of the nuclei, there are high chances of losing part of the input DNA in aerosols or stuck to the pipette tip, which will negatively impact the coverage of the WGA. Pipette all the volumes with extreme care using a single channel pipette. Do NOT dip the pipette tip in the sample already present in the tube, drop the volumes from the top of the surface of the sample without breaking the surface tension, and do NOT pipette up and down the volume for mixing.

23. Probably the most time-efficient way of performing this reaction is to perform the 16 h incubation overnight. Thus, reactions started around 5:00–6:00 p.m. will be conveniently done by 9:00–10:00 a.m. the next day.

24. Phi29 enzyme $3' \rightarrow 5'$ exonuclease activity is a high risk for degrading the synthesized DNA when the reaction runs out of nucleotides, the incubation temperature increases, or the reaction volume is reduced due to evaporation. Thus, it is critical not to let the reaction go longer than 16 h. Certain protocols include a 95 °C, 3 min inactivation step that can be programmed in the thermocycler and hold the reaction at 4 °C [47] although we obtained substantial amplification yield by immediately following the 16 h incubation with the AMPure® XP beads clean up. See **Note 23** for appropriate timing.

25. Due to the branched nature of the WGA DNA, AMPure® XP beads frequently aggregate to generate clumps. The dilution performed in Step 11 is intended to reduce this, but you can proceed as normal even if the clumps have been formed. In the resuspension Step 17, pipette the sample up and down several times to reduce the aggregation of the clumps and facilitate DNA elution.

26. This Qubit quantification can be performed during the clean-up incubation Steps 12 and 15.

27. Related to **Note 20**, the WGA of negative controls ("PBS" and "Blank") typically yields half of two-thirds of the concentration of the positive ones. However, amplification in the negative controls is expected for Phi29 enzyme since it can use primers dimers as substrates for elongation and generate equally long amplicons. This activity has not been reported to have an adverse effect when actual input DNA is added to the reaction. Until the WGA Quality Control PCR of Step 37, no estimation of contamination and real amplification can be done.

28. Mung-Bean Nuclease is a very aggressive nuclease. Although specific for single strand DNA, low concentrations of DNA or extended reaction times can lead to total degradation of the DNA sample. Despite the hold at 4 °C, it is strongly recommended to take the reaction tube and proceed to the

AMPure® XP beads clean up immediately after the incubation time is finished.

29. If first attempts measuring the concentration with the 1:50 dilutions of the original WGA reactions of Step 20 with 10 μL were within the range of the Qubit High Sensitivity assay, typically 1:4 dilutions will work too. Otherwise, proceed to 1:8 dilutions.

30. Note that the WGA Quality Control PCR in other protocols is performed with the 1:50 dilutions of the original WGA reactions of Step 20 [47]. However, after the debranching reaction, the amount of DNA can be severely reduced, but the DNA can eventually become more accessible for PCR. We consider that doing the PCR in the final step, after debranching and before proceeding to RC-seq or any other enrichment/sequencing methods is more appropriate for choosing the most suitable samples.

31. WGA Quality Control PCR cycling protocol takes more than 2 h. Steps 43 and 44 to evaluate the performance of the debranching reaction can be done during this time.

32. After the WGA Quality Control PCR, several conclusions can be drawn from the results of the different controls. If the No-Template Control has any of the four bands, reagents for the WGA Quality Control PCR are contaminated by any of the amplicons and they must be depurated and reordered, and the PCR must be repeated before assessing the quality of the samples. The positive PCR control with 10 ng of human genomic DNA must present the four expected bands, otherwise it would be an indication of failure of the PCR itself. PCR reagents should be checked and replaced in case they are expired or degraded. Blank control must NOT present any band, otherwise Lysis or Neutralization buffers, or any of the reagents used in the WGA reaction could be contaminated with DNA. PCR reactions using aliquots from the washing PBS droplet used during single nuclei picking (described in Step 29 of the Prodedure in Sect. 3.1) as input must NOT present any band, otherwise the nuclei preparation must have been contaminated and nuclei extraction need to be repeated paying special attention to steps aiming for nuclei integrity preservation. If all the controls work as expected and very few neurons produced ≥3 bands in the WGA Quality Control PCR, please ensure that the tissue sample is in good condition, verify that it has not been thawed at any time from the moment of extraction, extract genomic DNA from bulk tissue and assess its integrity by gel electrophoresis, and repeat the nuclei preparation paying special attention to steps aiming for nuclei integrity preservation.

33. The failure in the detection of TSDs may pinpoint possible deletions in the insertion site or the presence of a very long TSD, but may also result from two artifactual L1-genome junction chimeras mapping close to each other.

34. Be aware that the L1 element may undergo inversion/deletions during its retrotransposition and the 5' end of the element can be in antisense respect to the consensus [16, 68].

35. Pay special attention to the Repeat Masker tab, which indicates repeated sequences in the target site area. Repeats may hinder the validation PCR due to nonspecific primer binding. Try to adjust the frame of the selection to include non-repeated sequences on both sides of the insertion point.

36. Some criteria for narrowing the number of insertions: presence of TPRT hallmarks; identification of the L1 sequence as belonging to the L1_Ta subfamily [13, 65, 66]; the presence of an A-rich region (in sense) within the target site together with a non-clean polyA tract in the captured 3' junction is a strong indicative of chimerism; a requirement of a clean polyA tract of ≥30 nt is a suggested starting criteria.

37. Copy-paste this sequence: GGATAGCATTGGG<u>AGATATAC CTAATGCT**AGATGACAC**</u>**ATTAGTGGGTGCAG**CG CACCAGCATGGCACATGTATACATATGTAACTAACC TGCACAATGTGCACATGTACCCTAAAACTTAGAGTA TAATAAAAAAAAAAAAAAAAAAAA. The primer suggested in **Note 38** is underlined, and the target of the RC-seq probe is in bold.

38. This shortened primer from the reported L1HsTailSP1A2 has an optimized melting temperature of 60 °C: AGATATAC CTAATGCTAGATGACAC [45].

39. Some advice if the software does not generate any primer: if the sequence is pyrimidine-rich, try to increase "Primer Size" parameter; progressively increase the "Product Size Range" by 50 nt increments.

40. Despite the specificity imposed by a 3' flanking genomic primer, the L1_3' primer can generate multiple undesired amplicons. Start with 25 cycles and increase this number only if positive but very faint bands are detected.

41. Although primers are designed for 60 °C, it is better to start with melting temperature of 58 °C since DMSO is included to increase the stringency of the annealing conditions. Decrease melting temperature or DMSO, or increase the input DNA if no bands are detected. On the contrary, increase temperature to 60 °C and/or increase DMSO contribution to 1.25 μL if you obtain a ladder of nonspecific bands.

42. Some recommended L1 primers for this attempt are provided in reference [50].

43. When analyzing the results of full panel of neurons, take in consideration that the length of the 3' polyA can vary across the neurons [48].

Acknowledgments

G.J.F. acknowledges the support of a CSL Centenary Fellowship. F.J.S-L. was supported by a postdoctoral fellowship from the Alfonso Martín Escudero Foundation (Spain) and the Peoples Programme (Marie Curie Actions) of the European Union's Seventh Framework Programme (FP7/2007-2013) under REA grant agreement No PIOF-GA-2013-623324.

We also acknowledge the significant contribution of Dr. Adam D. Ewing in TEBreak settings adjustment for RC-seq sequencing data analysis, and J. Samuel Jesuadian and Marie-Jeanne H. C. Kempen in technical assistance.

References

1. Lander ES, Linton LM, Birren B et al (2001) Initial sequencing and analysis of the human genome. Nature 409(6822):860–921. doi:10.1038/35057062

2. Lindblad-Toh K, Wade CM, Mikkelsen TS et al (2005) Genome sequence, comparative analysis and haplotype structure of the domestic dog. Nature 438(7069):803–819. doi:10.1038/nature04338

3. Mouse Genome Sequencing C, Waterston RH, Lindblad-Toh K et al (2002) Initial sequencing and comparative analysis of the mouse genome. Nature 420(6915):520–562. doi:10.1038/nature01262

4. Craig NL, Caraigie R, Gellert M et al (2002) Mobile DNA II. ASM Press, Washington, DC

5. Boeke JD, Garfinkel DJ, Styles CA et al (1985) Ty elements transpose through an RNA intermediate. Cell 40(3):491–500

6. Luan DD, Korman MH, Jakubczak JL et al (1993) Reverse transcription of R2Bm RNA is primed by a nick at the chromosomal target site: a mechanism for non-LTR retrotransposition. Cell 72(4):595–605

7. Batzer MA, Deininger PL (2002) Alu repeats and human genomic diversity. Nat Rev Genet 3(5):370–379. doi:10.1038/nrg798

8. Hancks DC, Kazazian HH Jr (2010) SVA retrotransposons: evolution and genetic instability. Semin Cancer Biol 20(4):234–245. doi:10.1016/j.semcancer.2010.04.001

9. Khan H, Smit A, Boissinot S (2006) Molecular evolution and tempo of amplification of human LINE-1 retrotransposons since the origin of primates. Genome Res 16(1):78–87. doi:10.1101/gr.4001406

10. Smit AF, Toth G, Riggs AD et al (1995) Ancestral, mammalian-wide subfamilies of LINE-1 repetitive sequences. J Mol Biol 246(3):401–417. doi:10.1006/jmbi.1994.0095

11. de Koning AP, Gu W, Castoe TA et al (2011) Repetitive elements may comprise over two-thirds of the human genome. PLoS Genet 7(12):e1002384. doi:10.1371/journal.pgen.1002384

12. Grimaldi G, Skowronski J, Singer MF (1984) Defining the beginning and end of KpnI family segments. EMBO J 3(8):1753–1759

13. Dombroski BA, Mathias SL, Nanthakumar E et al (1991) Isolation of an active human transposable element. Science 254(5039):1805–1808

14. Swergold GD (1990) Identification, characterization, and cell specificity of a human LINE-1 promoter. Mol Cell Biol 10(12):6718–6729

15. Garcia-Perez JL, Doucet AJ, Bucheton A et al (2007) Distinct mechanisms for trans-mediated mobilization of cellular RNAs by the LINE-1 reverse transcriptase. Genome Res 17(5):602–611. doi:10.1101/gr.5870107

16. Gilbert N, Lutz S, Morrish TA et al (2005) Multiple fates of L1 retrotransposition intermediates in cultured human cells. Mol Cell Biol 25(17):7780–7795. doi:10.1128/MCB.25.17.7780-7795.2005

17. Richardson SR, Salvador-Palomeque C, Faulkner GJ (2014) Diversity through duplication: whole-genome sequencing reveals novel gene retrocopies in the human population. BioEssays 36(5):475–481. doi:10.1002/bies.201300181

18. Esnault C, Maestre J, Heidmann T (2000) Human LINE retrotransposons generate processed pseudogenes. Nat Genet 24(4):363–367. doi:10.1038/74184

19. Dewannieux M, Esnault C, Heidmann T (2003) LINE-mediated retrotransposition of marked Alu sequences. Nat Genet 35(1):41–48. doi:10.1038/ng1223

20. Hancks DC, Goodier JL, Mandal PK et al (2011) Retrotransposition of marked SVA elements by human L1s in cultured cells. Hum Mol Genet 20(17):3386–3400. doi:10.1093/hmg/ddr245

21. Cost GJ, Feng Q, Jacquier A et al (2002) Human L1 element target-primed reverse transcription in vitro. EMBO J 21(21):5899–5910

22. Jurka J (1997) Sequence patterns indicate an enzymatic involvement in integration of mammalian retroposons. Proc Natl Acad Sci U S A 94(5):1872–1877

23. Venter JC, Adams MD, Myers EW et al (2001) The sequence of the human genome. Science 291(5507):1304–1351. doi:10.1126/science.1058040

24. Kazazian HH Jr, Wong C, Youssoufian H et al (1988) Haemophilia a resulting from de novo insertion of L1 sequences represents a novel mechanism for mutation in man. Nature 332(6160):164–166. doi:10.1038/332164a0

25. Hancks DC, Kazazian HH Jr (2012) Active human retrotransposons: variation and disease. Curr Opin Genet Dev 22(3):191–203. doi:10.1016/j.gde.2012.02.006

26. Hancks DC, Kazazian HH Jr (2016) Roles for retrotransposon insertions in human disease. Mob DNA 7:9. doi:10.1186/s13100-016-0065-9

27. Beck CR, Collier P, Macfarlane C et al (2010) LINE-1 retrotransposition activity in human genomes. Cell 141(7):1159–1170. doi:10.1016/j.cell.2010.05.021

28. Brouha B, Schustak J, Badge RM et al (2003) Hot L1s account for the bulk of retrotransposition in the human population. Proc Natl Acad Sci U S A 100(9):5280–5285. doi:10.1073/pnas.0831042100

29. Philippe C, Vargas-Landin DB, Doucet AJ et al (2016) Activation of individual L1 retrotransposon instances is restricted to cell-type dependent permissive loci. elife 5. doi:10.7554/eLife.13926

30. Sassaman DM, Dombroski BA, Moran JV et al (1997) Many human L1 elements are capable of retrotransposition. Nat Genet 16(1):37–43. doi:10.1038/ng0597-37

31. Moran JV, Holmes SE, Naas TP et al (1996) High frequency retrotransposition in cultured mammalian cells. Cell 87(5):917–927

32. Coufal NG, Garcia-Perez JL, Peng GE et al (2009) L1 retrotransposition in human neural progenitor cells. Nature 460(7259):1127–1131. doi:10.1038/nature08248

33. Kuwabara T, Hsieh J, Muotri A et al (2009) Wnt-mediated activation of NeuroD1 and retro-elements during adult neurogenesis. Nat Neurosci 12(9):1097–1105. doi:10.1038/nn.2360

34. Muotri AR, Chu VT, Marchetto MC et al (2005) Somatic mosaicism in neuronal precursor cells mediated by L1 retrotransposition. Nature 435(7044):903–910. doi:10.1038/nature03663

35. Muotri AR, Marchetto MC, Coufal NG et al (2010) L1 retrotransposition in neurons is modulated by MeCP2. Nature 468(7322):443–446. doi:10.1038/nature09544

36. Richardson SR, Morell S, Faulkner GJ (2014) L1 retrotransposons and somatic mosaicism in the brain. Annu Rev Genet 48:1–27. doi:10.1146/annurev-genet-120213-092412

37. Hozumi N, Tonegawa S (1976) Evidence for somatic rearrangement of immunoglobulin genes coding for variable and constant regions. Proc Natl Acad Sci U S A 73(10):3628–3632

38. Muramatsu M, Kinoshita K, Fagarasan S et al (2000) Class switch recombination and hypermutation require activation-induced cytidine deaminase (AID), a potential RNA editing enzyme. Cell 102(5):553–563

39. Beck CR, Garcia-Perez JL, Badge RM et al (2011) LINE-1 elements in structural variation and disease. Annu Rev Genomics Hum Genet 12:187–215. doi:10.1146/annurev-genom-082509-141802

40. Hulme AE, Kulpa DA, Garcia-Perez JL et al (2006) The impact of LINE-1 retrotransposition on the human genome. In: Lupski JS (ed) Genomic disorders: the genomic basis of disease. Humana Press, Totowa, NJ, pp 35–72

41. Baillie JK, Barnett MW, Upton KR et al (2011) Somatic retrotransposition alters the genetic landscape of the human brain. Nature 479(7374):534–537. doi:10.1038/nature10531

42. Shukla R, Upton KR, Munoz-Lopez M et al (2013) Endogenous retrotransposition activates oncogenic pathways in hepatocellular carcinoma. Cell 153(1):101–111. doi:10.1016/j.cell.2013.02.032

43. Badge RM, Alisch RS, Moran JV (2003) ATLAS: a system to selectively identify human-specific L1 insertions. Am J Hum Genet 72(4):823–838. doi:10.1086/373939

44. Erwin JA, Paquola AC, Singer T et al (2016) L1-associated genomic regions are deleted in somatic cells of the healthy human brain. Nat Neurosci 19(12):1583–1591. doi:10.1038/n.4388

45. Ewing AD, Kazazian HH Jr (2010) High-throughput sequencing reveals extensive variation in human-specific L1 content in individual human genomes. Genome Res 20(9):1262–1270. doi:10.1101/gr.106419.110

46. Witherspoon DJ, Xing J, Zhang Y et al (2010) Mobile element scanning (ME-scan) by targeted high-throughput sequencing. BMC Genomics 11:410. doi:10.1186/1471-2164-11-410

47. Evrony GD, Cai X, Lee E et al (2012) Single-neuron sequencing analysis of L1 retrotransposition and somatic mutation in the human brain. Cell 151(3):483–496. doi:10.1016/j.cell.2012.09.035

48. Evrony GD, Lee E, Mehta BK et al (2015) Cell lineage analysis in human brain using endogenous retroelements. Neuron 85(1):49–59. doi:10.1016/j.neuron.2014.12.028

49. Upton KR, Gerhardt DJ, Jesuadian JS et al (2015) Ubiquitous L1 mosaicism in hippocampal neurons. Cell 161(2):228–239. doi:10.1016/j.cell.2015.03.026

50. Sanchez-Luque FJ, Richardson SR, Faulkner GJ (2016) Retrotransposon capture sequencing (RC-Seq): a targeted, high-throughput approach to resolve somatic L1 retrotransposition in humans. Methods Mol Biol 1400:47–77. doi:10.1007/978-1-4939-3372-3_4

51. Jiang Y, Matevossian A, Huang HS et al (2008) Isolation of neuronal chromatin from brain tissue. BMC Neurosci 9:42. doi:10.1186/1471-2202-9-42

52. Okada S, Saiwai H, Kumamaru H et al (2011) Flow cytometric sorting of neuronal and glial nuclei from central nervous system tissue. J Cell Physiol 226(2):552–558. doi:10.1002/jcp.22365

53. Dean FB, Hosono S, Fang L et al (2002) Comprehensive human genome amplification using multiple displacement amplification. Proc Natl Acad Sci U S A 99(8):5261–5266. doi:10.1073/pnas.082089499

54. Zong C, Lu S, Chapman AR et al (2012) Genome-wide detection of single-nucleotide and copy-number variations of a single human cell. Science 338(6114):1622–1626. doi:10.1126/science.1229164

55. Carreira PE, Ewing AD, Li G et al (2016) Evidence for L1-associated DNA rearrange-ments and negligible L1 retrotransposition in glioblastoma multiforme. Mob DNA 7:21. doi:10.1186/s13100-016-0076-6

56. Klawitter S, Fuchs NV, Upton KR et al (2016) Reprogramming triggers endogenous L1 and Alu retrotransposition in human induced pluripotent stem cells. Nat Commun 7:10286. doi:10.1038/ncomms10286

57. Streva VA, Jordan VE, Linker S et al (2015) Sequencing, identification and mapping of primed L1 elements (SIMPLE) reveals significant variation in full length L1 elements between individuals. BMC Genomics 16:220. doi:10.1186/s12864-015-1374-y

58. Li H (2013) Aligning sequence reads, clone sequences and assembly contigs with BWA-MEM. http://arxiv.org/abs/1303.3997.

59. Langmead B, Salzberg SL (2012) Fast gapped-read alignment with bowtie 2. Nat Methods 9(4):357–359. doi:10.1038/nmeth.1923

60. Rahbari R, Badge RM (2016) Combining amplification typing of L1 active subfamilies (ATLAS) with high-throughput sequencing. Methods Mol Biol 1400:95–106. doi:10.1007/978-1-4939-3372-3_6

61. Scott EC, Gardner EJ, Masood A et al (2016) A hot L1 retrotransposon evades somatic repression and initiates human colorectal cancer. Genome Res 26(6):745–755. doi:10.1101/gr.201814.115

62. Gilbert N, Lutz-Prigge S, Moran JV (2002) Genomic deletions created upon LINE-1 retrotransposition. Cell 110(3):315–325

63. Morrish TA, Gilbert N, Myers JS et al (2002) DNA repair mediated by endonuclease-independent LINE-1 retrotransposition. Nat Genet 31(2):159–165. doi:10.1038/ng898

64. Jurka J (1998) Repeats in genomic DNA: mining and meaning. Curr Opin Struct Biol 8(3):333–337

65. Boissinot S, Chevret P, Furano AV (2000) L1 (LINE-1) retrotransposon evolution and amplification in recent human history. Mol Biol Evol 17(6):915–928

66. Ovchinnikov I, Rubin A, Swergold GD (2002) Tracing the LINEs of human evolution. Proc Natl Acad Sci U S A 99(16):10522–10527. doi:10.1073/pnas.152346799

67. Skerra A (1992) Phosphorothioate primers improve the amplification of DNA sequences by DNA polymerases with proofreading activity. Nucleic Acids Res 20(14):3551–3554

68. Ostertag EM, Kazazian HH Jr (2001) Twin priming: a proposed mechanism for the creation of inversions in L1 retrotransposition. Genome Res 11(12):2059–2065. doi:10.1101/gr.205701

Chapter 13

Single-Cell Whole Genome Amplification and Sequencing to Study Neuronal Mosaicism and Diversity

Patrick J. Reed, Meiyan Wang, Jennifer A. Erwin, Apuã C.M. Paquola, and Fred H. Gage

Abstract

Neuronal mosaicism describes the extent of intercellular genotypic diversity within a single human brain. This somatic variability is driven by numerous mechanisms including errors in DNA replication acquired throughout development and by the activity of endogenous retrotransposons. The study of retrotransposition in neuronal mosaicism may prove crucial to understanding the true complexity of normal and aberrant brain function. Specifically, numerous lines of evidence suggest that retrotransposition specific aspects of neuronal mosaicism may contribute to the unresolved etiology of many neurologic and neuropsychiatric disorders. Here, we describe the SLAV-Seq method, a recent advancement in the field over previous approaches used to study the diversity of LINE-1 based neuronal mosaicism at the single-cell level. We describe in detail, methodology for the isolation of single cells from bulk tissue by FACS, the amplification of single-cell genomic DNA by multiple displacement amplification (MDA), the targeted enrichment of LINE-1 somatic events, and the sequencing of the LINE-1 enriched library. Finally, we discuss methods for the quantification and analysis of the neuronal mosaicism identified by SLAV-Seq and some of the current technical limitations.

Key words SLAV-Seq, Single cell, Neuronal mosaicism, Somatic mosaicism, WGA, MDA, Retrotransposition, LINE-1

1 Introduction

The adult human brain comprises approximately 90 billion neurons [1]. It has long been thought that the genetic information of these cells, needed to establish the diverse repertoire of neuronal subtypes and orchestrate assembly into complex neural circuits, was invariant. However, significant research over the last decade has identified numerous mechanisms which create stochastic intercellular genomic diversity throughout human development. The totality of this diversity, somatic mosaicism, is the presence of

---

Patrick J. Reed and Meiyan Wang contributed equally to this work.

José María Frade and Fred H. Gage (eds.), *Genomic Mosaicism in Neurons and Other Cell Types*, Neuromethods, vol. 131, DOI 10.1007/978-1-4939-7280-7_13, © Springer Science+Business Media LLC 2017

distinct genotypes within the somatic cells of a single organism. Throughout human development, somatic mosaicism is mediated by the accumulated burden of single nucleotide variants (SNVs), copy number variants (CNVs), and structural variants (SVs) owing to the imperfect fidelity of DNA replication, recombination, and repair and to cellular exposure to free radicals, UV light, oxidative stress, and other mutagens [2]. Additionally, several lines of evidence have proven that LINE-1 transposable elements are not only expressed in the human brain but also actively transpose during neurogenesis [3–5]. As such, LINE-1 Mobile Element Insertions (MEIs) in the developing brain add a unique additional layer of diversity to neuronal mosaicism.

Transposable elements comprise nearly 45% of the human genome [6]. However, only a small percentage of these mobile elements are still capable of mobilization [7]. LINE-1 elements are 6 kb elements that encode open reading frame 1 protein (L1ORF1p), an RNA-binding protein [8], and L1ORF2p, a protein with endonuclease [9] and reverse transcriptase [10] activity. Human-specific LINE-1 (LINE-1Hs) retrotransposons comprise the only transposon family known to be autonomously active in humans, with approximately 100 active LINE-1Hs elements per individual [11]. Interestingly, recent evidence suggests that LINE-1Hs activity is higher in the neuronal cells of Schizophrenic individuals compared to neurotypical controls [12]. This could be indicative of a unique role within the brain and may contribute to the unresolved etiology of many neurologic and neuropsychiatric disorders [13].

Whole genome sequencing of bulk brain tissue is a powerful method that can be used to study many aspects of genetic variation, including neuronal mosaicism. However, WGS of bulk tissue is severely limited by sequencing depth in its sensitivity to detecting rare events. For this reason, the systematic study of neuronal mosaicism requires single-cell genome sequencing [14]. Single-cell sequencing vastly increases the sensitivity of detecting unique events but is limited by technical variability of single-cell whole genome amplification (WGA) methods, cost, and the throughput of processing single cells. As WGA methods continue to improve, single-cell-based sequencing methods are quickly becoming the gold standard in the study of somatic mosaicism. For the study of MEIs, WGA followed by targeted enrichment for LINE-1 events greatly increases the sensitivity of detecting unique insertions and increases the ability to resolve the correct structure of the integration event.

The aim of this chapter is to give the reader a solid understanding of the SLAV-Seq method, its utility in the field of neuronal mosaicism, advantages, current limitations, and future applications. First, we begin by reviewing general methods for the isolation of single cells from bulk tissue as well as methods for the amplification of single-cell genomic DNA. We will then discuss the protocol for SLAV-Seq in detail, the analysis of bulk and single-cell SLAV-Seq data, methods for the validation of findings, and future directions for the field.

1.1 Single-Cell Isolation and Whole Genome Amplification

Individual cells can be isolated from the tissue using micromanipulation, for example, pipetting [15], serial dilution [16, 17], or microwell dilution [18]. Although these methods are cheap and easy to apply, they are low-throughput, and susceptible to errors. Another approach, which is also classified as micromanipulation, is the optical tweezer technology [19]. This technology relies on laser beam to capture cells. Laser-capture microdissection [20, 21] provides a low-throughput way of isolating DNA from single cells in their native spatial context, but the quality of sequencing data derived from microdissected single cells has been relatively poor [14].

Single cells or nuclei isolation can also be achieved by flow sorting using fluorescence-activated cell sorting (FACS) [22]. The advantages of FACS-based cell isolation include high accuracy, high-throughput, cell type-specific, and unbiased single-cell isolation [23, 24]. However, FACS requires cells in suspension as starting materials. The native spatial information of the cells will be lost during the preparation. Recent progress in microfluidic devices has enabled new era of single-cell isolation [23, 25]. These devices allow the compartmentalization and controlled management of reactions by controlled liquid streaming. Microfluidic devices provide several advantages, including high throughput with less effort, reduced reagent costs, and improved accuracy in single-cell isolation [24]. Similar with FACS, microfluidic devices require cells in suspension as starting materials.

Given the current limitations of sequencing technology, following single-cell isolation, genomic DNA must be amplified. A high-fidelity, low-bias method for WGA from single cells is a huge challenge in the field because it requires faithful amplification from picograms of genomic DNA without the loss or distortion of any particular loci or alleles (Fig. 1). Unlike for DNA sequencing from bulk cell populations or for single-cell RNA sequencing, the initial copy number for single-cell DNA is limited. The first group of methods that attempted to amplify entire human genomes from single cells relies on PCR amplification with either common sequences or degenerate or random oligonucleotide priming [26]. The principle of Degenerate Oligonucleotide Primed PCR (DOP-PCR) employs oligonucleotides of partially degenerate sequence, with a random six-base sequence at the 3' end and a fixed sequence at the 5' end. For the initial amplification, low annealing temperature ($\sim$30 °C) ensures priming from multiple (e.g., $\sim$$10^6$ in human) sites within a given genome. Strand extension is then achieved at an elevated temperature. The second PCR stage favors amplicon replication. High stringency amplification with a primer

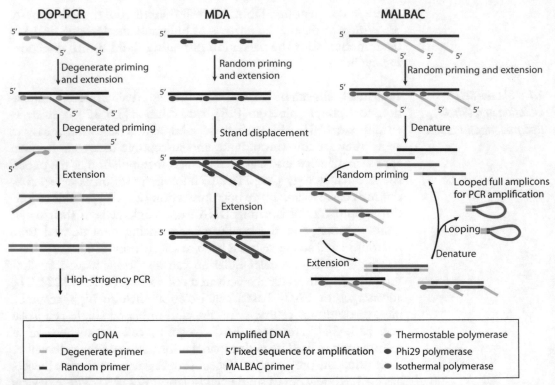

Fig. 1 Methods for whole genome amplification. Pure PCR methods include DOP-PCR, which has an initial amplification with low annealing temperature to ensure priming from multiple sites and a high stringency amplification to produce enough DNA for next-generation sequencing. Isothermal methods such as MDA use isothermal exponential amplification with DNA polymerase that has strand displacement activity. MALBAC has steps of temperature cycling to promote the looping of full amplicons, which may result in more uniform amplification

targeting the 5' fixed sequence at high annealing temperature aims to produce enough DNA for next-generation sequencing. DOP-PCR often yields low genome coverage, which is pertinent to the exponential amplification of PCR [27]. DOP-PCR uses thermostable polymerases, which have higher error rates than thermolabile polymerases, leading to more mutations generated from the amplification process. Although lacking whole genome coverage, DOP-PCR can be well suited for measuring CNVs on a large genomic scale with large bin sizes [28].

Multiple Displacement Amplification (MDA) was developed in 2001 by Lasken and coworkers [29] taking advantage of the unique properties of the DNA polymerase from bacteriophage Φ29 polymerase, which has high processivity and strong strand displacement activity [30]. The Φ29 DNA polymerase can amplify DNA isothermally at 30 °C and has a high replication fidelity because of its $3' \rightarrow 5'$ exonuclease activity and proofreading activity [27, 31]. DNA synthesis is primed by random hexamers. Amplification is achieved by a "hyperbranching" mechanism generating long DNA fragments (>10 kb). Of note, Φ29 polymerase activity results in the formation of a low level of chimeric sequences [32], which can be reduced with endonuclease treatment [33]. Nevertheless, MDA produces much higher genome coverage than DOP-PCR and lower error rate owing to the high fidelity of Φ29 DNA polymerase, which has an error rate of 1 in 10^6–10^7 [34]. However, like DOP-PCR, the exponential amplification process results in sequence-dependent bias, causing over-amplification in certain genomic regions and under-amplification in other regions [27].

Demands for unbiased single-cell WGA has inspired the development of new techniques. These include the development of the Multiple Annealing and Looping-Based Amplification Cycles method (MALBAC) by Zong et al. for single-cell WGA [35]. MALBAC primers have a common 27-nucleotide sequence at the 5′ end and eight random nucleotides at the 3′ end, which can hybridize to the DNA template at low annealing temperature. When the temperature is elevated, strand extension results in semiamplicons. The semiamplicons are then melted off from the templates. For the full amplicons with complementary ends, when the temperature is lowered (to 58 °C), the two ends hybridize to form hairpins, preventing their further amplification. After the step of looping full amplicons, single strand semi amplicons and genomic DNA are recycled as templates to produce additional semi amplicons and full amplicons. The quasi-linear amplification at these first few cycles is critical for avoiding the sequence-dependent bias exacerbated by exponential amplification [27, 35]. MALBAC uses a thermostable DNA polymerase with strand displacement activity [35]. Exponential amplification of the full amplicons by PCR is then followed, generating the amount of DNA required for next-generation sequencing. Please see Chap. 7 for further discussion of MALBAC.

1.2 SLAV-Seq

LINE-1 retrotransposition is known to create mosaicism by inserting LINE-1 sequences into new locations in the genome [36]. To identify somatic LINE-1 retrotransposition in single cells, several methods have been developed, including LINE-1-IP, RC-Scq (Chap. 12), and, most recently, SLAV-Seq [4, 37, 38]. These methods target the 5′ end or 3′ end of LINE-1 element, generating libraries enriched for LINE-1 and flanking sequences. As a method, SLAV-Seq outperforms other recently published methods in numerous ways (Fig. 2). LINE-1 capture library preparation using biotinylated primers increases enrichment purity. Paired-end sequencing increases alignment accuracy by allowing for discordant reads mapping to both reference genome and LINE-1 consensus sequence. Physical fragmentation of DNA facilitates unbiased targeting unlike the use of specific primer pools. In comparison with the other two methods, SLAV-Seq is capable of getting a much higher number of unique read starts per known non-reference germline loci, which leads to high confidence identification of somatic insertions [38].

		Erwin et al. 2016	Evrony et al. 2012	Upton et al. 2015
A	L1 targeting method	Illumina sequencing library preparation using biotynilated L1HS-specific capture primer.	Illumina sequencing library preparation using L1HS-specific primers.	Capture of L1-containing fragments after sequencing library preparation.
B	Library type	Paired-end	Single-end	Paired-end
C	Flanking end generation	Sonication	Amplification with a pool of 8 primers	Sonication
D	Median number of unique read start positions at KNRGL loci.	71 (heterozygous germline loci, table S6)	8-10 (Fig S4 of Evrony et al 2012)	1 (heterozygous germline loci, table S6)
E	DNA amplification	MDA	MDA	MALBAC
F	Main data analysis component (L1 prediction algorithm)	Random Forest classifier with 70 features.	Logistic regression with 4 features.	Researcher-defined rules identifying L1 3' junction sequences

Fig. 2 Comparison of recent LINE-1 targeted sequencing methods. Methods used to quantify LINE-1 activity in single cells and bulk tissue include SLAV-Seq (Erwin et al. [38]), RC-Seq (Upton et al. [37]), and LINE-1-IP (Evrony et al. [4]). These methods target the 5′ end or 3′ end of LINE-1 element, generating libraries enriched for LINE-1 and flanking sequences. These LINE-1 targeted enrichment assays differ significantly in their methodology. This is likely the primary reason for the conflicting results that have been published on rates and prevalence of LINE-1 activity

2   Materials

2.1   Nuclei Preparation

1. Sucrose.
2. KCl 1 M.
3. MgCl$_2$ 1 M.
4. Tris–HCl 1 M.
5. Dithiothreitol 1 M.
6. Roche complete™ Protease Inhibitor Cocktail 50×.
7. Iodixanol 60%.
8. Triton X-100 10%.
9. BSA (100× from NEB).
10. NeuN antibody.
11. DAPI.
12. 1 mL dounce homogenizer.
13. 40 μm Falcon™ Cell Strainers.

2.2   Single-Cell WGA by MDA

1. Dithiothreitol 1 M.
2. KOH 5 M.
3. Molecular grade water (Dnase/Rnase free). Reagents from GenomiPhi V2 Kit (GE Healthcare).
4. Sample buffer.

5. Reaction buffer.

6. Enzyme mix.

2.3 SLAV-Seq

1. Molecular grade water (Dnase/Rnase free).

2. AccuPrime Pfx SuperMix (Invitrogen).

3. Agencourt AMPure XP system (Beckman Coulter).

4. Dynabeads MyOne Streptavidin C1 (Life Technologies).

5. NEBNext dA-Tailing Module (New England Biolabs).

6. NEBNext End Repair Module (New England Biolabs).

7. T4 DNA Ligase (Rapid) (ENZYMATICS INC).

8. KAPA HiFi HotStart ReadyMix (2×) (kapa biosystems).

9. JE281L1_ACA: /5biosg/ATATACCTAATGCTAGATGAC
   AC*A
   (The asterisk denotes phosphorothioate linkages).

10. Custom asymmetric annealed adapters:

11. JED501+JED50×_lig: 5′-AATGATACGGCGACCACCGAG
    ATCTACACNNNNNNNNNACACTCTTTCCCTACACGAC
    GCTCTTCCGATC*T-3′
    Annealed to:
    JED50×_lig: /5Phos/GATCGGAAGAGCGTCGTGTAGGG
    AAAGAGTGT/3AmM/-3′
    (The asterisk denotes phosphorothioate linkages).

12. JE50*:   5′-CAAGCAGAAGACGGCATACGAGANNNNNN
    NGTGACTGGAGTTCAGACGTGTGCTCTTCCGA
    TCTNT AACTAACCTGCACAATGTGCAC-3′.

13. JE620:5′-AATGATACGGCGACCACCGAGATCTACAC-3′.

3 Methods

3.1 Nuclei Isolation for FACS

Prepare the following buffers.

NIM	
Reagent	Final concentration
Sucrose	0.25 M
KCl	25 mM
MgCl$_2$	5 mM
Tris–HCl	10 mM
DTT[a]	1 mM
Protease inhibitor[a]	1×
Water	–
[a]Add freshly before use	

ODN	
Reagent	Final concentration
Sucrose	0.25 M
KCl	150 mM
MgCl₂	30 mM
Tris–HCl	60 mM
Water	–

Nuclei storage buffer	
Reagent	Final concentration
Sucrose	5.7%
MgCl₂	5 mM
Tris–HCl	10 mM
DTT[a]	1 mM
Protease inhibitor[a]	1×
Water	–
[a]Add freshly before use	

ODN with 29% Iodixanol	
Reagent	Final concentration
Iodixanol	29%
ODN	

ODN with 50% Iodixanol	
Reagent	Final concentration
Iodixanol	50%
ODN	

Keep everything on ice.

1. Cut piece of sample on dry ice into tube to dounce.
2. Add 1 mL NIM +0.1% Triton X-100 mix per sample.
3. Dounce 10–15 strokes on ice. Start with loose 1–2 times then move to tight. Spin at $1000 \times g$, 8 min, 4 °C.
4. Aspirate supernatant and gently resuspend in 250 μL of NIM.

5. Strain through 40 µM filter.

6. Add 250 µL ODN with 50% Iodixanol to tube and mix well.

7. Add 500 µL ODN with 29% Iodixanol to a new tube. Then slowly layer 500 µL of NIM/iodixanol mixed sample on top of 29% iodixanol.

8. Spin at $13,500 \times g$ for 20 min at 4 °C.

9. Remove supernatant, leave as little as possible without disturbing pellet.

10. Add 50 µL nuclei storage buffer on top of nuclei. Leave on ice about 10 min for passive resuspension.

11. NeuN staining:

    (a) Add 1 µL of NeuN antibody to 50 µL of resuspended nuclei + 15 µL of BSA (100× from NEB).

    (b) Incubate at 4 °C in dark rotating for 1 h.

    (c) Add 450 µL of nuclei storage buffer to nuclei before sorting. Add 1:10,000 of DAPI.

3.2 Single-Cell WGA by MDA

Protocol adapted from [39] using GenomiPhi V2 DNA Amplification Kit (GE Healthcare).

*MDA kit is also available from Qiagen REPLI-g Single Cell Kit.

1. Prepare lysis buffer.

Component	Volume (µL)
DTT, 1 M	40
KOH, 5 M	40
H_2O	920

2. Aliquot 1.5 µL lysis buffer into each tube.

3. Place single-cell material into a PCR tube with 1.5 µL lysis buffer.

4. Incubate the samples on ice for 10 min.

5. Incubate the sample in a thermal cycler with the following condition.

Cycles	Temperature (°C)	Time
1	65	10 min
	4	Forever

6. Add 9 µL sample buffer to each sample.

7. Add 10 µL of reaction buffer and enzyme mix to each sample and mix well by pipetting.

Component	Volume per reaction (µL)
Reaction buffer	9
Enzyme mix	1

8. Incubate the sample in a thermal cycler with the following condition.

Cycles	Temperature (°C)	Time
1	30	5 h
	65	10 min
	4	Forever

9. Samples can be stored at −20 °C.

3.3 SLAV-Seq

1. Sonicate 10 µg of DNA to an average size of 500 bp by Covaris S2 sonicator with the following condition:
   5% duty cycle, intensity of 3200 cycles per burst, 80 s total time.

2. Purify DNA using Agencourt Ampure XP beads and resuspend DNA with the PCR master mix.

3. Perform LINE-1 capture hybridization and single cycle extension.

Component	Volume (µL)
Platinum Pfx DNA polymerase	19.5
JE281L1_ACA	0.5

4. Incubate in a thermal cycler with the following condition.

Temperature (°C)	Time
94	5 min
61.5	30 s
68	3 min
4	Forever

5. Clean up reaction with Agencourt Ampure XP beads.

6. Incubate DNA with 10 µL of streptavidin magnetic beads overnight at 4 °C.

7. Wash streptavidin magnetic beads three times with 200 µL 1× B&W buffer.

8. 2× B&W buffer composition

Component	Final concentration
Tris–HCl (pH 7.5)	10 mM
EDTA	1 mM
NaCl	2 M

9. Resuspend streptavidin magnetic beads directly in 10 μL NEBNext dA-Tailing Module.

Component	Volume (μL)
H_2O	8.5
NEBNext End Repair Reaction Buffer	1
NEBNext End Repair Enzyme mix	0.5

10. Incubate at room temperature for 30 min and then at 75 °C for 30 min to heat inactivate.

11. Wash once with 200 μL 1× B&W buffer.

12. Resuspend in 10 μL dA tailing.

Component	Volume (μL)
H_2O	8.5
NEBNext dA-Tailing Reaction Buffer	1
Klenow Fragment (3′ → 5'exo-)	0.5

13. Incubate for 30 min at 37 °C and then at 75 °C 30 min to heat inactivate.

14. Wash once with 200 μL 1× B&W buffer.

15. Resuspend in 10 μL ligation reaction.

Component	Volume (μL)
H_2O	4
50 μM annealed adapters	0.5
2× Rapid ligation buffer	5
T4 DNA ligase	0.5

16. Incubate at room temperature for 15 min.

17. Wash three times with 200 μL 1× B&W buffer.

18. Resuspend in 13 μL H_2O.

19. Add the amplification master mix.

Component	Volume (µL)
Kappa Hifi2× mix	15
10 µM JE620	0.8
10 µM JE50*	0.8

20. Incubate in a thermal cycler with the following condition.

Cycle	Temperature (°C)	Time
1	98	3 min
7	98	30 s
	54	30 s
	72	1 min
9	98	30 s
	68	30 s
	72	1 min
1	72	5 min
1	4	Hold

21. Purify DNA twice with Agencourt Ampure beads (ratio of Ampure beads to DNA = 0.8:1).

3.4 Quantifying Single-Cell Mosaicism

After enrichment and purification by SLAV-Seq, the amplified DNA is used to construct libraries for NGS. In NGS, the genomic DNA is sheared into millions of fragments, ranging from 35 to 400 bp in size for *en masse* amplification [40]. While several sequencing platforms are available, Illumina has emerged as the primary tool in most studies due to low cost per base, high-throughput, and low error rates [40–43]. Following amplification the subsequent steps prior to sequencing are dependent upon (1) the type of genetic variation being studied, and (2) the scope over which that variation is being studied. Scope of study can be split roughly into Whole Genome and Targeted sequencing. Types of genetic variation include SNVs, CNVs, MEIs, and complex Structural Variants.

LINE-1 Mobile Element Insertions generate a unique class of somatic variation which can be identified using both whole genome and targeted sequencing [37, 38, 44]. Unlike SNVs, CNVs, and most SVs which are a product of errors in DNA replication, recombination, and repair, MEIs are characterized by the use of retrotransposition machinery to generate novel events. Here, we will focus specifically on the detection of LINE-1 MEIs from targeted SLAV-Seq [38]. Alignment of SLAV-Seq data using BWA-MEM [45] produces clean peaks with minimal background that are quantified using MACS Model-based Analysis of ChIP-Seq [46]. These peaks represent the locations of full length, truncated, and

LINE-1-associated insertion events, both germline and somatic. Annotation of SLAV-Seq "peaks" with RepBase [47] and Repeat-Masker [48] identifies known and likely germline LINE-1 positions. Recent methods for quantifying non-germline events utilize the discordant and split reads produced by BWA-MEM to identify insertion junction sites. Local reassembly of reads around these sites can help resolve the actual sequence of the insertion. Recently, more sophisticated machine learning approaches have been shown to be very accurate in distinguishing somatic from germline LINE-1 events. Post alignment, the data is binned into 750-bp equally spaced windows across the reference genome. Within each window, 70 features are collected from each window to train a random forest classifier and predict insertion events [38].

3.5 Validating Neuronal Mosaicism

There are a significant number of factors which contribute to the error of single-cell sequencing analysis, and can naively be interpreted as true biological variation without proper validation. During single-cell isolation, the population of cells being isolated can be biased through selection based on size, viability, or propensity to enter the cell cycle. Further, unlike DNA sequencing from bulk tissue or even single-cell RNA analysis, the limited amount of input material for single-cell WGA can produce biased sequencing results, which during analysis generate both false-positive (FP) errors and false-negative (FN) errors. Artifacts of amplification include incorrect SNV calls, loss of coverage, decreased coverage uniformity, allelic imbalance, and other complex variants not present in the initial diploid genome. However, by far the largest source of WGA error is derived from allelic dropout events (ADO) at 10–50% of true mutation sites [35, 49, 50].

To minimize the inclusion of technical and analytical errors in final call sets, variant validation is crucial. The "gold standard" methods for validating single-cell variants are comparison of single-cell to bulk sequence data and PCR followed by Sanger-sequencing of bulk and scDNA. Assays including 3′ PCR, 5′ PCR, and flanking genomic PCR (Fig. 13.3) have been applied to validate somatic LINE-1 insertion candidates [4, 37, 38]. The 3′ PCR or 5′ PCR employs one primer complementary to the flanking genomic sequence and one primer complementary to the 3′ end or 5′ end of LINE-1, respectively. Flanking genomic PCR assay involves primers complementary to the 5′ and 3′ sequences flanking the insertions, such that the LINE-1 insertion generates a larger fragment for an insertion allele and a smaller fragment for the reference allele. In addition to PCR, significant information both for the validation of somatic events and for QC of WGA bias can be obtained from comparing single-cell to bulk sequence data. Levels of ADO, false-positive rates, coverage uniformity, and other forms of errors can be directly calculated.

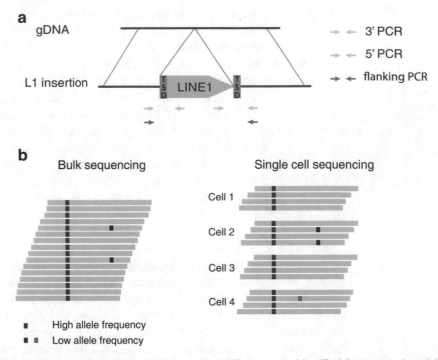

Fig. 3 Validation of somatic variation. (**a**) Somatic LINE-1 events identified by targeted enrichment and sequencing are validated by PCR of genomic DNA. New LINE-1 insertions are often truncated on the five prime or three prime end. Validation rates can vary drastically depending upon the region(s) being amplified. (**b**) The integration of results from bulk and single-cell-targeted enrichment assays better resolves the prevalence of low frequency somatic events than either assay alone. Rare events Identified in single cells (*blue* and *green*) can be properly characterized by their prevalence in bulk data as either originating in a sequenced cell (*green*) or originating in the progenitor of a sequenced cell (*blue*)

4　Conclusion

In this chapter we highlighted current methods for the study of single-cell neuronal mosaicism. Specifically, methods for the isolation of single cells from bulk tissue, the amplification of single-cell genomic DNA by MDA and MALBAC, and the targeted enrichment of LINE-1 sequences by SLAV-Seq. We are only now beginning to appreciate the extent of complexity that neuronal mosaicism may add to brain function. Specifically, numerous lines of evidence suggest that retrotransposition-specific aspects of neuronal mosaicism may contribute to the unresolved etiology of many neurological and neuropsychiatric disorders. It is thus crucial to understand the prevalence and diversity of mosaic retrotransposition at the single-cell level. To this end, large multi-institutional projects such as the Brain Somatic Mosaicism Network (BSMN) are currently undertaking this endeavor. The primary limitations to single-cell studies are whole genome amplification bias and sample processing throughput. Addressing these limitations is critical to the advancement of the field. In many ways analogous to current sequencing technology, it is conceivable that the number of single cells capable of being studied at once will soon drastically increase. We believe that the advances in single-cell sequencing-based technologies will bring us closer to understanding the prevalence and function of neuronal mosaicism.

References

1. Azevedo FAC, Carvalho LRB, Grinberg LT et al (2009) Equal numbers of neuronal and nonneuronal cells make the human brain an isometrically scaled-up primate brain. J Comp Neurol 513:532–541

2. De S (2011) Somatic mosaicism in healthy human tissues. Trends Genet 27:217–223

3. Muotri AR, Chu VT, Marchetto MCN et al (2005) Somatic mosaicism in neuronal precursor cells mediated by L1 retrotransposition. Nature 435:903–910

4. Evrony GD, Cai X, Lee E et al (2012) Single-neuron sequencing analysis of L1 retrotransposition and somatic mutation in the human brain. Cell 151:483–496

5. Coufal NG, Garcia-Perez JL, Peng GE et al (2009) L1 retrotransposition in human neural progenitor cells. Nature 460:1127–1131

6. Lander ES, Linton LM, Birren B et al (2001) Initial sequencing and analysis of the human genome. Nature 409:860–921

7. Cordaux R, Batzer MA (2009) The impact of retrotransposons on human genome evolution. Nat Rev Genet 10:691–703

8. Martin SL (1991) Ribonucleoprotein particles with LINE-1 RNA in mouse embryonal carcinoma cells. Mol Cell Biol 11:4804–4807

9. Feng Q, Moran JV, Kazazian HH et al (1996) Human L1 retrotransposon encodes a conserved endonuclease required for retrotransposition. Cell 87:905–916

10. Mathias SL, Scott AF, Kazazian HH Jr et al (1991) Reverse transcriptase encoded by a human transposable element. Science 254:1808–1810

11. Hancks DC, Kazazian HH (2012) Active human retrotransposons: variation and disease. Curr Opin Genet Dev 22:191–203

12. Bundo M, Toyoshima M, Okada Y et al (2014) Increased l1 retrotransposition in the neuronal genome in schizophrenia. Neuron 81:306–313

13. Singer T, McConnell MJ, Marchetto MCN et al (2010) LINE-1 retrotransposons: mediators of somatic variation in neuronal genomes? Trends Neurosci 33:345–354

14. Kalisky T, Blainey P, Quake SR (2011) Genomic analysis at the single-cell level. Annu Rev Genet 45:431–445

15. Kurimoto K, Yabuta Y, Ohinata Y et al (2007) Global single-cell cDNA amplification to provide a template for representative high-density oligonucleotide microarray analysis. Nat Protoc 2:739–752

16. Christopher Love J, Ronan JL, Grotenbreg GM et al (2006) A microengraving method for rapid selection of single cells producing antigen-specific antibodies. Nat Biotechnol 24:703–707

17. Choi JH, Ogunniyi AO, Du M et al (2010) Development and optimization of a process for automated recovery of single cells identified by microengraving. Biotechnol Prog 26:888–895

18. Rettig JR, Folch A (2005) Large-scale single-cell trapping and imaging using microwell arrays. Anal Chem 77:5628–5634

19. Zhang H, Liu K-K (2008) Optical tweezers for single cells. J R Soc Interface 5:671–690

20. Emmert-Buck MR, Bonner RF, Smith PD et al (1996) Laser capture microdissection. Science 274:998–1001

21. Espina V, Wulfkuhle JD, Calvert VS et al (2006) Laser-capture microdissection. Nat Protoc 1:586–603

22. Navin N, Kendall J, Troge J et al (2011) Tumour evolution inferred by single-cell sequencing. Nature 472:90–94

23. Dalerba P, Kalisky T, Sahoo D et al (2011) Single-cell dissection of transcriptional heterogeneity in human colon tumors. Nat Biotechnol 29:1120–1127

24. Shapiro E, Biezuner T, Linnarsson S (2013) Single-cell sequencing-based technologies will revolutionize whole-organism science. Nat Rev Genet 14:618–630

25. Guo MT, Rotem A, Heyman JA et al (2012) Droplet microfluidics for high-throughput biological assays. Lab Chip 12:2146–2155

26. Telenius H, Carter NP, Bebb CE et al (1992) Degenerate oligonucleotide-primed PCR: general amplification of target DNA by a single degenerate primer. Genomics 13:718–725

27. Huang L, Ma F, Chapman A et al (2015) Single-cell whole-genome amplification and sequencing: methodology and applications. Annu Rev Genomics Hum Genet 16:79–102

28. Navin NE (2014) Cancer genomics: one cell at a time. Genome Biol 15:452

29. Dean FB, Nelson JR, Giesler TL et al (2001) Rapid amplification of plasmid and phage DNA using phi 29 DNA polymerase and multiply-primed rolling circle amplification. Genome Res 11:1095–1099

30. Blanco L, Bernad A, Lázaro JM et al (1989) Highly efficient DNA synthesis by the phage phi 29 DNA polymerase. Symmetrical mode of DNA replication. J Biol Chem 264:8935–8940

31. Garmendia C, Bernad A, Esteban JA et al (1992) The bacteriophage phi 29 DNA polymerase, a proofreading enzyme. J Biol Chem 267:2594–2599

32. Lasken RS, Stockwell TB (2007) Mechanism of chimera formation during the multiple displacement amplification reaction. BMC Biotechnol 7:19

33. Zhang K, Martiny AC, Reppas NB et al (2006) Sequencing genomes from single cells by polymerase cloning. Nat Biotechnol 24:680–686

34. Esteban JA, Salas M, Blanco L (1993) Fidelity of phi 29 DNA polymerase. Comparison between protein-primed initiation and DNA polymerization. J Biol Chem 268:2719–2726

35. Zong C, Lu S, Chapman AR et al (2012) Genome-wide detection of single-nucleotide and copy-number variations of a single human cell. Science 338:1622–1626

36. Erwin JA, Marchetto MC, Gage FH (2014) Mobile DNA elements in the generation of diversity and complexity in the brain. Nat Rev Neurosci 15:497–506

37. Upton KR, Gerhardt DJ, Jesuadian JS et al (2015) Ubiquitous L1 mosaicism in hippocampal neurons. Cell 161:228–239

38. Erwin JA, Paquola ACM, Singer T et al (2016) L1-associated genomic regions are deleted in somatic cells of the healthy human brain. Nat Neurosci 19(12):1583–1591

39. Dean FB, Hosono S, Fang L et al (2002) Comprehensive human genome amplification using multiple displacement amplification. Proc Natl Acad Sci U S A 99:5261–5266

40. Ulahannan D, Kovac MB, Mulholland PJ et al (2013) Technical and implementation issues in using next-generation sequencing of cancers in clinical practice. Br J Cancer 109:827–835

41. Mardis ER (2008) Next-generation DNA sequencing methods. Annu Rev Genomics Hum Genet 9:387–402

42. Ross JS, Cronin M (2011) Whole cancer genome sequencing by next-generation methods. Am J Clin Pathol 136:527–539

43. Sun H-J, Chen J, Ni B et al (2015) Recent advances and current issues in single-cell sequencing of tumors. Cancer Lett 365:1–10

44. Lodato MA, Woodworth MB, Lee S et al (2015) Somatic mutation in single human neurons tracks developmental and transcriptional history. Science 350:94–98

45. Li H (2013) Aligning sequence reads, clone sequences and assembly contigs with BWA-MEM. https://arxiv.org/abs/1303.3997

46. Zhang Y, Liu T, Meyer CA et al (2008) Model-based analysis of ChIP-Seq (MACS). Genome Biol 9:R137

47. Jurka J, Kapitonov VV, Pavlicek A et al (2005) Repbase update, a database of eukaryotic repetitive elements. Cytogenet Genome Res 110:462–467

48. Smit AFA, Hubley R, Green P (1996), RepeatMasker Open-3.0

49. Hou Y, Song L, Zhu P et al (2012) Single-cell exome sequencing and monoclonal evolution of a JAK2-negative myeloproliferative neoplasm. Cell 148:873–885

50. Lasken RS (2007) Single-cell genomic sequencing using multiple displacement amplification. Curr Opin Microbiol 10:510–516

Part V

Genetic and Genomic Mosaicism in Aging and Disease

FISH Analysis of Aging-Associated Aneuploidy in Neurons and Nonneuronal Brain Cells

Grasiella A. Andriani and Cristina Montagna

Abstract

Aging is a ubiquitous complex process characterized by tissue degeneration and loss of cellular fitness. Genome instability (GIN) has long been implicated as a main causal factor in aging. The most severe form of genomic instability is whole chromosome instability (W-CIN), a state where dysfunction in chromosome segregation leads to whole chromosomes gains and losses. Aneuploidy is commonly linked to pathological states. It is a hallmark of spontaneous abortions and birth defects and it is observed virtually in every human tumor. There is mounting evidence that W-CIN increases with age, with the underlying hypothesis that some of the age-related loss of fitness phenotypes may be the result of W-CIN. Methodologically, the detection of stochastic W-CIN during the aging process poses unique challenges: aneuploid cells are scattered among diploid cells and, contrary to the cancer genome where aneuploidy is present in the background of massive ploidy changes, the number of aneuploid chromosome per cells is usually low (few per cell). Aging-associated aneuploidy is also largely stochastic or with limited clonal expansion. Therefore analysis at the single-cell level and the examination of a large number of cells is necessary. Here we describe a modification of the standard fluorescent in situ hybridization (FISH) protocol adapted for the detection of low-frequency mosaic aneuploidy in interphase cells isolated from the adult brain or within frozen tissue sections. This approach represents a straightforward method for the single-cell analysis of W-CIN in mammalian cells. It is based on the combination of four probes mapping to two different chromosomes and analysis of interphase cells, highly reducing false positives and enabling studying W-CIN also in post-mitotic tissues.

Key words FISH (Fluorescent in situ hybridization), Fluorophores, Aneuploidy, Interphase FISH, Genomic instability (GIN), Whole chromosome instability (W-CIN), Aging, Brain, NeuN+, NeuN−

1 Introduction

Cells that carry a chromosome number that deviates from a multiple of their karyotype are defined as aneuploid (an = not, eu = good, ploid = fold) [1] and they generally suffer from fitness disadvantages relative to their diploid counterparts [2].

Aneuploidy has been associated with pathological states and it is not tolerated at the systemic level. In fact, only three chromosomes (HSA 13, HSA 18 and HSA 21) are compatible with life in

José María Frade and Fred H. Gage (eds.), *Genomic Mosaicism in Neurons and Other Cell Types*, Neuromethods, vol. 131, DOI 10.1007/978-1-4939-7280-7_14, © Springer Science+Business Media LLC 2017

humans when present in the germline as trisomy. In these rare cases, survival of human germline trisomies occurs at the cost of fitness, resulting in developmental and cognitive defects. Yet, aneuploidy is also associated with uncontrolled cell proliferation. It is a hallmark of cancer, being found in two out of three tumors, and it has long been proposed as a mechanism to promote malignant transformation [3–5].

Work from our group and others has shown that aneuploidy is present in disease-free tissues and it is more common than what was previously presumed (summarized in [6, 7]). A variety of mammalian tissues undergo changes in ploidy during normative aging, such as the brain [8–11], liver [12, 13], lymphocytes [14, 15], oocytes [16], mouth mucosa [17], lungs, kidney, and heart [18], suggesting that mosaic aneuploidy is compatible with normal cellular functions and physiology. The link between aneuploidy and aging is strengthened further by the observations that aneuploidy can reduce the replicative life span in yeast [19] and induce progeroid features in vivo. Indeed, mice defective for components of the Spindle Assembly Checkpoint, a surveillance machinery that maintains fidelity of chromosome number, exhibit increased levels of aneuploidy, progeroid features, and shortened life span [20, 21]. Binucleation and aneuploidy generated by lack of vimentin phosphorylation in vivo is also associated with increased expression of senescence markers, lens cataract, and aging phenotypes in the skin [22, 23]. These findings suggest that somatic W-CIN is associated with aging and propose its causal role in the induction of senescence and age-associated phenotypes.

Aneuploidy is significantly increased in cerebral cortex of aged mammals [8–10], where more than 10% of all cells have a gain or a loss of at least one chromosome [8]. Our group, using the techniques described in this section, has shown that aneuploidy accumulates in a chromosome-specific fashion in the cortex with chromosomes 7, 18, and Y being the most affected in mice [8]. This phenomenon prompts the question of what are the consequences of mosaic aneuploidy in the aging brain. It has been suggested that accumulation of aneuploid brain cells could contribute to the gradual impairment of cognitive functions associated with aging [7, 24]. In a functional TP53 background, aneuploid cells can undergo apoptosis, resulting in depletion of brain cells [7, 25]. Alternatively, not-diploid cells can enter senescence and induce activation of the senescence-associated secretory phenotype (SASP), as we recently reported [26]. This observation is highly relevant to aging because the SASP contributes to age-related inflammation and tissue dysfunction [27, 28], suggesting that aneuploid cells accumulated during aging potentially play a role in the etiology of age-related diseases [7]. On the other hand, aneuploid cells can proliferate in the absence of TP53 [29], which could facilitate cancer initiation and higher brain tumor incidence at older age [30]. Nevertheless, aneuploid brain cells were shown to establish active neural circuitry and to be able to differentiate into neuronal and glial lineages [9, 31], suggesting that they could provide genetic diversity to the cerebral cortex. Likewise tetraploidy in the chick retina has been suggested as a mechanism to increase neuronal diversity [32]. We are only beginning to uncover this phenomenon, which is likely multifaceted. The brain, which controls a variety of biological functions, is a heterogeneous organ because it is composed of distinct subregions and several specialized cell types [33–35]. Thus, levels of aneuploidy may be variable between distinct brain areas, possibly resulting in a variety of functional consequences with a variable degree of severity. In support of this hypothesis, we have observed very low levels of aneuploidy in the mice cerebellum relative to matched cerebral cortex tissue [8, 36]. More studies are needed to evaluate the susceptibility of other brain regions to the accumulation of ploidy changes during aging and to begin understanding their respective outcomes.

The field of age-related aneuploidization has been progressing at a slow peace, likely due the technical challenges associated with its detection. Measuring aneuploidy during aging, in our view, poses unique difficulties that make its characterization more challenging than other systems where this phenomenon is more prominent (i.e., cancer). In aging studies, the frequency of aneuploid cells within a tissue is relatively low and the level of copy number changes is limited to one or few chromosomes per cell [8, 18, 37]. To overcome these technical difficulties, especially in the brain, we introduced a variety of modifications to routine FISH protocols to adapt this technique for the analysis of low-frequency, low-level changes in chromosome number. The approach described here benefits from the use of four-color interphase FISH, by labeling two different probes targeting the same chromosome at two distinct loci. This approach allows highly sensitive and accurate quantification of low-level chromosome number changes in disease-free tissues. By analyzing two chromosomes at once within the same nucleus, we can distinguish aneuploidy from polyploidy. The use of interphase FISH is important for the estimation of ploidy changes in the whole population of cells, including post-mitotic cells, which are the main cell types present in the brain.

A variety of cell subtype-specific markers have been extensively characterized for the identification of undifferentiated precursors, as well as fully differentiated cells from different brain areas [38–42] (Table 1). Combining approaches to isolate cell subtypes of interest with our custom FISH approach enable an unprecedented sensitivity and specificity to study mosaic aneuploidy in disease-free tissues. In this chapter we will provide the technical details to perform four-color interphase FISH in isolated nuclei and on frozen brain tissue sections.

Table 1
Common markers for identification of specific brain areas/cell types

Cells and cell type	Marker	Alias	Gene name	Cellular localization
Microglia	CD11	ITGAM	Integrin subunit Alpha M	Surface
Neural stem (NSC)	CD133	PROM1	Prominin 1	Surface
	Nestin	NES	Nestin	Intracellular
	SOX1, 2	SOX1, 2	SRY-Boxes 1, 2	Intracellular
Neural progenitor (NPC)	GLAST	SLC1A3	Solute carrier family 1 member 3	Surface
	Nestin	NES	Nestin	Intracellular
	S100B	S100B	S100 calcium binding protein B	Intracellular/nucleus
Type 1and type 2 astrocytes	GFAP	GFAP	Glial fibrillary acidic protein	Intracellular
	GLAST	SLC1A3	Solute carrier family 1 member 3	Surface
	GLT-1	SLC1A2	Solute carrier family 1 member 2	Surface
	S100B	S100B	S100 calcium binding protein B	Intracellular/nucleus
Oligodendrocytes precursors	Nestin	NES	Nestin	Intracellular
	NG2	CSPG4	Neural/glial antigen 2	Surface
	CD271	NGFR	Nerve growth factor receptor	Surface
Differentiated post-mitotic neurons	NeuN	NeuN	RNA binding protein, Fox-1 Homolog 3	Nuclear membrane
Other markers				
Oligodendrocytes, astrocytes precursors	A2B5		Ganglioside surface markers	Surface
Oligodendrocytes	O4		O-Antigens	Surface
For additional resources, please check:				
http://docs.abcam.com/pdf/neuroscience/neural-markers-guide-web.pdf				
https://www.rndsystems.com/research-area/neural-stem-cell-and-differentiation-markers				
https://www.labome.com/method/Neuronal-Cell-Markers.html				

2 Materials

2.1 Reagents

2.1.1 Labeling DNA Probes by Nick Translation

1. BAC clone DNA of interest (already purified by standard protocols).

2. Nick-dNTPs solution: 0.5 mM of each (dATP, dCTP, and dGTP) and 0.05 mM dTTP. Aliquot and store at −20 °C.

3. DNase I from bovine pancreas (Sigma #DN25).

   (a) DNase I stock solution: 1 mg/mL dissolved in 0.15 M NaCl/50% glycerol. Aliquot and store at −20 °C.

   (b) DNase I working solution: 3 μL of stock DNase I in 997 μL of ice-cold water. Prepare fresh every time.

4. DNA Polymerase I: 10 U/μL (ThermoFisher Scientific #EP0042).

5. 10× DNA Polymerase I buffer (supplied with DNA Polymerase I).

6. Modified dUTPs directly labeled: We routinely use dyes from Dyomics: DY-590-dUTP (#590–34), DY-495-dUTP (#495–34), DY-415-dUTP (#415–34), and DY-647-dUTP1 (#647–34) (*see* **Notes 1** and **2**).

7. DNase-free water, sterile.

8. β-Mercaptoethanol: 0.1 M in water. Aliquot and store at −20 °C.

9. Agarose, powder.

10. 1× Tris–acetate-EDTA Buffer (TAE).

11. 1 kb DNA Ladder.

2.1.2   Slide Preparation for FISH on Isolated Single Cells

1. KCl powder.

2. Methanol, absolute (*see* **Note 3**).

3. Acetic acid glacial (Sigma #537020).

4. Hypotonic solution: 0.075 M KCl in water. Pre-warm to 37 °C (*see* **Note 4**).

5. Fixative solution: Combine methanol and acetic acid at a 3:1 ratio (vol/vol) (*see* **Note 5**).

2.1.3   Slide Preparation for FISH on Frozen Tissue Sections

1. Liquid nitrogen or another snap freezing method (i.e., dry ice or isopentane).

2. Optimal cutting temperature (OCT) compound (VWR #25608–930).

3. Methanol, absolute (ice-cold) (*see* **Note 6**).

2.1.4   DNA Probe Precipitation and Preparation for Hybridization

1. Nick-translated DNA of interest.

2. Cot-1 DNA: 1 mg/mL (ThermoFisher Scientific #15279011 or #18440016) (*see* **Note 7**).

3. DNase-free water, sterile.

4. Sodium acetate (NaOAc): 3 M NaOAc in water, pH 5.2.

5. Ethanol, absolute.

6. Formamide, deionized (Sigma #F9037). Pre-warm to 37 °C.

7. 20× Saline-Sodium Citrate buffer (SSC).

8. Master Mix: 50% Dextran sulfate (Sigma #42867) in 2× SSC, pH 7. Pre-warm to 37 °C.

2.1.5   Slide Pretreatment

1. 2× SSC buffer.

2. PBS: 1×, pH 7.4, without calcium and $MgCl_2$ (ThermoFisher Scientific #10010).

3. 0.01 M HCl solution—Pre-warm 100 mL to 37 °C in a clean beaker.

4. Pepsin from porcine gastric mucosa (Sigma #P6887).

   (a) Pepsin stock solution: 100 mg/mL in sterile water (store aliquots at −20 °C).

   (b) Pepsin working solution (should be prepared fresh every time): Add 5–30 µL pepsin stock solution inside an empty and clean Coplin jar. Then add 100 mL of pre-warmed 0.01 M HCl into it, mixing very well with a glass pipette (*see* **Notes 8, 9,** and **10**). Keep working solution inside of a 37 °C water bath until use.

5. Pretreatment solution 1: 1× PBS/0.05 M MgCl$_2$ (5 mL of 1 M MgCl$_2$ and 95 mL of 1× PBS).

6. Pretreatment solution 2 (should be prepared fresh every time): 1% (vol/vol) formaldehyde in solution 1 (1.35 mL of 37% formaldehyde and 50 mL 1 × PBS/MgCl$_2$) (*see* **Note 11**).

7. Ethanol: 70, 90, and 100% (*see* **Note 12**).

2.1.6 Slide Denaturation and Hybridization

1. 20× SSC.

2. Deionized formamide (ThermoFisher Scientific #AM9342).

3. Denaturation solution: 70% (vol/vol) deionized formamide/2× SSC (70 mL deionized formamide, 10 mL of 20× SSC, and 20 mL of water). Adjust pH to 7.25 with 1 N HCl. Mix well and store as 1–2 mL aliquots at −20 °C (*see* **Note 13**).

4. Ethanol: 70, 90, and 100% (*see* **Note 14**).

5. Rubber cement.

2.1.7 Detection for FISH on Isolated Single Cells

1. Formamide (Fisher Scientific #BP228–100).

2. 20× SSC.

3. Distilled water.

4. Tween 20.

5. Wash Solution C1 (should be prepared fresh every time): 50% (vol/vol) formamide/2× SSC (100 mL of formamide, 20 mL 20× SSC, and 80 mL of water). Adjust pH to 7 with 1 N HCl. Pre-warm to 45 °C.

6. Wash Solution C2: 1× SSC (25 mL of 20× SSC and 475 mL of water). Pre-warm to 45 °C.

7. Wash Solution C3: 4× SSC/0.1%Tween 20 (200 mL of 20× SSC, 799 mL of water, and 1 mL of Tween 20). Pre-warm to 45 °C.

8. Ethanol: 70, 90, and 100%.

9. Mounting media: Antifade with DAPI (e.g., ProLong Gold—ThermoFisher Scientific #P36931 or VECTASHIELD—Vector laboratories #H-1200).

2.1.8   Detection for FISH on Frozen Tissue Sections

1. 20× SSC.

2. Distilled water.

3. Tween 20.

4. Wash Solution T1: 0.4× SSC (20 mL of 20× SSC and 980 mL of water). Pre-warm to 74 °C.

5. Wash Solution T2: 4× SSC/0.1%Tween 20 (200 mL of 20× SSC, 799 mL of water, and 1 mL of Tween 20).

6. Ethanol: 70, 90, and 100%.

7. Mounting media: Antifade with DAPI (e.g., ProLong Gold—ThermoFisher Scientific #P36931 or VECTASHIELD—Vector laboratories #H-1200).

2.2   Equipment

1. Vortex mixer.

2. Microcentrifuge (e.g., Sorvall legend micro 21, ThermoFisher Scientific).

3. Electrophoresis gel system and gel imager to allow visualization of Nick-translated DNA probes.

4. Benchtop water bath.

5. Benchtop orbital shaker (e.g., VWR standard orbital shaker model 3500, VWR international).

6. Thermotron humidity chamber (e.g., model CDS-5, Thermotron).

7. Drying oven (e.g., model 107,800, Boekel).

8. Slide warmer (e.g., ThermoBrite StatSpin, Abbott Molecular).

9. Thermomixer (e.g., VorTemp 56 shaking incubator, Labnet international).

10. Biological safety cabinet.

11. Dry bath incubator (e.g., Isotemp 125D, Fisher Scientific).

12. Microscope slides, glass (e.g., #12–544-7, Fisher Scientific). For frozen tissue sections we recommend the use of positively charged slides (EMS #71869–10).

13. Microscope cover glass: 18 mm × 18 mm (e.g., #2865–22, Corning) and 24 mm × 60 mm (e.g., #3322, ThermoFisher Scientific).

14. Diamond point marker (e.g., #750, ThermoFisher Scientific).

15. Inverted tissue culture microscope.

16. Temperature-controlled hybridization chamber (e.g., Slide Moat 240,000, Boekel).

17. Fluorescent microscope equipped with filters corresponding to the wavelength excitation/emission of the dUTPs used (see below) (e.g., Zeiss Axiovert 200 equipped with Chroma Technology specific filters). For four-color FISH, five filters are

necessary for visualization of the nucleus and hybridization signals in two chromosomes: DAPI/UV excitation (e.g., 350/470), DY-415-dUTP/Blue probe (e.g., 436/480), DY-495-dUTP/Green probe (e.g., 470/540), DY-590-dUTP/Red probe (e.g., 546/600), and DY-647-dUTP1/Yellow probe (e.g., 620/700).

3   Methods

Figure 1: Flowchart of procedures and indicative timeline

3.1   Selection and Labeling of DNA Probes by Nick Translation

1. DNA regions of interest can be chosen using publicly available tools and repositories containing the latest build of the genome of interest. We commonly use bacterial artificial chromosomes (BAC clones), which can be purchased through the BACPAC Resources Center at Children's Hospital Oakland Research Institute in Oakland, California, in the United States (https://bacpac.chori.org/about.htm). Clones corresponding to the region of interest can be visualized using the UCSC (University of California Santa Cruz) Genome Browser website (http://genome.ucsc.edu/) (*see* **Note 15**).

2. For four-color FISH we suggest the use of two probes for each chromosome selected (one in the *p* and the other in the *q* arm). Ideally, one subcentromeric and one distal probe mapping to the same chromosome should be selected and labeled with two different dyes (*see* Fig. 2 for a schematic representation of probe selection).

3. Culture of BACs and subsequent DNA purification can be carried on using standard protocols (Qiagen #12143 or Wako chemicals #PL-S2) (*see* **Note 16**). We recommend the addition of RNase A (e.g., Qiagen #19101) to the DNA preparation to avoid RNA contamination that may interfere with the nick translation reaction.

4. High molecular weight DNA should be visible on a 1.2% agarose gel (Fig. 3a).

5. All the following steps should be performed in the dark. For each DNA probe, prepare one Eppendorf tube containing:
   (a) 2 μg DNA (*see* **Note 17**).
   (b) 10 μL 10× DNA Polymerase I buffer.
   (c) 10 μL Nick-dNTPs solution.
   (d) 10 μL 0.1 M ß-mercaptoethanol.
   (e) 1.5–4 μL modified dUTP (1 mM) (*see* **Notes 18** and **19**).
   (f) X μL sterile water (the final total volume should be 100 μL).

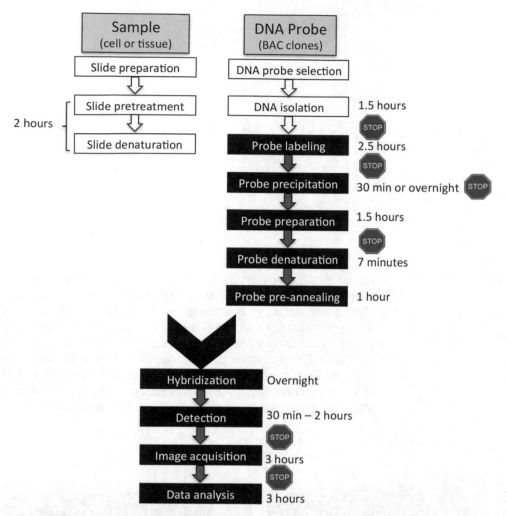

Fig. 1 Flowchart indicating the steps to carry on the four-color interphase FISH using isolated cells or frozen tissue sections. On the *left*, steps for sample processing (from slide preparation to denaturation). On the *right*, step-by-step procedures for generating locus-specific probes (from DNA probes selection to a ready-to-use FISH probes). Below the *black arrow*, the pre-annealed probe is hybridized to the denatured slide overnight, followed by detection, image acquisition, and data analysis. Steps described within *black boxes* indicate procedures that should be performed in the dark or protected from direct light. The "Stop" signs represent stages where the experiment can be paused: BAC DNA, nick-translated DNA, and DNA probe resuspended in deionized formamide and master mix can be stored at −20 °C until use. Precipitation of nick-translated DNA can also be done at −20 °C overnight or over the weekend. Detected slides can be stored at 4 °C protected from light and should be preferably imaged within 1 week. Once images are acquired, data analysis can be performed at convenience. Indicative time frames in hours required to carry on each step are also indicated

6. Vortex, centrifuge, and place tubes on ice.

7. Add 2 μL DNA Polymerase I (always add before DNase I).

8. Vortex, centrifuge, and place tubes on ice.

9. Add 6 μL of DNase I working solution.

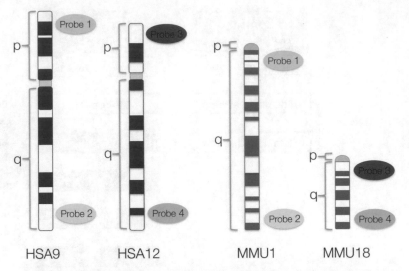

Fig. 2 FISH probes design. Ideograms of *Homo sapiens* Autosome (HSA, *dark blue*) and *Mus musculus* (MMU, *dark green*) chromosomes. For both species, the p and q chromosome arms are pinpointed as regions of interest. For human chromosomes we suggest to select one probe for each arm of the same chromosome (for example: *blue and yellow* probes for HSA9 and red and green probes for HSA12). Because murine chromosomes are all acrocentric, locus-specific probes for this species should be selected distal and proximal to the centromere of the same autosome (e.g., *blue* probe proximal and yellow probe distal on MMU1 and red probe proximal and green probe distal on MMU18)

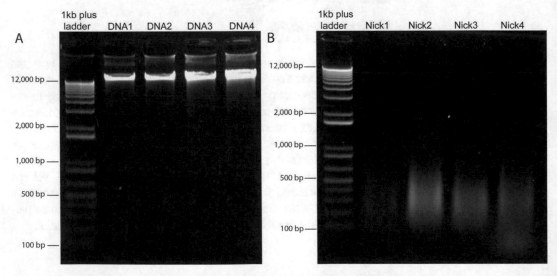

Fig. 3 Probes preparation. Representative gel images depicting (**a**) plasmid DNA purified from four different BAC clones, and (**b**) the same DNA samples after dye incorporation by Nick-translation. Note that the size of the bulk nick-translated DNA should be within 200–800 bp

10. Vortex and centrifuge.

11. Incubate protected from light at 15 °C for 1.5 h.

12. Stop the Nick translation by placing tubes at −20 °C (*see* **Note 20**).

13. Prepare gel electrophoresis (1.2% agarose in 1× TAE).

14. Run 2–5 μL of each sample and the same amount of 1 kb DNA Ladder (*see* **Note 21** and Fig. 3b).

15. Based on gel picture, estimate the amount of Nick-translated DNA necessary for each probe (*see* **Note 22**).

3.2 DNA Probe Precipitation and Preparation for Hybridization

1. Combine the four probes of interest with Cot-1 DNA in a single tube for probe precipitation (*see* **Notes 23** and **24**):

   (a) 200–500 ng of each Nick-translated DNA.

   (b) 60–80 μL Cot-1 DNA (1 mg/mL) of the desired species.

   Example:

   Probe—DY-415-dUTP: 20 μL.

   Probe—DY-495-dUTP: 30 μL.

   Probe—DY-590-dUTP: 25 μL.

   Probe—DY-647-dUTP1: 45 μL.

   Human Cot-1 DNA: 60 μL.

   Total volume: 180 μL.

2. Add 3 M NaOAc: The amount should be 1/10 of the total volume of the combined DNA (in this example: total volume = 180 μL, add 18 μL of NaOAc).

3. Add absolute ethanol: The amount should be 2.5× of the total volume of combined DNA + NaOAc (in this example: 180 μL + 18 μL = 198 μL and 198 μL × 2.5 = 495 μL, add 495 μL of absolute ethanol).

4. Vortex well.

5. Let the DNA precipitate at −20 °C overnight or at −80 °C for at least 30 min.

6. Centrifuge at 16,100 × *g* at 4 °C for 30 min.

7. Pour off supernatant. Remove most of the ethanol possible with a micropipette without disturbing the pellet and speed vac for 10–12 min with medium heat to dry the DNA pellet.

8. Add 6 μL of pre-warmed deionized formamide to the dry DNA pellet. Centrifuge briefly (*see* **Note 25**).

9. Incubate probes in a thermomixer for at least 30 min (37 °C, 140 rpm).

10. Add 6 μL of Master Mix, vortex, and centrifuge briefly.

11. Denature DNA probe at 80 °C for 7 min. Immediately after, place tubes in ice for 1 min and centrifuge briefly.

12. Pre-anneal the probe at 37 °C for 1–2 h.

13. The probe is now ready for hybridization onto denatured slides.

3.3 Four-Color Interphase FISH in Isolated Single Cells

3.3.1 Preparation of Slides

1. Single-cell suspensions of the brain can be obtained from grossly micro-dissected brain subregions [43]. It is beyond the scope of this book chapter to provide protocols for tissue dissection and digestion for single-cell suspensions specific for the brain. We recommend consulting specific literature and books dedicated to this process [44, 45].

2. For analysis of aneuploidy levels in specific cell subtypes, we recommend the combination of four-color FISH approach, described here, with a flow cytometric method to enrich for cell populations of interest (protocol detailed in Chap. 3 of this book). Table 1 lists some commonly used markers for identification of specific brain areas/cell types.

3. Single-cell aneuploidy studies could also be performed using cultured primary cells. In this case, cells need to be trypsinized to obtain single-cell suspensions and processed for hypotonic treatment followed by fixation as described below.

4. We suggest using a minimum of 10,000 cells resuspended in 200–500 μL of buffer (e.g., PBS, DPBS, HBSS, culture media). Pipette up and down to ensure no cell clumps are present).

5. Add 1 mL hypotonic solution drop by drop to the tube containing the cells in suspension, while agitating gently. Incubate at 37 °C for 15–25 min (*see* **Notes 26** and **27**).

6. Add approximately 1/100 of volume of hypotonic of fresh fixative solution to the tube and invert to mix (this step will stop the hypotonic process and prefix the cells).

7. Spin cell suspension at 217 × g for 10 min at RT.

8. Remove the supernatant carefully, leaving ~100 μL liquid, flick the tube to loosen the pellet and fully resuspend all cells.

9. Add 1 mL fresh fixative very slowly, gently flicking pellet (*see* **Note 28**).

10. Spin cell suspension at 217 × g for 10 min at RT.

11. Repeat steps 9–11 two more times.

12. Resuspend cells in 500 μL of fixative after last centrifugation step.

13. Set the Thermotron at 48% humidity and 24 °C temperature (*see* **Note 29**).

14. Drop ~20 μL of the fixed cell suspension onto a clean microscope slide and allow the slide to fully dry by evaporation inside the Thermotron.

15. View the slide with a 20× phase objective on a bright-field microscope to determine final cell density (*see* **Note 30**).

16. Sample slides can be heated for 30 min at 45 °C for same day use (*see* **Note 31**).

3.3.2 Pretreatment and Denaturation of Slides

1. Equilibrate slides in a Coplin jar containing 2× SSC for 5 min at RT, shaking.

2. Incubate slides in pepsin working solution at 37 °C for 5–10 min (*see* **Note 9**).

3. Wash slides 2× for 5 min each in 1× PBS at RT, shaking.

4. Wash slides 1× for 5 min in Pretreatment solution 1 at RT, shaking.

5. Wash slides 1× for 10 min in Pretreatment solution 2 at RT, not shaking.

6. Wash slides for 5 min in 1× PBS at RT, shaking.

7. Dehydrate slides in ethanol series: 70, 90, 100% ethanol, 3 min each.

8. Air-dry slides. Slides are now ready for denaturation.

9. Place 70% solution in ice for denaturation for at least 30 min.

10. Apply 120 μL of denaturation solution to a 24 × 60 mm coverslip. Touch the slide to the coverslip.

11. Denature slides at 72 °C on a slide warmer for 1 min and 30 s (*see* **Note 32**).

12. Immediately let coverslip slide off and place slide in ice-cold 70% ethanol for 3 min, followed by 90% ethanol and 100% ethanol for 3 min each at RT.

13. Air-dry slides. Slides are now ready for hybridization.

14. Visually inspect slides under a bright-field microscope to ensure that nuclei are not damaged. Select a good area for hybridization and mark it with a diamond point marker.

3.3.3 Hybridization of Slides and Detection for Single-Cell Aneuploidy Analyses

1. After pre-annealing (see above step 12 of Sect. 3.2), apply the DNA probe to the area selected for hybridization and cover with 18 × 18 mm coverslip. Seal coverslip with rubber cement (*see* **Note 33**).

2. Place slides in the hybridization chamber and hybridize at 37 °C overnight.

3. Remove the slides from the hybridization chamber.

4. Carefully remove the rubber cement from the slides, being careful not to drag the coverslip across them, thereby scratching the cells (*see* **Note 34**).

5. Place the slides into a Coplin jar containing Wash Solution C1 and wash slides while shaking at 45 °C for 5 min. Remove solution.

6. Repeat step 5 two more times, using fresh solution every time.

7. Wash slides 3× with Wash Solution C2, shaking at 45 °C for 5 min, using fresh solution every time (*see* **Note 35**).

8. Wash slides 1× with Wash Solution C3 for 5 min at 45 °C, shaking.

9. Dehydrate slides in ethanol series: 70, 90, 100% ethanol, 3 min each.

10. Air-dry slides in the dark.

11. Add enough antifade containing DAPI in the area selected for hybridization, careful to not touch it.

12. Mount slides with 18 × 18 mm coverslips and carefully remove any air bubbles (*see* **Notes 36** and **37**). Representative images for one diploid and one aneuploid nucleus are shown in Fig. 4.

3.4 Four-Color Interphase FISH on Frozen Tissue Sections

When performing interphase FISH on tissue sections, it is important to locate the brain areas and/or cell types of interest for aneuploidy analysis. Ideally, immunoFISH should provide the best approach to identify specific cell types for ploidy analysis. However, in our experience, the intensity of FISH signals is reduced when this assay is combined with immunofluorescence (IF), resulting in a high percentage of nuclei not being suited for quantification, and therefore eliminated due to ambiguous scoring or lack of FISH signals for one or both chromosomes. Moreover, age-related accumulation of aneuploidy, at least in the cerebral cortex, is observed both at low frequency (~4.5% cells aneuploid for chromosome 18) and at low copy number (up to 5 copies) [8]. Therefore, a highly sensitive method that enables accurate ploidy quantification is required.

Thus, we have adopted a pipeline to work with serial tissue sections, in which we first locate the regions of interest by immunofluorescence and/or immunohistochemistry, and then perform four-color FISH in a slide that is adjacent to the one used for IF. Distinct defined brain sub-areas such as the cerebral cortex (Fig. 5), cerebellum, and hippocampus can also be "grossly" identified in hematoxylin and eosin stained tissue, with the support of a histopathologist or with the aid of brain atlas and/or histology books (*see* **Notes 38**). It should be noted that this approach is correlative, lacking the information regarding the specific cell type visualized using FISH at a cost of accurate ploidy measurements for two chromosomes.

For ploidy studies in which the identity of the cell is important and highly desired, we suggest performing four-color interphase FISH in tissue dissociated cells or nuclei, which can be separated into distinct populations with the use of antibodies and flow cytometric approaches. With this method, the ploidy of cell subtypes of interest can be determined with higher confidence.

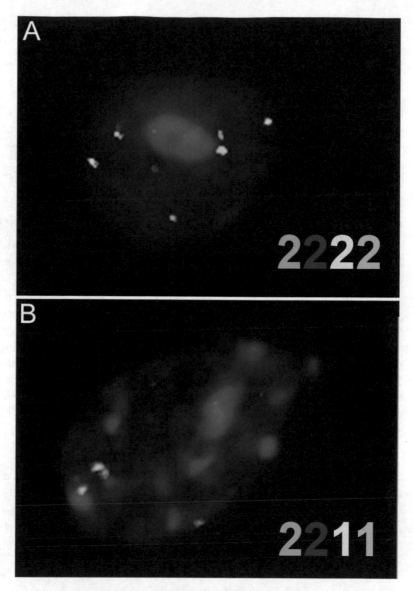

Fig. 4 Image acquisition on isolated single cells. Representative images of (**a**) one diploid nucleus where two copies of each autosome are visible (*green and red* for MMU18 and *yellow and blue* for MMU1); and (**b**) one aneuploid nucleus where two copies of MMU18 (*green and red*) and one copy of MMU1 (*yellow and blue*) are visible

3.4.1 Preparation of Slides

1. After tissue isolation, embed brain samples in OCT and perform snap freeze according to standard protocols [46].

2. Perform cryosectioning of the brain according to the desired orientation [47]. The thickness of the sections can range from 5 to 12 μm (*see* **Note 39**).

3. Keep slides at −80 °C until use.

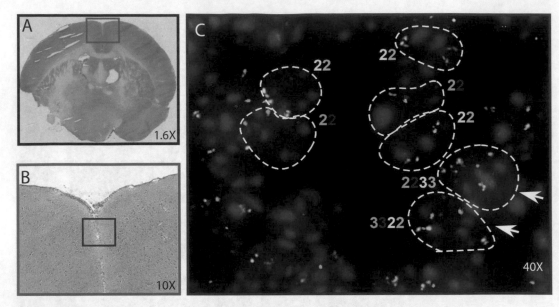

Fig. 5 Image acquisition on tissue sections, four colors analysis. Representative images of a mouse brain frozen section (**a, b**) stained with hematoxylin and eosin, and (**c**) hybridized using the four-color interphase FISH. (**a**) The *green rectangle* indicates the area of the cerebral cortex of interest in which FISH was performed. (**b**) Zoom-in of the area of the cerebral cortex highlighted in (**a**). The *red square* indicates the area of the cerebral cortex in which FISH was performed. (**c**) High magnification of the area of the cerebral cortex highlighted in (**b**) showing some nuclei containing hybridization signals for both chromosomes MMU1 and MMU18 (*white arrows*), while others only hybridize for probes mapping to MMU1 (*yellow and blue*) or MMU18 only (*green and red*)

3.4.2 Pretreatment and Denaturation of Slides

1. Place a Coplin jar containing absolute methanol in ice for at least 30 min.

2. Remove slides from the −80 °C and let them air dry at RT for 10–15 min.

3. Fix slides in ice-cold methanol for 10 min (*see* **Note 40**).

4. Equilibrate slides in a Coplin jar containing 2× SSC for 5 min at RT, shaking.

5. Incubate slides in pepsin working solution at 37 °C for 6–15 min (*see* **Note 9**).

6. Wash slides 2× for 5 min each in 1× PBS at RT, shaking.

7. Wash slides 1× for 5 min in Pretreatment solution 1 at RT, shaking.

8. Wash slides 1× for 10 min in Pretreatment solution 2 at RT, not shaking.

9. Wash slides for 5 min in 1× PBS at RT, shaking.

10. Dehydrate slides in ethanol series: 70, 90, 100% ethanol, 3 min each.

11. Air-dry slides. Slides are now ready for denaturation.

12. Place 70% solution in ice for denaturation for at least 30 min.

13. Apply 120 µL of denaturation solution to a 24 × 60 mm coverslip. Touch the slide to the coverslip.

14. Denature slides at 72 °C on a slide warmer for 1 min and 30 s (*see* **Note 32**).

15. Immediately let coverslip slide off and place slide in ice-cold 70% ethanol for 3 min, followed by 90% ethanol and 100% ethanol for 3 min each at RT.

16. Air-dry slides. Slides are now ready for hybridization.

3.4.3 Hybridization of Slides and Detection for Single-Cell Aneuploidy Analyses

1. After pre-annealing (see above step 12 of Sect. 3.2), apply the DNA probe to the tissue section and cover with 18 × 18 mm coverslip. Seal coverslip with rubber cement (*see* **Note 33**).

2. Place slides in the hybridization chamber and hybridize at 37 °C overnight.

3. Remove the slides from the hybridization.

4. Carefully remove the rubber cement from the slides, being careful not to drag the coverslip across them, thereby scratching the tissue (*see* **Note 41**).

5. Place the slides into a Coplin jar containing Wash Solution T1 and wash slides while shaking at 74 °C for 2 min. Remove solution.

6. Add Wash Solution T2 to Coplin jar and wash slides while shaking at RT for 3 min.

7. Dehydrate slides in ethanol series: 70, 90, 100% ethanol, 3 min each.

8. Air-dry slides in the dark.

9. Add enough antifade containing DAPI on top of the tissue section, careful to not touch it.

13. Mount slides with 18 × 18 mm coverslips and carefully remove any air bubbles (*see* **Notes 36** and **37**). Representative images for four-color and two-color FISH in frozen brain sections are shown in Fig. 6 (*see* **Notes 42**).

3.5 Image Acquisition and Data Analysis (see Note 43)

1. The precise procedure for image acquisition will vary depending on the microscope and software available. We have extensive experience using the Zeiss Axiovert 200 inverted fluorescence microscope with fine focusing oil immersion lens (×40, NA 1.3 oil and ×60, NA 1.35 oil). The microscope is equipped with a high-resolution CCD Camera Hall 100 and the images are acquired using the FISHView application of the Applied Spectral Imaging software (*see* **Notes 44**).

2. By incorporating the modified dUTPs to the DNAs of interest, four-color interphase FISH images are collected using the

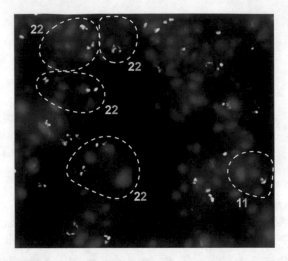

Fig. 6 Image acquisition on tissue sections, two colors analysis. Representative image of a mouse brain frozen section hybridized for four-color FISH but only showing signals for MMU1 (see Note **42**). Images are acquired in all channels but the analysis of ploidy was performed individually for each chromosome, thus increasing the amount of cells that can be analyzed in a single hybridization

following filters: 470 nm (for DAPI), 480 nm (for DY-415-dUTP), 540 nm (for DY-495-dUTP), 600 nm (for DY-590-dUTP), and 700 nm (for DY-647-dUTP1).

3. Images must be randomly acquired and samples should be blindly scored for hybridization signals. Initially we recommend the analysis of at least 200 nuclei per sample for an accurate estimation of ploidy. In tissues with very low levels of aneuploidy, it might be necessary to analyze a larger number of cells in order to observe ploidy changes (~500 cells). Appropriate statistical tools should be used computing the data collected on the initial 200 cells to estimate the sample size (i.e., biological replicates) and the number of nuclei required to obtain the desired statistical power.

4. For each microscopic field, multiple focal planes must be acquired within each channel to ensure that signals on different focal planes are included (*see* **Note 45**).

5. After acquisition, signals for each fluorophore are visually inspected and manually counted for each cell. Plotting of raw counts is summarized as shown in Table 2.

6. Diploid nuclei will have two signals for each probe (in our human chromosome example, two blue and two yellow signals for HSA9 and two red and two green signals for HSA12) (Figs. 2 and 4a). Likewise, polyploid nuclei can be identified by the presence of blue/yellow and red/green signals matching in number (i.e., four for tetraploid cells). Aneuploid nuclei are

Table 2
Template of cell scoring for four-color FISH

| Cell number | Number of signals | | | | Ploidy |
| | MMU 1 | | MMU 18 | | |
	Blue	Yellow	Red	Green	
1	2	2	2	2	Diploid
2	2	2	3	3	Aneuploid
3	4	4	4	4	Tetraploid
4	3	3	1	1	Aneuploid
5	5	5	5	5	>4n
...					
500					

From *left to right*, each column indicates, respectively, the cell analyzed, the number of signals detected in the *blue*, *yellow*, *red*, and *green* channels, and the ploidy determination based on the number of signals

Table 3
Template of summary of cell scoring

| Ploidy | Young | | Old | |
	Number of cells	Frequency (%)	Number of cells	Frequency (%)
2n	492	98.4	410	82
1n	0	0	1	0.2
3n	1	0.2	4	0.8
4n	3	0.6	21	4.2
>4n	0	0	7	1.4
Aneuploid	4	0.8	57	11.4
Total	500	100	500	100

For two samples analyzed (*young* and *old*), the table summarizes the number of cells found with any given ploidy (2n, 1n, 3n, 4n, >4n, or aneuploid), as well as their frequency (percentage of cells over the total number of cells analyzed)

considered when the numbers of signals for one chromosome do not match the other (Fig. 4b).

7. Once all nuclei have been inspected, data can be summarized as in Table 3.

4 Notes

1. Do not repeatedly freeze and thaw modified dUTPs. Store aliquots protected from light, according to manufacturer's directions, and carefully record expiration dates.

2. All steps involving the use of fluorophores (Nick translation, hybridization, and detection) should be performed in the dark.

3. Methanol is highly flammable. Use it only inside chemical fume hood and wear gloves while handling.

4. Autoclave solution after KCl solubilization. It can be stored at RT for at least 6 months.

5. Fixative should be freshly prepared every time and can be stored for 1–2 h before procedure at 4 °C.

6. Absolute methanol can be stored and reused for a couple of months.

7. Cot-1 DNA should match the genome of the species used for hybridization.

8. It is very important that the pepsin is added to an empty clean beaker first and not directly into the acid solution. Direct addition of the pepsin to the acid solution will cause the pepsin to precipitate and it will not dissolve properly into the acid solution.

9. The time of pepsin treatment and amount of pepsin stock solution to be used varies, depending on:

   (a) The amount of cytoplasm around single nuclei or thickness of tissue section.

   (b) The age of the slide.

      For fresh slides with single cells, we recommend 5 μL of pepsin for 5 min as a starting point. Slides with excess cytoplasm require longer treatment with pepsin and/or higher concentrations of stock pepsin ranging from 5 to 20 μL. Slides older than 6 months may also require more intensive pretreatment.

      For 5–12 μm frozen tissue sections, we recommend 12–30 μL of pepsin for 6–10 min as starting guides, respectively.

      For cell types that can only be dissociated/isolated as single nucleus (i.e., neurons), mild pepsin treatment (5 μL for 3 min) is desired, as cytoplasm is not abundant.

10. As an alternative to homemade pepsin solutions, commercial kits containing a mixture of proteases can also be used in tissue sections that are difficult to hybridize (Abbott Molecular #32–801,200).

11. Make Pretreatment Solution 2 fresh for each experiment, mix well, and store at room temperature until use. Formaldehyde is flammable, carcinogenic, and poisonous and it should be dispensed in a chemical fume hood and only handled when wearing gloves.

12. Ethanol solutions can be reused for a couple of weeks.

13. The use of deionized formamide is highly recommended as it contains fewer impurities than conventional formamide. Formamide is a known mutagen that causes eye and skin irritation. Wear gloves when handling and dispense under a chemical fume hood.

14. Precool 70% ethanol in ice for slide denaturation.

15. Select from the pull-down menu the desired organism and the most updated genome assembly. If BAC clones are not visible in the display window, make sure that the BAC End Pairs and the FISH clones tracks are selected (full display).

16. Purification of BAC clones using column-based methods produces DNA with higher 260/280 ratios.

17. The amount of DNA can be estimated with the aid of a high DNA mass ladder (ThermoFisher Scientific #10496016).

18. We suggest using DY-590-dUTP (#590–34) diluted 1:5 while all other modified dUTPs should be used without dilution.

19. The amount of modified dUTPs varies according to the fluorophore that is conjugated with: 1.5 μL of DY-415-dUTP and 2 μL of DY-495-dUTP and DY-647-dUTP1. For DY-590-dUTP, use 4 μL of a 1:5 diluted stock.

20. Stopping the nick translation by freezing the tubes at −20 °C is important to avoid further degradation of the DNA.

21. Ideally, the bulk of the digested DNA should be between 200–800 bp after Nick translation (see Fig. 3b). If the DNA fragments are too large, add more DNase and incubate at 15 °C for additional 10–30 min.

22. The amount of nick-translated DNA to be precipitated should be determined by the researcher based on the intensity of the nick-translated DNA in the gel. Generally, the amount varies between 15 and 60 μL for each probe. Probes nicked with DY-647-dUTP1 normally require more DNA for optimal signal.

23. The volume of the combined DNA should be at least 100 μL for optimal precipitation. If the volume is lower, add DNase-free water to the combined DNA to increase the final volume.

24. For aneuploidy analysis in mice cells/tissues, we use 80 μL of mouse Cot-1 DNA. For human cells/tissues, we use 60 μL of human Cot-1 DNA.

25. Add the deionized formamide on top of the pellet, quick spin, and place tube on shaker for DNA solubilization. Do not pipette up and down or vortex, as this may cause loss of DNA inside the tip or on tube sidewalls.

26. The amount of hypotonic solution to be added varies with the amount of cells. In general, it should be in excess relative to the volume of the cells. If more than 10,000 cells are being prepared, increase the amount of hypotonic solution accordingly (up to 10 mL). Invert the tube for complete mix when all the hypotonic solution has been added. If less than 10,000 cells are being prepared, scale down the hypotonic solution accordingly, as long as the volume is higher than 500 μL.

27. Detection of age-related aneuploidy is performed in interphase cells; therefore, hypotonic exposure is strictly not required. However, in our experience, performing hypotonic treatment reduces the amount of cytoplasm around the nuclei and improves hybridization signal. Of note, hypotonic treatment is not required when analyzing cell types that can only be isolated as single nucleus (i.e., neurons).

28. The amount of fixative to be added should be the same of the volume of hypotonic solution added previously.

29. Only interphase cells will be visible at this stage, we do not expect the presence of metaphase chromosomes. The integrity of chromatin structure is dependent on the evaporation rate of the fixative, as determined by the percent humidity, temperature, and success of the hypotonic procedure. Interphase nuclei should appear light gray in color and should not show shiny edges or have a bright halo around them. We recommend 48% humidity based on our equipment. If nuclei do not look as expected, try reducing the percentage of humidity gradually until the desired appearance is obtained.

30. If visual inspection of the slides reveals overcrowded cells that overlap with each other, add more fixative to dilute the preparation. If, on the other hand, the cells are scattered, the preparation can be centrifuged and the pellet resuspended in a smaller volume of fixative.

31. For better FISH performance we suggest aging slides in a 37 °C drying oven for 3–7 days. This process yields optimal results.

32. The denaturation time and temperature depend on the age of the slide, the species, and cell type. For example, mouse preparations usually require lower temperature and reduced denaturing time.

33. Make sure that no air bubbles are present by gently applying pressure to the coverslip (e.g., with forceps) to allow them to

escape from the edge. It is important that the coverslip is completely sealed with rubber cement to avoid drying of the probe, which would ultimately result in high background.

34. In case the coverslip does not slide off easily, dip the slide with the coverslip in a Coplin jar containing Wash Solution C3 pre-warmed to 45 °C. This will loosen the hybridization mix and aid the removal of the coverslip.

35. This protocol uses probes directly labeled with modified dUTP; therefore, the preferred solution to be used is Wash Solution C2 (1× SSC). If high background is observed, we recommend using 0.1× SSC at 45 °C instead.

36. Let the slides sit at room temperature for at least 30 min before visualization to allow the DAPI to uniformly stain the chromatin.

37. We recommend the use of nail polish to seal the coverslip to the slide. It avoids sliding of the coverslip on top of the sample, facilitates the cleaning of the coverslip, and prevents microbiotic contamination. However, if it is anticipated that the cells/tissue may be required for other analyses, do not seal the coverslip. Slides can be washed 3× for 5 min each in Wash Solution C3 (same as Wash Solution T2) pre-warmed to 45 °C in order to remove the antifade prior to another assay.

38. Here we list some useful online resources for brain atlas:

    (a) http://www.mbl.org/mbl_main/atlas.html—High-resolution images collected in a collaborative project between Robert W. Williams (University of Tennessee) and Glenn D. Rosen (Beth Israel Deaconess Medical Center, Boston).

    (b) http://www.hms.harvard.edu/research/brain/atlas. html—High resolution brain atlas project supervised by Dr. Richard Sidman (Harvard Medical School and Beth Israel Deaconess Medical Center, Boston).

    (c) http://brainmaps.org/index.php—Interactive, multi-resolution brain atlas project supervised by Dr. W. Martin Usrey (UC Davis).

39. In our experience, 12 μm sections preserve tissue structure and integrity better. However they can be more challenging to score low-level aneuploidy due to the presence of hybridization signals in other focal planes. The use of a confocal microscope may attenuate this problem.

40. The OCT compound is water soluble and it will separate from the tissue after methanol fixation, but it will remain on the slide. Make sure to remove it while the slide is still wet to ensure good hybridization.

41. In case the coverslip does not slide off easily, dip the slide with the coverslip in a Coplin jar containing Wash Solution T1 prewarmed to 74 °C. This will loosen the hybridization mix and aid the removal of the coverslip.

42. In our experience, analysis of four-color FISH in tissue sections is possible; however, the number of cells containing matching hybridization signals for all probes can be low (~10% of total cells). For example, in the nuclei highlighted by the arrows in Fig. 5c, signals for all four colors are clearly visible, while other neighboring cells carry strong signals for only two probes. In order to score the ploidy of more nuclei per experiment, we recommend hybridizing the tissue with the four probes but perform the counting separated for each chromosome (two-color FISH) (Table 4 and Fig. 6). This approach allows the inclusion of cells that would otherwise be discarded from the quantification. However, it also limits the assessment of ploidy to a single chromosome per nucleus.

43. Four-color FISH-hybridized slides can be stored at 4 °C until analysis. Because signal intensity diminishes with time, we recommend image acquisition to be carried on within 1 week from detection.

Table 4
Template of cell scoring for two-color FISH

| | Number of signals | | | | Number of signals | | |
| | MMU 1 | | | | MMU 18 | | |
Cell number	Blue	Yellow	Ploidy	Cell number	Red	Green	Ploidy
1	2	2	Diploid	a	2	2	Diploid
2	2	2	Diploid	b	3	3	Triploid
3	4	4	Tetraploid	c	4	4	Tetraploid
4	3	3	Triploid	d	1	1	Monosomic
5	5	5	>4n	e	5	5	>4n
...				...			
500				500			

In order to score more cells per experiment, tissue sections are hybridized for four-color FISH but the analysis is performed for each chromosome individually. We recommend labeling the cells in a different way for each chromosome (i.e., numbers for MMU1 and letters for MMU18) in order to differentiate the countings within a single image

From *left to right*, each column indicates, respectively, the cell analyzed for MMU1, the number of signals detected for MMU1, and respective ploidy; the cell analyzed for MMU18, the number of signals detected for MMU18, and respective ploidy. Note that in this type of analysis, any cell with ploidy different of 2n could be scored as aneuploid

44. ASI—GenASIs FISHView (http://www.spectral-imaging. com/applications/cytogenetics/fish/fishview).

45. This is especially important for FISH on tissue sections, in which we suggest acquiring five focal planes spaced ~1 μm and merging the resulting images. For isolated mammalian cells, acquisition in other focal planes is only necessary when the hybridization signals in adjacent cells are out of focus.

5   Conclusions

The techniques described in this chapter were designed to overcome the challenges of detecting somatic low-frequency and low copy number changes in chromosome content associated with normative aging. These approaches were proven reproducible and highly sensitive for the analysis of low-level aneuploidy in the brain [8, 36].

Few limitations are however associated with this approach, and these are a direct consequence of the analytical methods we adopted to quantify ploidy changes. Because we specifically measure chromosome copies by the presence or absence of two signals for each autosome (respectively red/green and aqua/yellow), our approach, while highly sensitive, is unable to detect structural aneuploidy and it is likely to underestimate aneuploidy. For instance, breakage between the two probes mapping to the same chromosome may result in the absence of one of the signals, which would result in the exclusion of this particular cell from the analysis. Therefore, when matching signals are actually lacking for a particular chromosome, we are unable to distinguish it from an artifact due to poor hybridization. Thus, our four-color interphase FISH protocol probably underestimates the complexity of the genomic instability (GIN) present in a sample.

Additionally, because we are only analyzing two chromosomes at a time within a single cell, we might overestimate the frequency of polyploid cells at a cost of aneuploid cells. Nuclei could be scored as tetraploid, for example, because both chromosomes analyzed contained four copies. However, we cannot rule out the possibility that chromosomes other than the ones tested may be present in numbers different from four. Likewise diploid cells could still be aneuploid for chromosomes other than the two tested.

More recent studies, using single-cell sequencing, have questioned the aneuploidy levels observed by interphase FISH in the mammalian brain, reporting remarkably lower rates of not-diploid cells [48, 49]. Yet, multiple somatic copy number variations have been detected by independent groups in these cells [37, 50], suggesting that GIN is a feature of normal mammalian neurons and nonneuronal cells. It is possible that the levels of whole-chromosome gains and losses may have been overestimated by

FISH studies and aneuploidy levels are actually lower than previously anticipated. Like FISH analysis, single-cell sequencing also has its own limitations. While it allows for the analysis of ploidy of all chromosomes at once, it is limited in the measurement of ploidy changes, since this cellular state produces a balanced genome content. In addition, generally the number of cells analyzed by single-cell sequencing is lower than what is feasible for FISH, reducing the power of detection of copy number changes occurring at low frequency. In this case, more single cells may need to be sequenced to determine the biological range of chromosome number variation. In our opinion the combined use of FISH and single-cell sequencing on the same sample should provide unprecedented sensitivity for the analysis of mosaic copy number changes.

References

1. Täckholm G (1922) Zytologische studien über die gattung rosa. Acta Horti Bergiani 7:97–381

2. Santaguida S, Amon A (2015) Short- and long-term effects of chromosome mis-segregation and aneuploidy. Nat Rev Mol Cell Biol 16(8):473–485. doi:10.1038/nrm4025

3. Holland AJ, Cleveland DW (2012) Losing balance: the origin and impact of aneuploidy in cancer. EMBO Rep 13(6):501–514. doi:10.1038/embor.2012.55

4. Siegel JJ, Amon A (2012) New insights into the troubles of aneuploidy. Annu Rev Cell Dev Biol 28:189–214. doi:10.1146/annurev-cellbio-101011-155807

5. Bakker B, van den Bos H, Lansdorp PM, Foijer F (2015) How to count chromosomes in a cell: an overview of current and novel technologies. BioEssays 37(5):570–577. doi:10.1002/bies.201400218

6. Faggioli F, Vijg J, Montagna C (2011) Chromosomal aneuploidy in the aging brain. Mech Ageing Dev 132(8–9):429–436. doi:10.1016/j.mad.2011.04.008

7. Andriani GA, Vijg J, Montagna C (2016) Mechanisms and consequences of aneuploidy and chromosome instability in the aging brain. Mech Ageing Dev. doi:10.1016/j.mad.2016.03.007

8. Faggioli F, Wang T, Vijg J, Montagna C (2012) Chromosome-specific accumulation of aneuploidy in the aging mouse brain. Hum Mol Genet 21(24):5246–5253. doi:10.1093/hmg/dds375

9. Rehen SK, McConnell MJ, Kaushal D, Kingsbury MA, Yang AH, Chun J (2001) Chromosomal variation in neurons of the developing and adult mammalian nervous system. Proc Natl Acad Sci U S A 98(23):13361–13366. doi:10.1073/pnas.231487398

10. Rehen SK, Yung YC, McCreight MP, Kaushal D, Yang AH, Almeida BS, Kingsbury MA, Cabral KM, McConnell MJ, Anliker B, Fontanoz M, Chun J (2005) Constitutional aneuploidy in the normal human brain. J Neurosci 25(9):2176–2180. doi:10.1523/JNEUROSCI.4560-04.2005

11. Yurov YB, Iourov IY, Monakhov VV, Soloviev IV, Vostrikov VM, Vorsanova SG (2005) The variation of aneuploidy frequency in the developing and adult human brain revealed by an interphase FISH study. J Histochem Cytochem 53(3):385–390. doi:10.1369/jhc.4A6430.2005

12. Duncan AW, Hanlon Newell AE, Smith L, Wilson EM, Olson SB, Thayer MJ, Strom SC, Grompe M (2012) Frequent aneuploidy among normal human hepatocytes. Gastroenterology 142(1):25–28. doi:10.1053/j.gastro.2011.10.029

13. Faggioli F, Vezzoni P, Montagna C (2011) Single-cell analysis of ploidy and centrosomes underscores the peculiarity of normal hepatocytes. PLoS One 6(10):e26080. doi:10.1371/journal.pone.0026080

14. Jacobs PA, Court Brown WM, Doll R (1961) Distribution of human chromosome counts in relation to age. Nature 191:1178–1180

15. Jacobs KB, Yeager M, Zhou W, Wacholder S, Wang Z, Rodriguez-Santiago B, Hutchinson A, Deng X, Liu C, Horner MJ, Cullen M, Epstein CG, Burdett L, Dean MC, Chatterjee N, Sampson J, Chung CC, Kovaks J, Gapstur SM, Stevens VL, Teras LT, Gaudet MM,

Albanes D, Weinstein SJ, Virtamo J, Taylor PR, Freedman ND, Abnet CC, Goldstein AM, Hu N, Yu K, Yuan JM, Liao L, Ding T, Qiao YL, Gao YT, Koh WP, Xiang YB, Tang ZZ, Fan JH, Aldrich MC, Amos C, Blot WJ, Bock CH, Gillanders EM, Harris CC, Haiman CA, Henderson BE, Kolonel LN, Le Marchand L, McNeill LH, Rybicki BA, Schwartz AG, Signorello LB, Spitz MR, Wiencke JK, Wrensch M, Wu X, Zanetti KA, Ziegler RG, Figueroa JD, Garcia-Closas M, Malats N, Marenne G, Prokunina-Olsson L, Baris D, Schwenn M, Johnson A, Landi MT, Goldin L, Consonni D, Bertazzi PA, Rotunno M, Rajaraman P, Andersson U, Beane Freeman LE, Berg CD, Buring JE, Butler MA, Carreon T, Feychting M, Ahlbom A, Gaziano JM, Giles GG, Hallmans G, Hankinson SE, Hartge P, Henriksson R, Inskip PD, Johansen C, Landgren A, McKean-Cowdin R, Michaud DS, Melin BS, Peters U, Ruder AM, Sesso HD, Severi G, Shu XO, Visvanathan K, White E, Wolk A, Zeleniuch-Jacquotte A, Zheng W, Silverman DT, Kogevinas M, Gonzalez JR, Villa O, Li D, Duell EJ, Risch HA, Olson SH, Kooperberg C, Wolpin BM, Jiao L, Hassan M, Wheeler W, Arslan AA, Bueno-de-Mesquita HB, Fuchs CS, Gallinger S, Gross MD, Holly EA, Klein AP, LaCroix A, Mandelson MT, Petersen G, Boutron-Ruault MC, Bracci PM, Canzian F, Chang K, Cotterchio M, Giovannucci EL, Goggins M, Hoffman Bolton JA, Jenab M, Khaw KT, Krogh V, Kurtz RC, McWilliams RR, Mendelsohn JB, Rabe KG, Riboli E, Tjonneland A, Tobias GS, Trichopoulos D, Elena JW, Yu H, Amundadottir L, Stolzenberg-Solomon RZ, Kraft P, Schumacher F, Stram D, Savage SA, Mirabello L, Andrulis IL, Wunder JS, Patino Garcia A, Sierrasesumaga L, Barkauskas DA, Gorlick RG, Purdue M, Chow WH, Moore LE, Schwartz KL, Davis FG, Hsing AW, Berndt SI, Black A, Wentzensen N, Brinton LA, Lissowska J, Peplonska B, McGlynn KA, Cook MB, Graubard BI, Kratz CP, Greene MH, Erickson RL, Hunter DJ, Thomas G, Hoover RN, Real FX, Fraumeni JF Jr, Caporaso NE, Tucker M, Rothman N, Perez-Jurado LA, Chanock SJ (2012) Detectable clonal mosaicism and its relationship to aging and cancer. Nat Genet 44(6):651–658. doi:10.1038/ng.2270

16. Jones KT (2008) Meiosis in oocytes: predisposition to aneuploidy and its increased incidence with age. Hum Reprod Update 14(2):143–158. doi:10.1093/humupd/dmm043

17. Thomas P, Fenech M (2008) Chromosome 17 and 21 aneuploidy in buccal cells is increased with ageing and in Alzheimer's disease. Mutagenesis 23(1):57–65. doi:10.1093/mutage/gem044

18. Baker DJ, Dawlaty MM, Wijshake T, Jeganathan KB, Malureanu L, van Ree JH, Crespo-Diaz R, Reyes S, Seaburg L, Shapiro V, Behfar A, Terzic A, van de Sluis B, van Deursen JM (2013) Increased expression of BubR1 protects against aneuploidy and cancer and extends healthy lifespan. Nat Cell Biol 15(1):96–102. doi:10.1038/ncb2643

19. Sunshine AB, Ong GT, Nickerson DP, Carr D, Murakami CJ, Wasko BM, Shemorry A, Merz AJ, Kaeberlein M, Dunham MJ (2016) Aneuploidy shortens replicative lifespan in Saccharomyces cerevisiae. Aging Cell 15(2):317–324. doi:10.1111/acel.12443

20. Baker DJ, Jeganathan KB, Cameron JD, Thompson M, Juneja S, Kopecka A, Kumar R, Jenkins RB, de Groen PC, Roche P, van Deursen JM (2004) BubR1 insufficiency causes early onset of aging-associated phenotypes and infertility in mice. Nat Genet 36(7):744–749. doi:10.1038/ng1382

21. Baker DJ, Jeganathan KB, Malureanu L, Perez-Terzic C, Terzic A, van Deursen JM (2006) Early aging-associated phenotypes in Bub3/Rae1 haploinsufficient mice. J Cell Biol 172(4):529–540. doi:10.1083/jcb.200507081

22. Matsuyama M, Tanaka H, Inoko A, Goto H, Yonemura S, Kobori K, Hayashi Y, Kondo E, Itohara S, Izawa I, Inagaki M (2013) Defect of mitotic vimentin phosphorylation causes microophthalmia and cataract via aneuploidy and senescence in lens epithelial cells. J Biol Chem 288(50):35626–35635. doi:10.1074/jbc.M113.514737

23. Tanaka H, Goto H, Inoko A, Makihara H, Enomoto A, Horimoto K, Matsuyama M, Kurita K, Izawa I, Inagaki M (2015) Cytokinetic failure-induced Tetraploidy develops into aneuploidy, triggering skin aging in Phosphovimentin-deficient mice. J Biol Chem 290(21):12984–12998. doi:10.1074/jbc.M114.633891

24. Glisky EL (2007) Changes in cognitive function in human aging. In: Riddle DR (ed) Brain aging: models, methods, and mechanisms. Frontiers in Neuroscience, Boca Raton, FL

25. Peterson SE, Yang AH, Bushman DM, Westra JW, Yung YC, Barral S, Mutoh T, Rehen SK, Chun J (2012) Aneuploid cells are differentially susceptible to caspase-mediated death during embryonic cerebral cortical development. J Neurosci 32(46):16213–16222. doi:10.1523/JNEUROSCI.3706-12.2012

26. Andriani GA, Almeida VP, Faggioli F, Mauro M, Tsai WL, Santambrogio L, Maslov A, Gadina M, Campisi J, Vijg J, Montagna C (2016) Whole chromosome instability induces senescence and promotes SASP. Sci Rep 6:35218. doi:10.1038/Srep35218

27. Franceschi C, Campisi J (2014) Chronic inflammation (inflammaging) and its potential contribution to age-associated diseases. J Gerontol A: Biol 69:S4–S9. doi:10.1093/gerona/glu057

28. Bhatia-Dey N, Kanherkar RR, Stair SE, Makarev EO, Csoka AB (2016) Cellular senescence as the causal nexus of aging. Front Genet 7:13

29. Thompson SL, Compton DA (2010) Proliferation of aneuploid human cells is limited by a p53-dependent mechanism. J Cell Biol 188(3):369–381. doi:10.1083/jcb.200905057

30. Flowers A (2000) Brain tumors in the older person. Cancer Control 7(6):523–538

31. Kaushal D, Contos JJA, Treuner K, Yang AH, Kingsbury MA, Rehen SK, McConnell MJ, Okabe M, Barlow C, Chun J (2003) Alteration of gene expression by chromosome loss in the postnatal mouse brain. J Neurosci 23(13):5599–5606

32. Lopez-Sanchez N, Ovejero-Benito MC, Borreguero L, Frade JM (2011) Control of neuronal ploidy during vertebrate development. Results Probl Cell Differ 53:547–563. doi:10.1007/978-3-642-19065-0_22

33. Darmanis S, Sloan SA, Zhang Y, Enge M, Caneda C, Shuer LM, Gephart MGH, Barres BA, Quake SR (2015) A survey of human brain transcriptome diversity at the single cell level. Proc Natl Acad Sci U S A 112(23):7285–7290. doi:10.1073/pnas.1507125112

34. Rose CR, Kirchhoff F (2015) Glial heterogeneity: the increasing complexity of the brain. e-Neuroforum 6(3):59–62. doi:10.1007/s13295-015-0012-0

35. McCaffrey JB (2015) The brain's heterogeneous functional landscape. Philos Sci 82(5):1010–1022. doi:10.1086/683436

36. Andriani GA, Faggioli F, Baker D, Dolle ME, Sellers RS, Hebert JM, Van Steeg H, Hoeijmakers J, Vijg J, Montagna C (2016) Whole chromosome aneuploidy in the brain of Bub1bH/H and Ercc1−/Delta7 mice. Hum Mol Genet 25(4):755–765. doi:10.1093/hmg/ddv612

37. McConnell MJ, Lindberg MR, Brennand KJ, Piper JC, Voet T, Cowing-Zitron C, Shumilina S, Lasken RS, Vermeesch JR, Hall IM, Gage FH (2013) Mosaic copy number variation in human neurons. Science 342(6158):632–637. doi:10.1126/science.1243472

38. Regan CM (1988) Neuronal and glial markers of the central nervous system. Experientia 44(8):695–697. doi:10.1007/bf01941031

39. Redwine JM, Evans CF (2002) Markers of central nervous system glia and neurons in vivo during normal and pathological conditions. Curr Top Microbiol Immunol 265:119–140

40. Yuan SH, Martin J, Elia J, Flippin J, Paramban RI, Hefferan MP, Vidal JG, Mu Y, Killian RL, Israel MA, Emre N, Marsala S, Marsala M, Gage FH, Goldstein LSB, Carson CT (2011) Cell-surface marker signatures for the isolation of neural stem cells, glia and neurons derived from human pluripotent stem cells. PLoS One 6(3):e17540. doi:10.1371/journal.pone.0017540

41. Lopez-Sanchez N, Frade JM (2015) Flow cytometric analysis of DNA synthesis and apoptosis in central nervous system using fresh cell nuclei. Methods Mol Biol 1254:33–42. doi:10.1007/978-1-4939-2152-2_3

42. Slaninova I, Lopez-Sanchez N, Sebrlova K, Vymazal O, Frade JM, Taborska E (2016) Introduction of macarpine as a novel cell-permeant DNA dye for live cell imaging and flow cytometry sorting. Biol Cell 108(1):1–18. doi:10.1111/boc.201500047

43. Chiu K, Lau WM, Lau HT, So KF, Chang RCC (2007) Micro-dissection of rat brain for RNA or Protein extraction from specific brain region. J Vis Exp 7:269

44. Freshney RI (2010) Cell Separation. In: Culture of animal cells. Wiley, Hoboken, NJ, pp 227–237. doi:10.1002/9780470649367.ch14

45. Matevossian A, Akbarian S (2008) Neuronal nuclei isolation from human postmortem brain tissue. J Vis Exp 20:914

46. Peters SR (2010) Embedding of tissue for frozen section. In: Peters SR (ed) A practical guide to frozen section technique. Springer, New York, NY, pp 37–74. doi:10.1007/978-1-4419-1234-3_3

47. Peters SR (2010) Cutting the frozen section. In: Peters SR (ed) A practical guide to frozen section technique. Springer, New York, NY, pp 75–95. doi:10.1007/978-1-4419-1234-3_4

48. Knouse KA, Wu J, Whittaker CA, Amon A (2014) Single cell sequencing reveals low levels of aneuploidy across mammalian tissues. Proc Natl Acad Sci U S A 111(37):13409–13414. doi:10.1073/pnas.1415287111

49. van den Bos H, Spierings DC, Taudt AS, Bakker B, Porubsky D, Falconer E, Novoa C, Halsema N, Kazemier HG, Hoekstra-Wakker K, Guryev V, den Dunnen WF, Foijer F, Tatche MC, Boddeke HW, Lansdorp PM (2016) Single-cell whole genome sequencing reveals no evidence for common aneuploidy in normal and Alzheimer's disease neurons. Genome Biol 17(1):116. doi:10.1186/s13059-016-0976-2

50. Cai X, Evrony GD, Lehmann HS, Elhosary PC, Mehta BK, Poduri A, Walsh CA (2014) Single-cell, genome-wide sequencing identifies clonal somatic copy-number variation in the human brain. Cell Rep 8(5):1280–1289. doi:10.1016/j.celrep.2014.07.043

Chapter 15

Genomic Analysis and In Vivo Functional Validation of Brain Somatic Mutations Leading to Focal Cortical Malformations

Jae Seok Lim and Jeong Ho Lee

Abstract

Focal cortical malformation (FCM), such as focal cortical dysplasia (FCD) and hemimegalencephaly (HME), is a major developmental brain malformation in the cerebral cortex leading to intractable epilepsy. The sporadic occurrence of most FCM and histologic characteristics of surgically resected brain tissue showing scattered dysmorphic cells suggest that FCM might be caused by a somatic mutation in an area affecting brain development. Indeed, recent genomic studies of these conditions have shown that low-frequency somatic mutations in PI3K-AKT-mTOR pathway genes are a major genetic cause of FCM. In addition, functional validation using an in vivo disease model not only confirmed the causality of the identified somatic mutations but also helped to reveal their molecular genetic mechanisms. Here, we highlight the key points to be considered regarding the application of sequencing methods and bioinformatics analysis to identify brain somatic mutations with a low allelic frequency in FCM patients. In addition, we describe the generation of an in vivo disease model recapitulating the pathologic phenotype of FCM such as dysmorphic neurons, migration defects, and electrographic seizures. Our goal is to provide guidelines for the analysis of sequencing data and functional validation using a disease model of FCM caused by somatic mutations.

Key words Focal cortical malformation, Brain somatic mutation, PI3K-AKT-mTOR pathway, Intractable epilepsy, Bioinformatics analysis, Low-frequency somatic mutation, In utero electroporation, In vivo disease-modeling

1 Introduction: Brain Somatic Mutation in Focal Cortical Malformation

A somatic mutation is a genetic variation acquired by somatic cells, which is neither inherited from a parent nor passed to offspring. Somatic mutations can occur in any dividing or nondividing somatic cells due to DNA replication errors or environmental factors such as UV light, smoking, and other carcinogens [1]. Recent studies have revealed that the mutational burden in somatic cells is quite high, and estimated mutation rates suggest that every cell division creates a somatic point mutation and structural variation,

José María Frade and Fred H. Gage (eds.), *Genomic Mosaicism in Neurons and Other Cell Types*, Neuromethods, vol. 131, DOI 10.1007/978-1-4939-7280-7_15, © Springer Science+Business Media LLC 2017

which may or may not have an effect on the resulting phenotype [2]. Although somatic mutations are well known to be the cause of most cancers [3–5], recent studies have shown that they are also causative for non-neoplastic skin disorders such as McCune–Albright syndrome, Sturge–Weber syndrome, and Proteus syndrome [6–8]. Somatic mutations can occur in the human brain. During the 4–24-week gestation period, the developing brain generates 10^{10} neurons from a few neural precursor cells. The division rate during this period is approximately 10^5 times per minute, which is faster than the division that occurs in any other organ and even faster than cancer [9]. This suggests that the human brain has a higher chance of acquiring somatic mutations during its early developmental period compared with any other period.

Recently, increasing evidence supports the importance of somatic mutation as a major genetic cause of sporadic neurodevelopmental disorders. Malformations of cortical development (MCD) include various structural abnormalities of the cerebral cortex that arise during the formation of the cortical plate [10]. Development of the cerebral cortex involves complex processes including neural progenitor cell proliferation, migration, and organization. Abnormalities in any step of these processes may cause various clinical phenotypes such as cortical layer disruption and an enlarged brain [11, 12]. Among these MCD, focal cortical malformation (FCM) including focal cortical dysplasia (FCD) and hemimegalencephaly (HME) share a pathologic phenotype characterized by disorganized or absent cortical lamination, loss of radial neuronal orientation, and the presence of large and dysmorphic neurons in the affected brain region [13]. Most FCMs occur sporadically without familial history and present focal brain lesions [14], implying that somatic mutations in key genes that regulate neuronal cell growth and migration in affected brains are the underlying disease mechanism. Recently, we and other groups have reported the disease-causing somatic mutation in PI3K-AKT-mTOR pathway genes in hemimegalencephaly (HME), which is characterized by enlargement and extensive malformation of an entire cerebral hemisphere. These mutations were found in surgically resected tissue from the affected brain of HME patients, not in their peripheral blood, consistent with somatic mosaicism in the brain. The frequency of the mutated allele in the affected brain of HME has been reported to be as low as 8%, implying that low-level somatic mosaicism in the affected brain plays an important role in the development of cortical malformation in sporadic neurodevelopmental disorder. More recently, brain somatic mutations in *MTOR* were identified in up to 25% of patients with FCD, which is a major cause of medically refractory epilepsy [15, 16]. Interestingly, the brain MRIs of FCD patients are occasionally normal, and histological examination has revealed that a small subset of neurons show a dysmorphic or enlarged phenotype in affected tissue, thereby suggesting the presence of a somatic mutation in a small fraction of neurons. Notably, the mutated allele frequency in FCD was reported to be as low as ~1%, as validated by deep sequencing with various experimental replications [15]. Moreover, an in vivo mouse model of FCD carrying the identified somatic mutation with a low allele frequency was able to recapitulate the clinical and pathological symptoms found in patients. These studies provide direct evidence that a low-level somatic mutation in the affected brain region is sufficient to cause a neurodevelopmental disorder [15].

1.1 Detecting Brain Somatic Mutations in Human Samples

Recent advances in next-generation sequencing technology provide great opportunities to quantify the mutational burden and assess low-level somatic mutations. Deep sequencing is frequently used to detect somatic mutations in noncancerous disorders such as cortical malformation disease [7, 15]. Although the accurate detection of somatic mutations with low allelic frequency is still challenging, mosaicism levels as low as 1% can be distinguished from sequencing errors by using ultra-deep sequencing combined with various experimental validation methods, including laser capture microdissection (LCM) for enriching mutated cells and mass spectrometry assays such as a single allele base extension reaction assay (SABER) [7, 15, 17]. To accurately detect brain somatic mutations in FCM, researchers should be concerned with several key experimental steps, including (a) extraction of quality-controlled genomic DNA (gDNA) from matched brain-blood (or saliva) tissue, (b) generation of raw sequencing data and exclusion of any contamination (e.g., plasmid vectors and other tissue samples), (c) selection of aligner and variant callers, and (d) validation sequencing of the identified mutations.

1.2 In Vivo Functional Validation of Identified Somatic Mutations Using in Utero Electroporation

In general, the detection of somatic mutations that are specific to a patient's tissues does not indicate a causal relationship between the identified mutations and disease phenotypes. Hence, it is necessary to examine the biological consequence of the mutations. In vitro studies such as mutagenesis followed by various imaging and biochemical experiments in cultured cells have been widely used to screen the biological functions of candidate variants and elucidate their molecular mechanism at the cellular or subcellular level. In studies of FCM, in vivo functional validation of somatic mutations is also important because it is necessary to test whether low-level somatic mutations in the focal brain area actually lead to behavioral changes like epilepsy. In this chapter, we would like to focus more on in vivo modeling of brain somatic mutation than in vitro modeling commonly used in many other molecular biology and genetic studies. In utero electroporation has emerged as an effective tool for modeling somatic mutations by manipulating gene expression in the brain at specific time points and among particular cell populations [18, 19]. One of the most useful features of in utero electroporation is that it can recapitulate phenotypes observed in focal cortical malformations, including dysmorphic neurons, migration defects, and seizures. If a novel disease-causing somatic mutation is identified in a neurodevelopmental disorder, this tool can be applied to generate a disease-specific mouse model of the clinical symptoms of patients.

2   Materials

2.1   Extraction and Assessment of Genomic DNA

To extract gDNA, the QIAamp DNA kit (Qiagen, USA) and QIAamp DNA FFPE tissue kit (Qiagen, USA) were used according to the procedure recommended by the manufacturer. The Quant-iT™ PicoGreen® dsDNA Assay Kit (Invitrogen, USA) were used to quantify the concentration of gDNA extracted from patient tissues. Assessing the gDNA quality can be performed using the fragment analyzer (Bioanalyzer 2100, Agilent, USA), which reads the fragment signal up to 12 kbp. The fragment analyzer assesses DNA quality by calculating a Genomic Quality Number (GQN). To handle a degraded sample such as FFPE tissue, it is highly recommended to evaluate the quality of the gDNA using the fragment analyzer.

2.2   Establishing a Computing Environment

Building a computing environment is an essential part of analyzing the NGS data from a patient sample. In theory, any computer with sufficient RAM, hard drive space, and CPU power can be used for analysis. In general, we recommend the following computing hardware: (1) a minimum of 8–16 Gb of RAM (better to have more than 32+ Gb); (2) 2 Tb of disk space (better to have 10 Tb of space); (3) a fast CPU (better to have at least eight cores, the more the better). For example, in our previous study [15], computing hardware consisting of 24 Gb of RAM, 4 TB of disk space, and eight core CPU was sufficient for analyzing deep WES data (average depth > 300×) from ten paired blood-brain samples. Most software for NGS is developed for Linux/UNIX style operating systems. Linux systems are the most compatible with academic software, and it is easier to install the analysis software on Linux than any other operating system.

2.3   Software for Detecting Somatic Mutations

Various pipelines or protocols have been used to analyze the NGS data. However, the best analysis pipeline for detecting low-level single nucleotide variants (SNVs) remains elusive. Thus, we will provide a brief introduction to the software used to identify SNVs in FCM. Bioinformatics analysis for detecting somatic SNVs in WES data consists of three steps: (1) preprocessing, (2) variant discovery, and (3) variant evaluation. In the preprocessing step to align the raw files (*.fastq) to the reference human genome, we used the "Best Practices" workflow suggested by the Broad

Institute consisting of Burrows-Wheeler Aligner (BWA) and the Genome Analysis Toolkit (GATK) and the generated aligned files (*.bam), which can be utilized for the subsequent analysis [20, 21]. The BWA and GATK software can be downloaded at official websites [22, 23]. In the variant discovery step for detecting candidate brain-specific SNVs, we utilized both the Virmid [24] and MuTect [25] algorithms to jointly analyze blood-brain paired WES data. The Virmid and MuTect are available at official websites [26, 27]. All variants from two algorithms were annotated by functional prediction using the snpEFF program for variant evaluation [28]. The general quality and contamination issue of raw data can be assessed by FastQC [29] and Vecuum [30], respectively.

2.4 Primer Design for Targeted Amplicon Sequencing

Primers including adapter sequence are required to generate the amplicon library. Primers (usually 20–24 bp) targeting specific sites were designed using the web-based Primer3 software. The PCR fragment size was determined according to the sequencing length of NGS. For example, for a sequencing length of 150 bp using paired-end sequencing, a PCR fragment size <300 bp is suitable for covering all of the target fragment. The sequence of the target primer is then added to the adapter sequence. The Truseq universal and index adapter sequence can be found at official website [31]. The primer composition is as follows: (1) sense, 5'- Truseq universal adapter sequence + forward primer sequence of the target site-3'; (2) antisense, 5'-Truseq index adapter sequence + reverse primer sequence of the target site-3'.

2.5 Tissue Slide Preparation for Laser Capture Microdissection (LCM)

Patient or mouse brain tissue was fixed in freshly prepared 4% paraformaldehyde phosphate-buffered saline overnight, cryoprotected overnight in 30% buffered sucrose, used to generate gelatin-embedded tissue blocks (7.5% gelatin in 10% sucrose/PB), and then stored at −80 °C. Sections cut using a cryostat were collected and placed on a glass slide (Superfrost* Plus Micro Slide, Fisher Scientific, USA). To capture the nuclei in laser pulse catapult (LPC) mode, we recommend a section thickness <10 μm. For FFPE slides, deparaffinization and rehydration were performed. A heat-induced retrieval process was then performed using the deparaffinized FFPE slides with citrate buffer (10 mM sodium citrate, pH 6.0) to enhance the antibody staining intensity. For immunofluorescence staining, sections cut using a cryostat or processed FFPE slides were blocked in PBS-GT (0.2% gelatin and 0.2% Triton X-100 in PBS) for 1 h at RT and stained with the following antibodies: rabbit antibody against phosphorylated S6 ribosomal protein (Ser240/Ser244) (1:100 dilution; 5364, Cell Signaling Technology) and mouse antibody against NeuN (1:100 dilution; MAB377, Millipore). Slides were then washed in PBS and stained with the following secondary antibodies: Alexa Fluor 488-conjugated goat anti-mouse antibody (1:200 dilution; A21422,

Invitrogen) and Alexa Fluor 555-conjugated goat anti-rabbit antibody (1:200 dilution; A11008, Invitrogen). DAPI solution (Thermofisher Scientific, USA) was used for nuclear staining.

2.6 Preparation for in Utero Electroporation (IUE)

Plasmids for IUE were purified using the EndoFree PlasmidKit (Qiagen, USA) according to the manufacture's protocol. The final concentration of the DNA should be higher than 1 μg/μL. Higher concentrations of DNA produce brighter fluorescence. Then 1/10 volume of 1% Fastgreen (Sigma, USA) was mixed with purified plasmids as tracer. For injection of prepared DNA solution, 1-mm diameter glass capillary tubes (Drummond Scientific, USA) were pulled with the micropipette puller (P-97, Sutter Instruments, USA). The following parameters are recommended: pressure, 500; heat, 600; pull, 30; velocity, 40; time, 1. Then, pulled capillary tubes were cut with forceps at 0.5–1.0 cm from the end of the capillary tubes. For anesthesia pregnant mouse, isoflurane vaporizer (Harvard apparatus, USA) with oxygen supplier was used. The following surgical instruments were required: fine forceps × 2, surgery scissors × 1, ring forceps, needle holder, vicryl suture, silk suture. Embryo was pinched with a forceps-type electrode (CUY650P3, Nepa gene, Japan), and electrical pulses are applied with an electroporator (BTX-Harvard apparatus, USA). All electroporations were performed with five 50 ms pulses of 35–50 V at 950 ms intervals.

3 Methods

3.1 Extraction of High-Quality Genomic DNA from Patient Brain Tissues

Brain tissue samples are commonly stored as freshly frozen tissue or formalin-fixed paraffin-embedded (FFPE) blocks to preserve specimens for a longer period for retrospective experiments such as immunohistochemistry. DNA was extracted using the QIAamp DNA extraction kit (Qiagen, USA) according to a procedure recommended by the manufacturer. Briefly, 10–25 mg of freshly frozen brain tissue was treated with 180 μL ATL buffer and incubated with 20 μL proteinase K at 56 °C overnight. After tissue lysis, 200 μL AL buffer was added to 200 μL ethanol, and the lysate was transferred to the column. After washing, 20–100 μL AE buffer was added and the eluted genomic DNA was collected. For extraction of DNA from FFPE tissue, additional procedures including paraffin removal and heating were performed to reverse formalin cross-linking of nucleic acids. Briefly, FFPE tissue sections (eight sections, 5–10 μm thick) were collected, and 1 mL of xylene was then added and mixed vigorously for paraffin removal. For lysis, the pellet was treated with 180 μL ATL buffer and incubated with 20 μL proteinase K at 56 °C for 1 h. The lysate was then incubated at 90 °C for 1 h to reverse formalin cross-linking. The subsequent procedures were the same as the protocol used for freshly frozen tissue.

Library preparation of NGS-based sequencing includes random fragmentation of gDNA and the addition of library-specific adapter sequences to the flanking ends. Considering the substantial loss of gDNA during these processes and additional use for validation sequencing, microgram ranges of gDNA should be extracted from FCM patient tissues such as brain, blood, or saliva. Although minute amounts of gDNA can be amplified by excessive targeted PCR or whole-genome amplification (WGA), specific artifacts such as PCR-induced base substitution or chimeric sequences jeopardize the discrimination of low-frequency somatic mutations due to erroneous variants [32–34]. Therefore, the extraction of a sufficient amount of gDNA is required for further analysis as well as the avoidance of an unnecessary genome amplification step. Regarding the accurate quantification of gDNA, the Picogreen assay (e.g., QuBit: Invitrogen) is preferred to UV absorbance measurements (e.g., Nanodrop). Picogreen is a widely used intercalating dye that specifically binds to double-stranded DNA (dsDNA), and thus protein, RNA, and contaminants do not interfere with the measurements. Several studies have shown that gDNA quantification with the Picogreen assay is more accurate than UV absorbance measurements, which have a tendency to overestimate the DNA concentration due to the presence of RNA and other contaminants that are commonly found in gDNA preparations [35–38]. Thus, estimation of the gDNA quantity using fluorescence intercalating dye is highly recommended. Greater than 1 µg of total gDNA assessed using the Picogreen assay is sufficient to perform the subsequent library preparation for deep sequencing.

To check the gDNA quality, gel electrophoresis is able to reveal the condition of the DNA, such as impurities, RNA contamination, and DNA degradation. Impurities, such as detergents or proteins, can be detected by nonspecific signal near the loading well. RNA, which interferes with 260 nm readings, is often visible at the bottom of a gel. A ladder or smear below a band of interest may indicate nicking or degradation of the DNA. Partially degraded gDNA often found in FFPE samples can be used as input material to satisfy QC criteria. In FFPE samples, formalin decreases the PCR efficiency due to protein cross-linking, and the degradation of nucleic acids increases during storage depending on the pH value of the fixative [39]. Nevertheless, several recent studies have reported successful NGS analysis using FFPE specimens, which satisfies the recommended criteria suggested by the manufacturers' guideline (e.g., Illumina). According to this guideline, a DNA input of 100–300 ng is recommended for FFPE samples with a genomic quality number (GQN) value >0.3, and low-quality FFPE samples with GQN values <0.3 are not recommended [40]. Assessment of the sample quality can be handled efficiently through the use of the Genomic Quality Number (GQN), assigning a value between 0 and 10, which is determined based on the extent by which the band intensity exceeds the threshold size designated by the researcher. For example, if 90% or 10% of the band intensity in a single gel electrophoresis lane is located above the threshold (usually 10 kbp), the GQN value of this sample is 9 or 1, respectively. Therefore, the above criteria established using a GQN value of 0.3 indicate a minimum requirement of 3% intact total input DNA for successful library preparation of the FFPE sample. The band intensity observed by gel electrophoresis and the GQN value can be measured simply by using the fragment size analyzer (e.g., Bioanalyzer, Agilent, USA).

3.2 Consideration of the Sequencing Depth to Detect low-Frequency Somatic Mutations

Prior to the NGS era, Sanger sequencing was widely used to detect or validate somatic variants, but it is difficult to detect low-level somatic mutations with less than 10–20% allelic frequency in a given tissue sample due to sequencing noise or low resolution (Fig. 1). In the NGS era, deep targeted sequencing over a 1000× read depth is able to detect low-level SNVs even at levels as low as ~1% (Fig. 1). However, in most genetic studies using NGS, samples are sequenced at read depths of 100–150× for whole exomes and 30–60× for whole genomes [41, 42], thereby limiting the potential to detect low-allelic-fraction SNVs in the sample, especially those at less than 5% (Fig. 2). The sensitivity and specificity of calling low-level SNVs depend on several factors, including the depth of the sequence coverage in the affected and matched normal sample, selection of variant callers, sequencing error rate, expected allelic fraction of the mutation, and thresholds used to declare a mutation [25, 43–47]. These various factors contribute to the determination of the sequencing depth to accurately detect low-level SNVs. Nonetheless, it is widely accepted that higher coverage samples are sequenced with greater sensitivity for detecting mutations with low allelic frequencies (Fig. 3). Based on calculations of sensitivity using the mutant allele fraction and the tumor sequencing depth using MuTect, a read depth of >200× is sufficient for detecting an allele frequency of 5 with 99% probability (Fig. 3). Indeed, in our experience, identifying brain somatic mutations with a low frequency in FCM patients suggest that an approximately 300× coverage of deep WES is suitable for detecting ~5% allele frequency SNVs in matched brain and blood (or salvia) samples [15]. Practically, the manufacturer provides an online coverage calculator [48] that calculates the sequencing output needed to reach the desired coverage for a given experiment based on the Lander/Waterman eq. [49]. Because some reads are not mapped to target regions due to the nature of the capture method (off-target read), the on-target coverage is consistently lower than the calculated coverage [50, 51]. Therefore, it is plausible to design a sequencing depth more than approximately 1.5–2 times the expected depth.

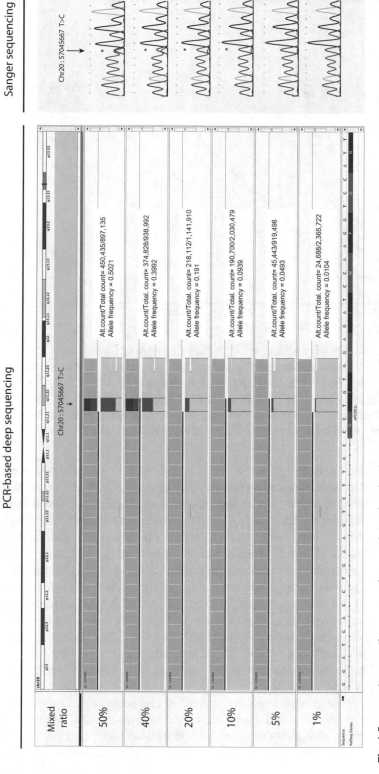

Fig. 1 Deep next-generation sequencing can detect low-level somatic mutations. To generate the test samples with various mutated allele frequency, two genomic DNA containing different genotypes at the Chr20:57,045,667 locus were mixed at various ratios as shown above. This site was amplified by PCR using a targeted primer and sequenced by Sanger or PCR-based amplicon sequencing, respectively. Arrows indicate the target locus determined by Sanger sequencing or integrated genomic view (collapsed mode) of the aligned sequencing file. Asterisks indicate the discernible peak of altered allele observed in Sanger sequencing. The observed allelic frequency of deep sequencing is calculated by dividing the altered allele count by the total allele count. The result shows that low-level SNVs that are not detected by Sanger sequencing can be revealed using PCR-based deep sequencing. The following abbreviation is used: *Alt* altered

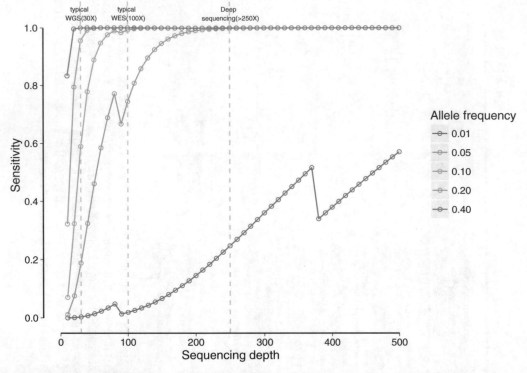

Fig. 2 Sensitivity of the MuTect algorithm according to the allele frequency and sequencing depth. Using a sequencing depth of 30× or 100× sequencing depth (typical depth of the whole genome or exome sequencing), MuTect showed an 18.8% or 74.3% sensitivity for an allele frequency of 5%, respectively. The sensitivity increased to 99.8% with deeper sequencing (250×) in the 5% allele frequency [25]. This result suggests that the appropriate depth of sequencing is critical for determining sensitivity. Therefore, to detect low-level SNV presented as low as 5%, deep sequencing with over 250× reads depth is recommended

3.3 Bioinformatics Analysis of Brain Somatic Mutations

After acquiring the raw data (e.g., fastq file) from deep sequencing, it is necessary to perform quality control (QC) processes to determine whether raw sequencing data are suitable for the subsequent bioinformatics analysis. Several studies and tools have reported that a low base quality, contamination with primer/adapter sequences, and biases in base composition can have harmful effects on downstream analytic processes [52–54]. The initial steps in the QC process typically involve the evaluation of the intrinsic quality of the raw reads using metrics produced by the sequencing platform (e.g., base quality scores) or calculated directly from the raw reads (e.g., base composition). FastQC [29] is one of the most popular tools for the QC of raw sequence reads. The FastQC tool generates representative key metrics that characterize the raw data quality, including the Phred score distribution per base and per base sequence contents, as well as the sequence duplication level. The Phred score distribution per base shows an overview of the range of quality values with a box whisker plot across all sequencing reads at each position in the FastQ file. In general, this metric is a

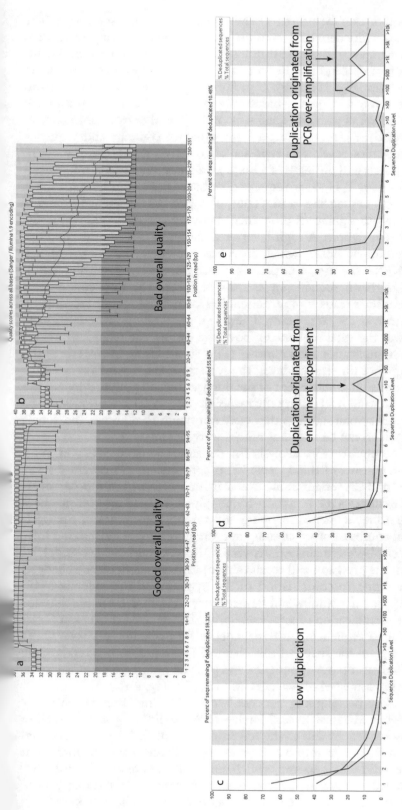

Fig. 3 Representative plots of FastQC assessing the quality of the raw data. (**a**, **b**) The Phred score distribution per base shows an overview of the range of quality values with box whisker plots across all bases of the sequencing reads at each position in the FastQ file. In general, a Phred score 25 is accepted as the cutoff value for determining the overall quality of the data. The data containing good quality sequences show the high Phred score (**a**). In contrast, the base quality of bad data steeply decreases as the run progresses, and thus the box whisker plot descends into the *orange/red* area at the end of a sequencing read (**b**). (**c**) This is an example of library with a low level of duplication. The *blue and red lines* simply show the percentages of the raw and deduplicated libraries, respectively. In this library, percentage of remaining reads after removing the duplication is 59.32%. (**d**) This duplication plot presents an example of raw read from target-enrichment experiment such as panel sequencing targeting small subset of genes. A group of sequences with intermediate duplication (ten duplications) may encompass some repeat sequences of specific genes. Considering the percentage of remaining reads, this intermediated duplication in target-enrichment sequencing may not significantly affect the total duplication level. (**e**) This plot is a representative result of PCR over-amplification. Most of raw reads show a high duplication level (>100 duplications). This PCR over-amplification severely affects the percentage of remaining reads after deduplication, thus eventually lowering the sequencing depth

good marker for evaluating the overall data quality at a glance. The base quality derived from most platforms will gradually decrease as the run progresses, and thus it is common to observe the box whisker plot descend into the orange/red area at the end of a sequencing read (Fig. 15.3a, b). If the base quality near the end of a read is severely low, quality-based clipping should be considered to clip the base call with low quality from the 3' end or from both ends of a read [55]. Next, the per base sequence contents and sequence duplication level are a representative metric for assessing the randomness and diversity of a library. In a random library, four nucleotides (A, T, G, C) are equally distributed across the entire position of the sequencing read. Thus, the per base sequence contents reveal no significant differences between the nucleotide contents of each sample. A high level of duplication is likely due to PCR over-amplification, which increases the total amount of library DNA or the use of targeted capture for enriching specific genes. High-depth sequencing can also increase the duplication level because it leads to the generation of multiple reads starting by chance at the same point. These high duplication levels eventually lower the total sequencing coverage during the mapping process with duplication read removal (Fig. 3c–e). Using the metric of the sequence duplication level, the researcher can obtain a reasonable impression of the extent of the duplication and potential loss of the sequencing read during the mapping process. This tool was used to report several statistics that were not mentioned above, the details of which can be found at official website [29].

Bioinformatics analysis for detecting somatic SNVs in WES data consists of three steps: (1) preprocessing; (2) variant discovery; and (3) variant evaluation (Fig. 4a). Although the Broad institute suggests the mutation-calling pipeline GATK as best practices regarding sequence mapping and variant identification, the optimal combination of aligning and variant calling for detecting low-level SNVs remains controversial. Recent studies have reported that the precision and recall value of variant calling can be changed according to combination of the mapper and the caller [44–47, 56]. Further studies will be necessary to optimize the pipeline for the identification of low-level SNVs. The detailed discussion about the best sequencing analysis pipeline for detecting low-level SNVs is beyond the scope of this chapter. Therefore, in this section, we will focus on our experience and provide brief guidelines for identifying low-level SNVs in FCM [15, 57] (Fig. 4b).

For the aligning process, we utilized the Burrows-Wheeler Aligner-Maximal Exact Matches (BWA-MEM) algorithm to align raw sequence reads to the reference genome. BWA is a widely used read alignment tool based on Burrows-Wheeler transformation of the reference genome, which not only minimizes the memory needed to store the reference but also allows for rapid mapping of short read sequences [58]. Other read aligners based on

Burrows-Wheeler transformation, such as Bowtie and SOAP2, utilize different strategies for mismatches [58–60]. BWA has two different align modes: ALN and MEM. BWA-ALN is the original alignment algorithm of the BWA designed to align reads up to 100 bp in length. The subsequently released BWA-MEM is designed to align reads between 70 bp and 1 Mbp long, and it allows for more flexible read clipping to identify the maximal match for the reference sequence. In comparison to BWA-ALN mode, BWA-MEM rescued a larger number of vector-contaminated reads by active read clipping (Fig. 5). Therefore, we utilized BWA-MEM to align and discriminate contamination from somatic SNVs.

Accurate calling of somatic SNVs with a low allelic frequency has been considered as one of the major challenges in genetics. Early genetic studies using NGS have relied on independent variant calling of each sample in comparison to the reference genome (e.g., hg19), followed by the subtraction of variants in the normal sample from those in the diseased sample to obtain disease-associated somatic SNVs [61]. However, this method of calling SNVs in a single sample was not suitable for detecting variants with low allelic fractions. To overcome this challenge, a number of tools with enhanced accuracy based on different statistical models have been developed. These tools directly compare a tumor-normal samples at each variant locus [24, 25, 62, 63]. Although each of these tools has its own merit, the relative advantages of selecting one algorithm over another are not clear. Therefore, in most studies, a single algorithm is commonly used to call somatic SNVs, and predicted mutations are validated by an orthogonal method such as Sanger sequencing [64, 65]. However, the use of multiple algorithms compensates for algorithm-specific limitations and improves confidence by removing many of the false-positive results produced by data artifacts [15, 43, 56]. We used this approach, thereby combining both MuTect and Virmid algorithms to predict the true SNVs in the FCM sample. We were then able to identify the overlapped SNVs in both algorithms and validate them using other orthogonal methods such as deep amplicon sequencing and laser capture microdissection (LCM) (Fig. 4b). MuTect and Virmid are designed to detect somatic mutations with a low allelic fraction in genetically heterogeneous samples, whereas they are based on different probability models emerging from independent approaches. Selection of the somatic SNV reproducibly called from different detection algorithms helps to increase the validation rate.

The use of multiple algorithms is not sufficient to prioritize the SNV and provides a small subset of candidate SNVs for subsequent functional analysis. Thus, further annotation with a functional prediction score such as SIFT [66], PolyPhen [67], and GERP [68] is necessary. For example, SNVs with a phastCons score and PolyPhen score ≤ 0.5 are unlikely to be disease-causing due to the low conservation and protein damaging scores, respectively [15].

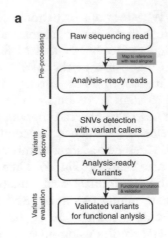

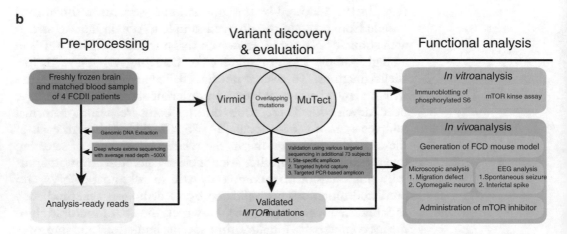

Fig. 4 Schematic presentation of the analysis pipeline for identifying somatic SNVs causative for FCDII. (**a**) Raw sequencing data (e.g., fastq file) were mapped to the reference genomic sequence (e.g., hg19) with short read aligner. Analysis-ready reads were analyzed to detect somatic SNVs using variant callers. The analysis-ready variants were annotated with functional scores such as PolyPhen and prioritized using predetermined criteria. To remove false-positive variants, various validation sequencing should be performed. Finally, these validated variants are ready to be used for in vitro or in vivo functional studies. (**b**) Deep whole exome sequencing (WES) with >500× read depth was performed in matched brain-blood samples from four FCDII patients. The raw sequencing data were analyzed to discover somatic variants in affected brain samples. Overlapping mutations in both the Virmid and MuTect algorithms were selected as candidate single nucleotide variants (SNVs). Subsequently, we performed deep sequencing of the *MTOR* gene with various sequencing platforms in a large FCDII cohort comprising an additional 73 patients. Finally, the validated *MTOR* mutations were subjected to in vitro and in vivo functional analysis. The image of panel B reprinted with permission from [57]

Although various functional annotation methods are useful for prioritizing causal variants, current approaches have some limitations. First, each annotation method has its own metric, and these metrics are hardly comparable, making it difficult to assess the relative importance of variants. Second, there is no consensus for selecting the annotation methods among many of them and determining the cutoff value for filtering. Recently, to overcome these

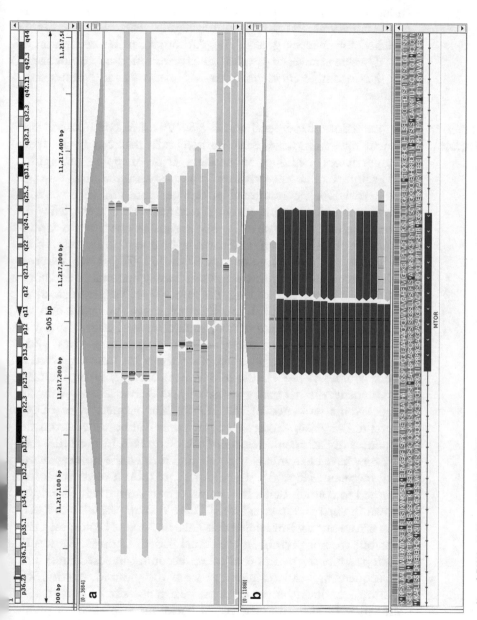

Fig. 5 Alignment results of BWA-ALN and BWA-MEM on the same sample contaminated with the mutant plasmid. (**a**) BWA-ALN discarded most of the contaminated reads (clipped at the exon junction). As a result, it lost mapping signatures of the vector contamination at exon junctions. However, unclipped reads containing the intentionally inserted mutation were fully retained (reads with the red mutation mark). BWA-ALN alignment makes false variants in the mutant plasmid undetectable. (**b**) Unlike BWA-ALN, BWA-MEM successfully rescued most of the contaminated reads in *MTOR* by active read clipping, which is represented by a discrete increase in the read depth at exon junctions. An obvious distortion of the mutated allele in the clipped reads could be observed. All images were reprinted with permission from [30]

limitations, a general framework known as the Combined Annotation-Dependent Depletion (CADD) value was developed as a comprehensive tool that takes into account the results of many known functional prediction scores [69]. CADD is helpful for discriminating causative mutations from non-causative mutations. Indeed, a recent epilepsy study suggested that the group of de novo mutations predicted to be highly deleterious is highly enriched for epilepsy genes [70]. Although further evaluation of CADD value is necessary, we expect that this method compensates for the incompleteness and bias of many existing annotation methods.

3.4 Determination of Tissue and Vector Contamination

The detection of low-level somatic SNVs using NGS-based deep sequencing accompanies false-positive calls that can result from various factors including sequencing error, mapping ambiguity, and imprecise calling algorithms. In addition, even a small number of external DNA contamination can generate critical erroneous calls that mimic genuine somatic variants with a low allelic frequency. Indeed, the cause of genetic anomalies observed in various sequencing data has been confirmed to be due to unexpected contamination of the cell line or viral DNA [71–75]. Thus, external DNA contamination must be evaluated using systematic and computational approaches during the process of variant detection. Cross-contamination of human samples can be evaluated by ContEst, which is a tool used to estimate the level of cross-individual contamination in next-generation sequencing data [76]. External contamination with plasmid vectors has been reported in several studies [77–80]. Plasmid vectors are highly problematic contaminants due to the presence of cDNA inserts harboring the target exonic sequence of the gDNA. The sequences of cDNA inserts in the plasmid vector are not easily distinguished from the sequences of the exonic region of sample gDNA in the NGS data after they have been mixed together and formed the homogeneous read fragment (Fig. 6). In addition, cDNA inserts are often designed to contain deliberate mutations to study their functional roles at in vitro and in vivo levels. These mutant plasmids can generate erroneous variant alleles that have important biological functions but are not present in the actual patient samples. A recently developed bioinformatics tool called Vecuum can search and filter out sequencing reads originating from the plasmid vector [30]. Vecuum identifies the presence and existence site of vector contamination by searching for vector backbone sequences in public vector databases such as UniVec and Addgene. The Vecuum isolate-contaminated reads from the sample read using the intron-less feature of cDNA inserts shows split or unevenly mapped reads at exon junctions (Fig. 6). Finally, Vecuum reports a list of the genomic position of plasmid contamination and false variants based on the statistical analysis and provides cleaned mapping reads by

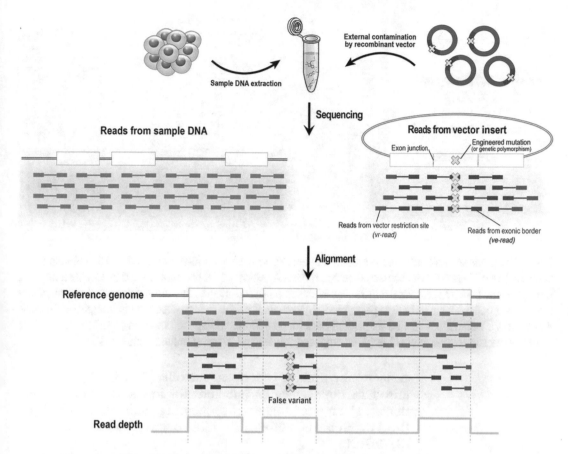

Fig. 6 Schematic representation of erroneous calls originated from the vector contamination. Sequencing data from a contaminated sample contains DNA sequencing reads originating from both the tissue sample and contaminated recombinant vector. *Grey, blue, and red* reads represent sequence from the tissue sample, vector backbone, and coding region of the target gene, respectively. Due to mapping of reads from both samples and vector inserts, engineered mutations in the vector can be observed as low-frequency variants (*yellow marks*). Since recombinant inserts do not include intronic sequences in general, unique mapping patterns are generated by vector-originating reads at exon junctions (clipped and discordant reads), providing a clue to identify the false variants. All images were reprinted with permission from [30]

filtering out the contaminating read from the original sequencing data. Thus, we highly recommend utilizing this tool to identify and filter out contamination sequence from the plasmid vector, especially in laboratories that simultaneously handle plasmid vectors and sample gDNA to generate sequencing data.

3.5 Validation Sequencing Methods of Low-Level Somatic Mutations

False-positive and erroneous variants can arise from various steps of sequencing experiments, including sample and library preparation, sequencing, and data analysis. These errors can be misinterpreted as somatic variants. Therefore, validation sequencing of predicted candidate variants is essential due to stochastic fluctuation and the systematic bias of the sequencing experiment. Recent studies suggest that cross-sequencing platform replication is effective to

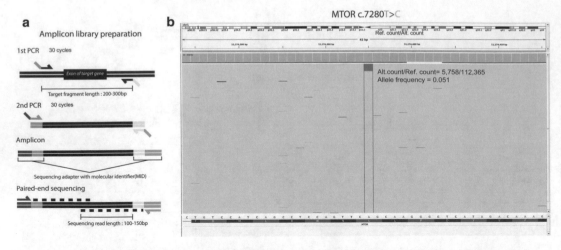

Fig. 7 The process of site-specific amplicon sequencing for validating low-level SNVs. (**a**) Schematic figure showing the PCR-based amplicon sequencing process. A two-step PCR protocol is used to amplify amplicons from genomic DNA that contained a molecular identifier (MID) surrounding the target site. Paired-end sequencing of the amplicon libraries is then performed. (**b**) Variants are visually inspected via the Integrative Genomic Viewer (IGV). Variants are labeled as colored bars (*brown*) in the "collapsed downsampling" mode of IGV to visualize the sequencing reads of targeted amplicon sequencing covering mutated sites of *MTOR*

exclude false-positive and erroneous calls [7, 15, 81]. Here, we introduce three validation methods for low-level somatic mutations: (1) targeted amplicon sequencing with ultra-high depth (read depth > 100,000×), (2) LCM to enrich mutated cells, and (3) SABER.

Unique sequencing artifacts can arise according to the library preparation methods, such as hybrid capture and PCR-based amplicon sequencing [82, 83]. Because most WES is based on the hybrid capture method, validation based on PCR-based amplicon sequencing can discern artifacts arising from library preparation of the hybrid capture method. In addition, site-specific target sequencing using the PCR-based amplicon can provide an extremely high sequencing depth that improves the sequencing accuracy. To generate a library of targeted amplicon sequencing, a two-step PCR protocol is used to amplify amplicons from genomic DNA containing a sequencing adapter and a molecular identifier (MID) surrounding the target region. The adapter and MID sequence can be provided by the manufacturer's protocol. This library is then sequenced on a Miseq or Hiseq sequencer (Illumina, USA) with a high depth (read depth of >100,000×) (Fig. 7a). Using a deep sequencing strategy to discriminate false-positive calls from true SNV, we successfully validated low-level somatic mutations present at ~5% (Fig. 7b).

Current NGS technologies have their own sequencing and imaging errors that mainly originate from the processing of the fluorescent or electrical signal. Each sequencing platform, such as Illumina, Iontorrent, or Pacbio, is known to have its own

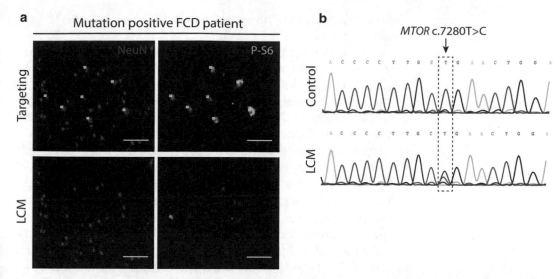

Fig. 8 Laser capture microdissection (LCM) enriches the *MTOR* c.7280 T > C mutant allele in FCD patients. (**a**) NeuN-positive cytomegalic neurons with the increased S6 phosphorylation (Major readout of mTOR activation) were labeled with yellow dot ("Targeting"). Then, the labeled neurons were microdissected ("LCM"). (**b**) Sanger sequencing revealed an enrichment for the *MTOR* c.7280 T > C mutant allele in ~20 of micro-dissected cyto-megalic neurons. The control indicates the use of bulk genomic DNA extracted from brain samples of patients without the enrichment. Scale bars, 100 μm. All images were reprinted with permission from [15]

intrinsic error rate ranging from approximately 0.1–1% [84, 85]. Validation sequencing data using orthogonal technologies such as LCM and SABER can also increase confidence that the identified mutations are true somatic SNVs without errors. Interestingly, FCM samples present dysmorphic enlarged neurons caused by somatic mutations in mTOR pathway genes that regulate neuronal cell growth [15, 86, 87]. Such dysmorphic neurons permit the discrimination of mutation-carrying cells from normal cells. LCM can be used to dissect dysmorphic neurons from brain tissue slices and collect nuclei from the neuron carrying the mutation [88–90]. Using a prepared tissue slide, phosphorylated S6-immunoreactive neurons with an enlarged soma size (n = ~20 per case) were targeted and microdissected using the PALM laser capture system (Carl Zeiss, Germany). They were then collected in AdhesiveCap (Carl Zeiss, Germany) using laser pulse catapult (LPC) mode (Fig. 8a). Genomic DNA was extracted from the collected neurons using the QiAamp microkit LCM tissue protocol (Qiagen, USA), and the target regions were amplified by PCR using targeted primer and high-fidelity DNA polymerase. The amplified PCR product was then purified using a DNA-binding column. The purified PCR product was sequenced using the Sanger method. Using the LCM method, we were able to enrich for the mutation-carrying cells and detect mutant allele peaks in dysmorphic neurons (Fig. 8b). This result indicates that dysmorphic neuron of affected brain in FCM patients harbor the somatic SNVs identified in WES.

SABER is used to restrict primer extension to the mutant-specific allele. It improves the detection sensitivity of the mutation [17, 91]. In the SABER reaction, only a primer annealed to a mutant allele undergoes single-base extension, leading to a mass difference between the original primer and the primer containing the extended allele. In combination with MALDI-TOF mass spectrometry, it can detect a mutation frequency of approximately 0.5% by measuring the mass difference of the extension product [91]. The assay was designed using MassARRAY Assay Design 4.0 software (Sequenom, USA). A schematic diagram of the assay is shown in Fig. 9a. After PCR amplification of the target site, single-base primer extension is performed according to the published protocol [17]. The final products are then spotted on a SpectroChip II (Sequenom, USA) and analyzed using a Compact Mass Spectrometer and MassARRAY Workstation software (Sequenom). Using the SABER method, low-level SNVs were successfully amplified and detected in FCM patients carrying somatic SNVs identified in WES (Fig. 9b).

In summary, somatic SNVs identified in WES of FCM patients can be validated by one of these three orthogonal methods, including PCR-based amplicon sequencing and an enrichment experiment using LCM and SABER. The reproduced results strongly indicate that identified somatic SNVs were in fact present in the patient samples. Together with a computational analysis using NGS technology, wise use of validation methods will provide confidence for the presence of low-level somatic SNVs in FCM patients.

3.6 Generation of an In Vivo Mouse Model Carrying Brain Somatic Mutations

Using advanced sequencing technology and bioinformatics tools, several studies have successfully identified mutations in specific genes related to FCM [15, 87, 92–94]. However, the determination and validation of the biological function of putative mutated genes has been much more challenging. Functional scoring methods based on a computational analysis of structural changes and chemical dissimilarity of the mutated protein have been widely used to assess the pathogenicity or damaging effects of candidate mutations [66, 67, 95]. However, a precise understanding of the biological role of putative genes harboring a mutation should be achieved through in vitro or in vivo experiments. Mutagenesis followed by relevant in vitro biochemical or imaging experiments has been widely accepted for the functional validation of the identified mutations. Recently, in vitro experiments utilizing CRISPR technology or human induced pluripotent stem cells (hiPSCs) have been applied to elucidate the pathologic role of the identified human mutation in the neurodevelopmental disorder. Although in vitro model systems are useful for understanding the molecular

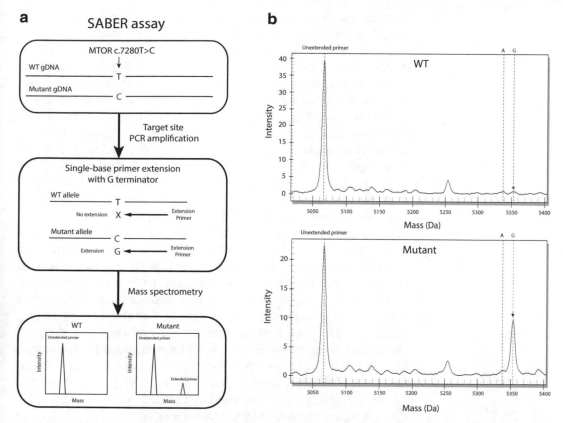

Fig. 9 Schematic presentation and experimental result of the SABER assay for detecting the *MTOR* c.7280 T > C mutation. (**a**) The genomic DNA (gDNA) was first amplified by PCR. The PCR products were then subjected to a single base primer extension. The SABER reaction only included terminators for the mutated nucleotide ("G"), and not the terminators for the wild-type nucleotide("A"). The primer extension products were analyzed using matrix-assisted laser desorption ionization-time of flight mass spectroscopy (MALDI-TOF MS). The mass difference in the single allele between unextended (*black peak*) and extended primers (*red peak*) can be detected. (**b**) An arrow at 5353 Da indicates the detection of the *MTOR* c.7280 T > C mutation. An asterisk indicates a nonspecific background peak. WT, wild type

function of mutated genes, there are obvious limitations of examining cortical layering, behavioral traits (e.g., seizure), or electrographic signatures (e.g., EEG). Therefore, to reveal the biological and pathological functions of the identified somatic variant in the context of the disease phenotype, an in vivo model of somatic brain mutations is necessary.

In utero electroporation (IUE) has been used as an effective tool for manipulating gene expression in a particular cell population of the brain or at specific time points [18, 19]. IUE selectively introduces mammalian-expressing plasmids in the neural progenitor of the target area. Various subpopulations of neural progenitor cells can be manipulated, depending on the stage of the embryo or the location of where the electrical pulse is administered [96]. Therefore, this IUE technique is particularly useful for investigating molecular mechanisms of how mutated genes regulate cortical development, including migration and regional patterning [97, 98].

IUE involves multiple surgical steps from anesthesia to surgical closure. First, timed pregnant mice (usually E12–E15) are anesthetized with isoflurane (0.4 L/min of oxygen and isoflurane vaporizer gauge 3 during surgery). Laparotomy is performed to obtain access to the uterine horns. Using pulled glass capillaries, 2–3 μg of endotoxin-free plasmids with 0.1% of Fast Green (Sigma, USA) are injected into a lateral ventricle of each embryo. A pulse of current is then applied to the head of the embryo by discharging 30–50 V with the electroporator (BTX-Harvard apparatus, USA) to introduce plasmid into neural progenitor cells. The uterine horns are then placed back into the abdominal cavity, and the incision window is sutured using an absorbable fiber such as vicryl (Fig. 10a). Embryonic mice can be screened using a fluorescence reporter such as GFP and harvested for further analysis at later stages. Since the body temperature of the mouse gradually decreases with the exposure time of the uterus and the effect of anesthetization, the mouse should be kept warm on a warmer plate. The duration of surgery critically affects the survival rate of the embryo, and thus the abdominal cavity of the mouse should not be open for more than 30 min [18]. The positioning of the electrodes determines the cortical region of DNA administration. In addition, cortical layer or cell type targeting can be achieved by the timing of electroporation. For example, plasmid DNA electroporated at E12 or E15 is expressed in deep layer neurons (cortical layer 5) or upper layer neurons (cortical layer 2/3), respectively. The electroporation at E15 induces the expression of the introduced plasmid in pyramidal neurons of cortical layers 2/3 but not in glial cells such as astrocytes [99]. Using IUE, we successfully introduced a DNA construct containing the *MTOR* activating mutation, which was confirmed by in vitro immunoblot analysis of pS6 protein, a well-known downstream marker of mTOR activation. Among the many neuropathological features of FCM, three important hallmarks of FCM are migration defects, dysmorphic neurons, and electrographic seizure. Thus, a good mouse model of FCM should display these features. First, we measured the cortical radial migration of reporter-positive neurons at E18 and observed that brain sections expressing the mutant construct showed a significant decrease in reporter-positive cells in the upper layer, consistent with the migration defect (Fig. 10b). After the birth of embryonic mice expressing the mutant construct, we observed that the soma sizes of reporter-positive neurons were markedly increased in the affected cortical regions (Fig. 10c). In addition, this mouse model displayed spontaneous seizures with epileptic discharge after seizure onset. Surprisingly, the behavioral seizures and dysmorphic neurons in our model mouse were almost completely rescued by rapamycin administration, a clinically approved mTOR inhibitor (Fig. 10c, d).

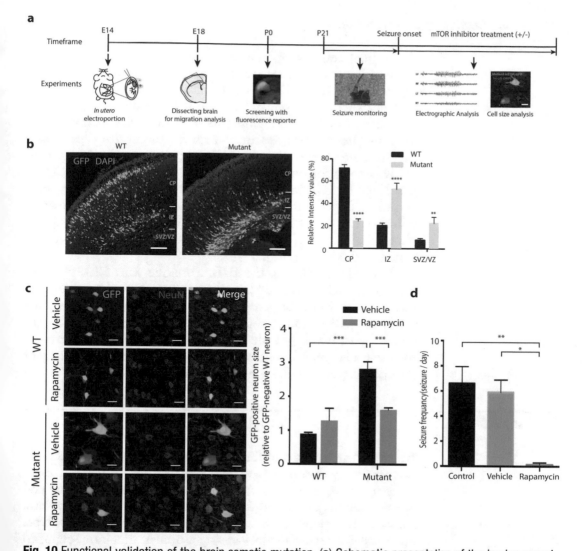

Fig. 10 Functional validation of the brain somatic mutation. (**a**) Schematic presentation of the in vivo experiment to validate the function of identified mutation according to the timeframe. Embryonic mice were electroporated at E14, and then their brains were harvested after 4 day of development (E18) for migration analysis. Next, we selected properly delivered mouse pups at birth (P0), which exhibited GFP signal in the electroporated cortical region. The mice were monitored by video-recording until the tonic-clonic seizures were observed (seizure onset). Then, mice with seizures were then video-EEG monitored to characterize the spontaneous seizure and cell size according to administration of the mTOR inhibitor (rapamycin). EEG signals were recorded from four epidural electrodes located on the left frontal lobe (LF), right frontal lobe (RF), left temporal lobe (LT), and right temporal lobe (RT). Scale bars, 20 μm. (**b**) In utero electroporation of *MTOR* mutants vectors disrupts neuronal migration in the developing mouse neocortex. The images show coronal sections of mouse brains 4 day after electroporation at E14 with wild-type or mutant *MTOR* vectors. The bar charts correspond to the relative fluorescence intensities reflecting the distribution of electroporated cells within the cortex. ** $p < 0.01$, and **** $p < 0.0001$ (relative to wild type, $n = 6-8$, two-way ANOVA with Bonferroni's multiple comparison test). Error bars, s.e.m. CP, cortical plate; IZ, intermediate zone; SVZ/VZ, subventricular and ventricular zone. Scale bars, 100 μm. (**c**) NeuN staining of GFP-positive cells showed an increased soma size of GFP-positive neurons in the affected cortical regions of the mice carrying the mutation. The increased soma size was rescued by rapamycin treatment. *** $p < 0.001$ (relative to GFP negative neurons, $n = 20-263$ per each group. Student's t-test). Scale bars, 20 μm. Error bars, s.e.m. (**d**) The seizure frequency in mice expressing the mutant vector was dramatically reduced by rapamycin treatment. * $p < 0.05$ and ** $p < 0.01$ ($n = 7-17$ for each group, one-way ANOVA with Bonferroni's post test). All images were reprinted with permission from [15]

Animal models of human brain disorders are frequently evaluated according to three criteria: construct, face, and predictive validity [100–102]. An adequate mouse model of a brain disorder should recapitulate the underlying disease mechanism (construct validity), the symptoms of human disease (face validity), and the effective treatment with the target drug (predictive validity). From this perspective, our mouse model is a proper model system for understanding FCM. Our mouse model (1) is generated by IUE introducing the mutant gene into neural progenitor cells recapitulating the development of FCM (construct validity); (2) shows the neuropathological symptoms of patients (face validity); and (3) are rescued by an inhibitor of the target gene (predictive validity).

Recently, combining the IUE technique with CRISPR technology via HDR-mediated genome editing allows for a more precise introduction of a genetic variant onto the human genome [103, 104]. Therefore, we are confident that a model mouse generated by IUE will provide an opportunity to better understand the molecular genetic mechanism of FCM and the connectivity of the epileptogenic brain leading to intractable epilepsy.

4   Conclusion

Recently, the extent and functional role of somatic mutations in noncancerous disorders have received increased attention, especially in FCM. The advancement of sequencing technology and in vivo modeling tools provide a means to investigate the low-level somatic mutation in the context of the disease phenotype. Specifically, a mouse model carrying a human mutation identified by deep sequencing with various replications and validations made it possible to elucidate the pathological role of the somatic mutation in FCM and to provide a potential molecular target for drug treatment.

The combination of deep sequencing, various replication methods, and bioinformatics analysis for detecting somatic SNVs minimize erroneous calls or false variants and provide true somatic SNVs for functional analysis. In our experience, we can successfully detect disease-causing somatic SNVs by utilizing (1) a BWA-MEM algorithm to achieve the maximal match of short read fragments to the reference sequence; (2) consensus approach selecting the overlapping somatic SNVs from different calling algorithms; and (3) validation methods including ultra-high depth reads (depth > 100,000×), LCM to enrich for mutated cells, and the single allele base extension reaction assay. Nevertheless, further studies that optimize the pipeline of bioinformatics analysis and selection of experimental validations with different technologies will be necessary to provide an accurate guideline for detecting true somatic mutations that are related to the disease phenotype.

Regarding the mouse model, IUE is a very powerful tool for modeling FCM. The most useful features of IUE are that it recapitulates the focal nature of the cortical malformation disorder by using temporal and spatial control of plasmid expression in the embryonic brain. Indeed, the mouse model of FCD revealed a very small fraction of reporter-positive cells expressing mutated mTOR, accounting for ~1% of the total neurons. However, imprecise spatial control of gene expression due to manual positioning of the electrode is the most challenging problem related to in utero electroporation. More accurate spatial control can be achieved using a specific promoter sequence that is selectively expressed in target brain areas or triple electrodes, increasing the spatial resolution of the electrical pulse. In future, combining IUE with CRISPR technology allows the generation of a model mouse carrying the same genetic variants of the human disease.

In conclusion, we anticipate that our translational approach to detection of the disease-causing mutation to generate a mouse model recapitulating the disease phenotype will advance our understanding of the genetics, pathogenic mechanism, and therapeutic potential of low-level somatic mutations in FCM. Moreover, this approach has the potential to be extended to other genetic disorders that occur sporadically and provide new insight into the functional role of brain somatic mutations in humans.

Acknowledgments

This work was supported by a grant of the Korean Health Technology R&D Project, Ministry of Health & Welfare, Republic of Korea (H15C3143, HI13C0208, and H16C0415), Citizens United for Research in Epilepsy, the Brain Research Program through the National Research Foundation of Korea (NRF) funded by the Ministry of Science, ICT & Future Planning (2013M3C7A1056564), and the KAIST Future Systems Healthcare Project from the Ministry of Science, ICT and Future Planning. The authors declare that they have no competing interests.

References

1. Lupski JR (2013) Genome mosaicism-one human, multiple genomes. Science 341:358–359. doi:10.1126/science.1239503
2. Lynch M (2010) Evolution of the mutation rate. Trends Genet 26:345–352. doi:10.1016/j.tig.2010.05.003
3. Bozic I, Antal T, Ohtsuki H et al (2010) Accumulation of driver and passenger mutations during tumor progression. Proc Natl Acad Sci U S A 107:18545–18550. doi:10.1073/pnas.1010978107
4. Poduri A, Evrony GD, Cai X, Walsh CA (2013) Somatic mutation, genomic variation, and neurological disease. Science 341:1237758. doi:10.1126/science.1237758
5. Kennedy SR, Loeb LA, Herr AJ (2012) Somatic mutations in aging, cancer and neurodegeneration. Mech Ageing Dev 133:118–126. doi:10.1016/j.mad.2011.10.009
6. Weinstein LS, Shenker A, Gejman PV et al (1991) Activating mutations of the stimulatory G protein in the McCune-Albright

syndrome. N Engl J Med 325:1688–1695. doi:10.1056/NEJM199112123252403

7. Shirley MD, Tang H, Gallione CJ et al (2013) Sturge–weber syndrome and port-wine stains caused by somatic mutation in GNAQ. N Engl J Med 368:1971–1979. doi:10.1056/NEJMoa1213507

8. Lindhurst MJ, Sapp JC, Teer JK et al (2011) A mosaic activating mutation in AKT1 associated with the Proteus syndrome. N Engl J Med 365:611–619. doi:10.1056/NEJMoa1104017

9. Insel TR (2014) Brain somatic mutations: the dark matter of psychiatric genetics? Mol Psychiatry 19:156–158. doi:10.1038/mp.2013.168

10. Barkovich AJ, Guerrini R, Kuzniecky RI et al (2012) A developmental and genetic classification for malformations of cortical development: update 2012. Brain 135:1348–1369. doi:10.1093/brain/aws019

11. Guerrini R (2005) Genetic malformations of the cerebral cortex and epilepsy. Epilepsia 46(Suppl 1):32–37. doi:10.1111/j.0013-9580.2005.461010.x

12. Pang T, Atefy R, Sheen V (2008) Malformations of cortical development. Neurologist 14:181–191. doi:10.1097/NRL.0b013e31816606b9

13. Wong M, Crino PB (2010) mTOR and epileptogenesis in developmental brain malformations. Epilepsia 51:72–72. doi:10.1111/j.1528-1167.2010.02858.x

14. Salamon N (2005) Contralateral hemimicrencephaly and clinical-pathological correlations in children with hemimegalencephaly. Brain 129:352–365. doi:10.1093/brain/awh681

15. Lim JS, Kim W-I, Kang HC et al (2015) Brain somatic mutations in MTOR cause focal cortical dysplasia type II leading to intractable epilepsy. Nat Med 21:395–400. doi:10.1038/nm.3824

16. Nakashima M, Saitsu H, Takei N et al (2015) Somatic mutations in the MTOR gene cause focal cortical dysplasia type IIb. Ann Neurol 78:375–386. doi:10.1002/ana.24444

17. Sakai K, Horiike A, Irwin DL et al (2013) Detection of epidermal growth factor receptor T790M mutation in plasma DNA from patients refractory to epidermal growth factor receptor tyrosine kinase inhibitor. Cancer Sci 104:1198–1204. doi:10.1111/cas.12211

18. Tabata H, Nakajima K (2001) Efficient in utero gene transfer system to the developing mouse brain using electroporation: visualization of neuronal migration in the developing cortex. Neuroscience 103:865–872

19. Tabata H, Nakajima K (2008) Labeling embryonic mouse central nervous system cells by in uteroelec-

troporation. Develop Growth Differ 50:507–511. doi:10.1111/j.1440-169X.2008.01043.x

20. McKenna A, Hanna M, Banks E et al (2010) The genome analysis toolkit: a MapReduce framework for analyzing next-generation DNA sequencing data. Genome Res 20:1297–1303. doi:10.1101/gr.107524.110

21. DePristo MA, Banks E, Poplin R et al (2011) A framework for variation discovery and genotyping using next-generation DNA sequencing data. Nat Genet 43:491–498. doi:10.1038/ng.806

22. McKenna A, Hanna M, Banks E, et al (2011) The genome analysis toolkit: a MapReduce framework for analyzing next-generation DNA sequencing data. https://software.broadinstitute.org/gatk/. Accessed 27 Dec 2016

23. Li H, Durbin R (2009) Fast and accurate short read alignment with Burrows-Wheeler transform. http://bio-bwa.sourceforge.net. Accessed 27 Dec 2016

24. Kim S, Jeong K, Bhutani K et al (2013) Virmid: accurate detection of somatic mutations with sample impurity inference. Genome Biol 14:R90. doi:10.1186/gb-2013-14-8-r90

25. Cibulskis K, Lawrence MS, Carter SL et al (2013) Sensitive detection of somatic point mutations in impure and heterogeneous cancer samples. Nat Biotechnol 31:213–219. doi:10.1038/nbt.2514

26. Kim S, Jeong K, Bhutani K, et al (2013) Virmid: accurate detection of somatic mutations with sample impurity inference. https://sourceforge.net/projects/virmid/. Accessed 27 Dec 2016

27. Cibulskis K, Lawrence MS, Carter SL, et al (2013) Sensitive detection of somatic point mutations in impure and heterogeneous cancer samples. http://archive.broadinstitute.org/cancer/cga/mutect. Accessed 27 Dec 2016

28. Cingolani P, Platts A, Le LW et al (2012) A program for annotating and predicting the effects of single nucleotide polymorphisms, SnpEff. Flying 6:80–92. doi:10.4161/fly.19695

29. Andrews S (2016) FastQC: a quality control tool for high throughput sequence data. http://www.bioinformatics.babraham.ac.uk/projects/fastqc/. Accessed 27 Dec 2016

30. Kim J, Maeng JH, Lim JS et al (2016) Vecuum: identification and filtration of false somatic variants caused by recombinant vector contamination. Bioinformatics 32:3072–3080. doi:10.1093/bioinformatics/btw383

31. Illumina (2016) Illumina adapter sequences document. http://support.illumina.com/

downloads/illumina-customer-sequence-letter.html. Accessed 27 Dec 2016

32. Acinas SG, Sarma- Rupavtarm R, Klepac-Ceraj V, Polz MF (2005) PCR-induced sequence artifacts and bias: insights from comparison of two 16S rRNA clone libraries constructed from the same sample. Appl Environ Microbiol 71:8966–8969. doi:10.1128/AEM.71.12.8966-8969.2005

33. Parkinson NJ, Maslau S, Ferneyhough B et al (2012) Preparation of high-quality next-generation sequencing libraries from picogram quantities of target DNA. Genome Res 22:125–133. doi:10.1101/gr.124016.111

34. Lasken RS, Stockwell TB (2007) Mechanism of chimera formation during the multiple displacement amplification reaction. BMC Biotechnol 7:19. doi:10.1186/1472-6750-7-19

35. Robin JD, Ludlow AT, LaRanger R et al (2016) Comparison of DNA quantification methods for next generation sequencing. Sci Rep 6:1–10. doi:10.1038/srep24067

36. Bhat S, Curach N, Mostyn T et al (2010) Comparison of methods for accurate quantification of DNA mass concentration with traceability to the international system of units. Anal Chem 82:7185–7192. doi:10.1021/ac100845m

37. Simbolo M, Gottardi M, Corbo V et al (2013) DNA qualification workflow for next generation sequencing of Histopathological samples. PLoS One 8:e62692–e62698. doi:10.1371/journal.pone.0062692

38. O' Neill M, McMillan ND, Smith SRP et al (2011) Performance studies on the transmitted light drop Analyser. J Phys Conf Ser 307:012035–012037. doi:10.1088/1742-6596/307/1/012035

39. Gilbert MTP, Haselkorn T, Bunce M et al (2007) The isolation of nucleic acids from fixed, paraffin-embedded tissues–which methods are useful when? PLoS One 2:e537–e512. doi:10.1371/journal.pone.0000537

40. Illumina (2016) Evaluating DNA quality from FFPE samples. 1–4.

41. Consortium TICG, committee E, committee EAP et al (2010) International network of cancer genome projects. Nature 464:993–998. doi:10.1038/nature08987

42. Weinstein JN, Collisson EA, Mills GB et al (2013) The cancer genome atlas pan-cancer analysis project. Nat Genet 45:1113–1120. doi:10.1038/ng.2764

43. Goode DL, Hunter SM, Doyle MA et al (2012) A simple consensus approach improves somatic mutation prediction accuracy. Genome Med 5:90–90. doi:10.1186/gm494

44. Xu H, DiCarlo J, Satya RV et al (2014) Comparison of somatic mutation calling methods in amplicon and whole exome

sequence data. BMC Genomics 15:244. doi:10.1186/1471-2164-15-244

45. Wang Q, Jia P, Li F et al (2013) Detecting somatic point mutations in cancer genome sequencing data: a comparison of mutation callers. Genome Med 5:91. doi:10.1186/gm495

46. Alioto TS, Buchhalter I, Derdak S et al (2015) A comprehensive assessment of somatic mutation detection in cancer using whole-genome sequencing. Nat Commun 6:1–13. doi:10.1038/ncomms10001

47. Roberts ND, Kortschak RD, Parker WT et al (2013) A comparative analysis of algorithms for somatic SNV detection in cancer. Bioinformatics 29:2223–2230. doi:10.1093/bioinformatics/btt375

48. Illumina (2016) Sequencing coverage calculator. http://support.illumina.com/downloads/sequencing_coverage_calculator.html. Accessed 27 Dec 2016

49. Sims D, Sudbery I, Ilott NE et al (2014) Sequencing depth and coverage: keyconsiderations in genomic analyses. Nat Rev Genet 15:121–132. doi:10.1038/nrg3642

50. Ng SB, Buckingham KJ, Lee C et al (2010) Exome sequencing identifies the cause of a mendelian disorder. Nat Genet 42:30–35. doi:10.1038/ng.499

51. Choi M, Scholl UI, Ji W et al (2009) Genetic diagnosis by whole exome capture and massively parallel DNA sequencing. Proc Natl Acad Sci 106:19096–19101. doi:10.1073/pnas.0910672106

52. Leggett RM, Ramirez- Gonzalez RH, Clavijo BJ et al (2013) Sequencing quality assessment tools to enable data-driven informatics for high throughput genomics. Front Genet 4:288. doi:10.3389/fgene.2013.00288

53. Patel RK, Jain M (2011) NGS QC toolkit: a toolkit for quality control of next generation sequencing data. PLoS One 7:e30619–e30619. doi:10.1371/journal.pone.0030619

54. Trivedi UH, Cézard T, Bridgett S et al (2014) Quality control of next-generation sequencing data without a reference. Front Genet 5:111. doi:10.3389/fgene.2014.00111

55. Smeds L, Künstner A (2010) ConDeTri—a content dependent read trimmer for Illumina data. PLoS One 6:e26314–e26314. doi:10.1371/journal.pone.0026314

56. Kim SY, Speed TP (2013) Comparing somatic mutation-callers: beyond Venn diagrams. BMC Bioinformatics 14:189. doi:10.1186/1471-2105-14-189

57. Lim JS, Lee JH (2016) Brain somatic mutations in MTOR leading to focal cortical dysplasia. BMB Rep 49:71–72. doi:10.5483/BMBRep.2016.49.2.010

58. Li H, Durbin R (2009) Fast and accurate short read alignment with Burrows-Wheeler

transform. Bioinformatics 25:1754–1760. doi:10.1093/bioinformatics/btp324

59. Langmead B, Trapnell C, Pop M, Salzberg SL (2009) Ultrafast and memory-efficient alignment of short DNA sequences to the human genome. Genome Biol 10:R25. doi:10.1186/gb-2009-10-3-r25

60. Li R, Yu C, Li Y et al (2009) SOAP2: an improved ultrafast tool for short read alignment. Bioinformatics 25:1966–1967. doi:10.1093/bioinformatics/btp336

61. Pleasance ED, Cheetham RK, Stephens PJ et al (2010) A comprehensive catalogue of somatic mutations from a human cancer genome. Nature 463:191–196. doi:10.1038/nature08658

62. Roth A, Ding J, Morin R et al (2012) JointSNVMix: a probabilistic model for accurate detection of somatic mutations in normal/tumour paired next-generation sequencing data. Bioinformatics 28:907–913. doi:10.1093/bioinformatics/bts053

63. Saunders CT, Wong WSW, Swamy S et al (2012) Strelka: accurate somatic small-variant calling from sequenced tumor-normal sample pairs. Bioinformatics 28:1811–1817. doi:10.1093/bioinformatics/bts271

64. Le Gallo M, O' Hara AJ, Rudd ML et al (2012) Exome sequencing of serous endometrial tumors identifies recurrent somatic mutations in chromatin-remodeling and ubiquitin ligase complex genes. Nat Genet 44:1310–1315. doi:10.1038/ng.2455

65. Shah SP, Roth A, Goya R et al (2012) The clonal and mutational evolution spectrum of primary triple-negative breast cancers. Nature 486:395–399. doi:10.1038/nature10933

66. Ng PC (2003) SIFT: predicting amino acid changes that affect protein function. Nucleic Acids Res 31:3812–3814. doi:10.1093/nar/gkg509

67. Adzhubei IA, Schmidt S, Peshkin L et al (2010) A method and server for predicting damaging missense mutations. Nat Methods 7:248–249. doi:10.1038/nmeth0410-248

68. Cooper GM, Stone EA, Asimenos G et al (2005) Distribution and intensity of constraint in mammalian genomic sequence. Genome Res 15:901–913. doi:10.1101/gr.3577405

69. Kircher M, Witten DM, Jain P et al (2014) A general framework for estimating the relative pathogenicity of human genetic variants. Nat Genet 46:310–315. doi:10.1038/ng.2892

70. Consortium E, Project EPG (2013) De novo mutations in epileptic encephalopathies. Nature 501(7466):217–221. doi:10.1038/nature12439

71. Xu B, Zhi N, Hu G et al (2013) Hybrid DNA virus in Chinese patients with seronegative hepatitis discovered by deep sequencing. Proc Natl Acad Sci 110:10264–10269. doi:10.1073/pnas.1303744110

72. Naccache SN, Hackett J, Delwart EL, Chiu CY (2014) Concerns over the origin of NIH-CQV, a novel virus discovered in Chinese patients with seronegative hepatitis. Proc Natl Acad Sci 111:E976–E976. doi:10.1073/pnas.1317064111

73. Hué S, Gray ER, Gall A et al (2009) Disease-associated XMRV sequences are consistent with laboratory contamination. Retrovirology 7:111–111. doi:10.1186/1742-4690-7-111

74. Kjartansdóttir KR, Friis- Nielsen J, Asplund M et al (2015) Traces of ATCV-1 associated with laboratory component contamination. Proc Natl Acad Sci 112:E925–E926. doi:10.1073/pnas.1423756112

75. Cantalupo PG, Katz JP, Pipas JM (2015) HeLa nucleic acid contamination in the cancer genome atlas leads to the misidentification of human papillomavirus 18. J Virol 89:4051–4057. doi:10.1128/JVI.03365-14

76. Cibulskis K, McKenna A, Fennell T et al (2011) ContEst: estimating cross-contamination of human samples in next-generation sequencing data. Bioinformatics 27:2601–2602. doi:10.1093/bioinformatics/btr446

77. Tao ZY, Sui X, Jun C et al (2015) Vector sequence contamination of the plasmodium vivax sequence database in PlasmoDB and in silico correction of 26 parasite sequences. Parasit Vectors 8:318. doi:10.1186/s13071-015-0927-x

78. Tang KW, Mahabadi BA, Samuelsson T et al (2013) The landscape of viral expression and host gene fusion and adaptation in human cancer. Nat Commun 4:2513. doi:10.1038/ncomms3513

79. López- Ríos F, Illei PB, Rusch V, Ladanyi M (2004) Evidence against a role for SV40 infection in human mesotheliomas and high risk of false-positive PCR results owing to presence of SV40 sequences in common laboratory plasmids. Lancet 364:1157–1166

80. Borst A, Box ATA, Fluit AC (2004) False-positive results and contamination in nucleic acid amplification assays: suggestions for a prevent and destroy strategy. Eur J Clin Microbiol Infect Dis 23:289–299. doi:10.1007/s10096-004-1100-1

81. Robasky K, Lewis NE, Church GM (2013) The role of replicates for error mitigation in next-generation sequencing. Nat Rev Genet 15:56–62. doi:10.1038/nrg3655

82. Costello M, Pugh TJ, Fennell TJ et al (2013) Discovery and characterization of artifactual mutations in deep coverage targeted capture sequencing data due to oxidative DNA dam-

age during sample preparation. Nucleic Acids Res 41:e67–e67. doi:10.1093/nar/gks1443

83. Schirmer M, Ijaz UZ, D' Amore R et al (2015) Insight into biases and sequencing errors for amplicon sequencing with the Illumina MiSeq platform. Nucleic Acids Res 43:e37–e37. doi:10.1093/nar/gku1341

84. GLENN TC (2011) Field guide to next-generation DNA sequencers. Mol Ecol Resour 11:759–769. doi:10.1111/j.1755-0998.2011.03024.x

85. Fox EJ, Reid- Bayliss KS, Emond MJ, Loeb LA (2014) Accuracy of next generation sequencing platforms. Next Gener Seq Appl. doi:10.4172/jngsa.1000106

86. Crino PB (2011) mTOR: a pathogenic signaling pathway in developmental brain malformations. Trends Mol Med 17:734–742. doi:10.1016/j.molmed.2011.07.008

87. Lee JH, Huynh M, Silhavy JL et al (2012) De novo somatic mutations in components of the PI3K-AKT3-mTOR pathway cause hemimegalencephaly. Nat Genet 44:941–945. doi:10.1038/ng.2329

88. Espina V, Wulfkuhle JD, Calvert VS et al (2006) Laser-capture microdissection. Nat Protoc 1:586–603. doi:10.1038/nprot.2006.85

89. Lutz HL, Marra NJ, Grewe F et al (2016) Laser capture microdissection microscopy and genome sequencing of the avian malaria parasite, plasmodium relictum. Parasitol Res 115:4503–4510. doi:10.1007/s00436-016-5237-5

90. Shapiro E, Biezuner T, Linnarsson S (2013) Single-cell sequencing-based technologies will revolutionize whole-organism science. Nat Rev Genet 14:618–630. doi:10.1038/nrg3542

91. Ding CM, Chiu R, Lau TK et al (2004) MS analysis of single-nucleotide differences in circulating nucleic acids: application to noninvasive prenatal diagnosis. Proc Natl Acad Sci U S A 101:10762–10767. doi:10.1073/pnas.0403962101

92. Poduri A, Evrony GD, Cai X et al (2012) Somatic activation of AKT3 causes hemispheric developmental brain malformations. Neuron 74:41–48. doi:10.1016/j.neuron.2012.03.010

93. Mirzaa GM, Campbell CD, Solovieff N et al (2016) Association of MTORMutations with developmental brain disorders, including megalencephaly, focal cortical dysplasia, and pigmentary mosaicism. JAMA Neurol 73(7): 836–845. doi:10.1001/jamaneurol.2016.0363

94. Jamuar SS, Lam A-TN, Kircher M et al (2014) Somatic mutations in cerebral cortical malformations. N Engl J Med 371:733–743. doi:10.1056/NEJMoa1314432

95. Davydov EV, Goode DL, Sirota M et al (2010) Identifying a high fraction of the human genome to be under selective constraint using GERP. PLoS Comp Biol 6:e1001025–e1001013. doi:10.1371/journal.pcbi.1001025

96. Maschio MD, Ghezzi D, Bony G et al (2012) High-performance and site-directed in utero electroporation by a triple-electrode probe. Nat Commun 3:960–911. doi:10.1038/ncomms1961

97. Takahashi M, Sato K, Nomura T, Osumi N (2002) Manipulating gene expressions by electroporation in the developing brain of mammalian embryos. Differentiation 70:155–162. doi:10.1046/j.1432-0436.2002.700405.x

98. Fukuchi-Shimogori T (2001) Neocortex patterning by the secreted signaling molecule FGF8. Science 294:1071–1074. doi:10.1126/science.1064252

99. Molyneaux BJ, Arlotta P, Menezes JRL, Macklis JD (2007) Neuronal subtype specification in the cerebral cortex. Nat Rev Neurosci 8:427–437. doi:10.1038/nrn2151

100. Belzung C, Lemoine M (2011) Criteria of validity for animal models of psychiatric disorders: focus on anxiety disorders and depression. Biol Mood Anxiety Disord 1:9. doi:10.1186/2045-5380-1-9

101. Willner P (1984) The validity of animal models of depression. Psychopharmacology 83:1–16. doi:10.1007/BF00427414

102. Grone BP, Baraban SC (2015) Animal models in epilepsy research: legacies and new directions. Nat Neurosci 18:339–343. doi:10.1038/nn.3934

103. Mikuni T, Nishiyama J, Sun Y et al (2016) High-throughput, high-resolution mapping of protein localization in mammalian brain by in vivo genome editing. Cell 165:1803–1817. doi:10.1016/j.cell.2016.04.044

104. Kalebic N, Taverna E, Tavano S et al (2016) CRISPR/Cas9-induced disruption of gene expression in mouse embryonic brain and single neural stem cells in vivo. EMBO Rep 17:338–348. doi:10.15252/embr.201541715

Chapter 16

Using Fluorescence In Situ Hybridization (FISH) Analysis to Measure Chromosome Instability and Mosaic Aneuploidy in Neurodegenerative Diseases

Julbert Caneus, Antoneta Granic, Heidi J. Chial, and Huntington Potter

Abstract

Chromosome instability is a form of genomic instability that leads to cells with an abnormal number of chromosomes, defined as aneuploidy. Aneuploidy that results from chromosome instability can be complete or mosaic, depending on whether all or only some of the cells that make up an organism have an abnormal number of chromosomes. Aneuploidy is associated with many human conditions, such as cancer and Down syndrome (DS, trisomy 21), and it has more recently become a focus of investigation in neurodegenerative diseases, including Alzheimer's disease (AD), Niemann-Pick C1 (NPC), and frontotemporal lobar degeneration (FTLD). In these disorders, aneuploid cells in affected brain regions appear to contribute significantly to apoptosis and neurodegeneration, and may thus underlie the associated cognitive deficits. Herein, we describe the methods that our laboratory has developed to analyze the frequency of chromosome instability (i.e., mosaic aneuploidy) in AD, NPC, and FTLD and associated cell death. Our goal is to provide the reader with guidelines for using these methods and to offer insights into their utility and potential limitations.

Key words Neurodegenerative disease, Alzheimer's disease (AD), Frontotemporal lobar degeneration (FTLD), Frontotemporal dementia (FTD), Niemann-Pick C1 (NPC), Mosaic aneuploidy, Sporadic disease, Fluorescent in situ hybridization (FISH), Metaphase chromosome spread, Apoptosis, Single-cell sequencing

1 Introduction: Chromosomal Instability and Neurodegenerative Disease

Alzheimer's disease (AD) is the most common age-associated cognitive disorder. Pathologically, AD is characterized by several distinct hallmarks, most notably the formation of extracellular amyloid deposits and intracellular neurofibrillary tangles. AD can be either late-onset sporadic (90% of reported cases with no identifiable causes), which usually affects people above the age of 65, or familial, which mostly leads to early-onset AD and is inherited in an autosomal dominant manner. Familial AD (FAD) is caused by mutations in the genes that encode the amyloid precursor protein

José María Frade and Fred H. Gage (eds.), *Genomic Mosaicism in Neurons and Other Cell Types*, Neuromethods, vol. 131, DOI 10.1007/978-1-4939-7280-7_16, © Springer Science+Business Media LLC 2017

(APP) or the Presenilin-1 and -2 (PSEN1 and PSEN2) proteins, or by duplication of the APP gene [1–7].

The APP gene resides on human chromosome 21 and encodes the APP transmembrane protein. Under normal conditions, APP is proteolytically cleaved by alpha (α)- and beta (β)- secretase (BACE) to produce soluble APP (sAPP) peptides, which are then released into the extracellular space [8]. Alternatively, APP can be cleaved by β-secretase, followed by gamma (γ)-secretase within the membrane to generate multiple different isoforms of beta-amyloid (Aβ) peptides, including Aβ-42 and Aβ-40, the main components of the amyloid plaques found in the brains of AD patients [9]. In AD, mutations in the Presenilin genes, predominantly in PSEN1, disrupt the catalytic activities of γ-secretase, leading to abnormal cleavage of the APP protein and a relative increase in the production of the Aβ-42 peptides [10, 11]. As a result, the ratio of Aβ-42/Aβ-40 has been reported to increase sharply in AD, leading to the formation of amyloid plaques in the AD brain [9].

The finding of mutations in the APP and PSEN genes that cause FAD indicated that the product of γ-secretase and BACE cleavage of APP, Aβ, must play a key role in AD pathogenesis. Indeed, the accumulation of Aβ-oligomers and their subsequent further polymerization and aggregation outside of cells to form senile plaques contributes to neuronal dysfunction and cell death, an integral part in the development of dementia in AD [12, 13]. Neuronal loss in the brain is important, because increasing neuronal loss correlates with the development of dementia from the early stage to late stage of AD [14, 15].

Aside from amyloid plaque formation, Aβ initiates numerous other neurotoxic events through its effects on downstream proteins or pathways. In particular, Aβ-42 has been shown to promote hyperphosphorylation of the microtubule-associated protein tau (MAPT) that binds to and stabilizes microtubules, whose stability and integrity are necessary for numerous cellular processes and for neuronal cell survival [16]. In AD, Tau phosphorylation and aggregation result in the formation of paired helical filaments and the accumulation of intracellular neurofibrillary tangles [17, 18].

Genomic instability, which includes both abnormal chromosome number (aneuploidy) and structural chromosome aberrations, has been an area of intense investigation for decades in cancer research due to the fact that somatic genomic instability, especially aneuploidy, is a hallmark of most cancerous cells [19–21]. Furthermore, many human developmental disorders, such as Down syndrome (DS, complete or mosaic trisomy 21) and Turner's syndrome (complete or partial monosomy X), arise due to chromosome mis-segregation during gametogenesis and the development of an aneuploid individual.

Because trisomy 21/DS invariably leads to AD pathology by age 30–40 and usually to AD dementia by age 60, due to the fact that the APP gene resides on chromosome 21, we proposed that both familial and sporadic AD might be caused by random chromosome mis-segregation events that occur over the course of an individual's lifetime, leading to mosaic aneuploidy that includes trisomy 21 [22]. We and others then tested this hypothesis **that AD is a mosaic form of DS**, and found that up to 10% of cells throughout the bodies of individuals with AD, including in the brain, were trisomic for chromosome 21, and that many other cells were aneuploid for other chromosomes [14, 23–31] (see Chaps. 1–3 and 5). Furthermore, many cells become tetraploid, especially in the cerebral cortex, during normal aging [32, 33] (see Chap. 4).

Our studies of mouse and cell models of AD revealed that mitotic spindle abnormalities and chromosome mis-segregation are caused by mutant forms of APP and presenilin (the active component of the γ-secretase enzyme that is required for cleaving amyloid Aβ peptide from APP) that cause FAD [1, 3]. Mechanistically, we found that chromosome mis-segregation and consequent aneuploidy was caused by the inhibition of certain microtubule-dependent motors, including Kinesin-5/Eg5, by the Aβ peptide [34]. Kinesin-5 is a double-ended motor protein that attaches to and moves along antiparallel microtubules toward their plus-ends, using the hydrolysis of ATP to produce the mechanical forces required to slide antiparallel microtubules past each other during cell division [35–37]. In addition to driving microtubule movement, Kinesin-5 also promotes microtubule polymerization and stability [38].

These studies resonate with other lines of evidence that have indicated that certain aspects of the cell cycle are dysregulated in the brains of individuals with AD, whether sporadic or familial [15, 39, 40]. Indeed, cell cycle defects and aneuploidy have also been found in other neurodegenerative diseases, besides AD, specifically Ataxia Telangectasia [41], Niemann-Pick C1 (NPC) [42], Lewy body disease [43], and frontotemporal lobar degeneration (FTLD) ([44–46]; Granic and Caneus et al., manuscript submitted), many of which, such as AD, are associated with atrophy of the affected brain regions as a result of neuronal dysfunction and subsequent cell death. Each of those disorders has distinct pathological and clinical characteristics caused by perturbations in different pathways, yet, despite their differences, AD, FTLD, and NPC share common features, including tauopathy, the presence of aneuploid brain cells, and cognitive impairment and/or cognitive decline associated with neuronal cell loss. Interestingly, normal aging can also lead to neuronal aneuploidy [25] (see Chap. 3).

Just as Aβ can inhibit Kinesin-5 to cause chromosome mis-segregation, Tau/MAPT and/or related microtubule-binding proteins are also vital for proper mitotic spindle assembly and cell division [47]. As a result, alterations in Tau function or expression (e.g., in FTLD patients with MAPT mutations or in *Drosophila* expressing human MAPT) have been shown to alter normal chromosome segregation, resulting in aneuploidy and subsequent neuronal dysfunction and cell death ([44–46]; Granic and Caneus et al., manuscript submitted). Indeed, studies have shown that neuronal loss and synapse loss are highly correlated with tau pathology and/or tangle levels in animal models of AD and in the brains of AD patients [17, 46, 48–51]. The fact that both Kinesin-5 and Tau/MAPT are microtubule-associated proteins that are also downstream targets of Aβ, and that mutations in MAPT can cause FTLD, combined with the finding that aneuploid cells that accumulate in AD and other neurodegenerative diseases are highly unstable and are prone to cell death, provides a potential common pathophysiology in which abnormal production and accumulation of different proteins can disrupt proper chromosome segregation, thereby resulting in aneuploidy.

Could the chromosome mis-segregation and aneuploidy in neurons and other cells that characterize many neurodegenerative diseases underlie or contribute to neurodegeneration? Certainly, genomic instability (aneuploidy) can stimulate cell death in many circumstances [52]. Specifically, aneuploid cells can either progress through mitosis to cytokinesis or arrest in the G1/2 phases. Cells arrested in the G1/2 phases of the cycle can subsequently become highly unstable and prone to degeneration, thereby causing them to undergo apoptosis [53–57]. Indeed, Thomas Arendt and colleagues (2010) presented convincing evidence that 90% of the neurodegeneration observed at autopsy of AD brains can be attributed to the generation and selective death of aneuploid neurons [14] (see Chap. 5), which, if extended to other neurodegenerative diseases, suggests that aneuploidy and consequent apoptosis underlie the cognitive deficits that these disorders share.

In sum, our results reveal an increased level of chromosomal aneuploidy, especially for chromosome 21, both in human subjects with neurodegenerative diseases and in their transgenic mouse and cell culture model counterparts. Because the methods we used to acquire these results can be applied to the study of other neurodegenerative diseases, we discuss them in detail below.

2   Materials

2.1   Single-Cell Suspensions of Brain Samples from Patients with AD, NPC, or FTLD

To investigate the frequency of genomic instability in neurodegenerative diseases, frozen samples of brain tissues from patients diagnosed with AD ([1]; Caneus and Granic et al., unpublished observation), NPC [42], or FTLD (Granic and Caneus et al., manuscript submitted) were obtained from the Mayo Clinic Brain Bank, Jacksonville, FL, the NICHD Brain and Tissue Bank for Developmental Disorders at the University of Maryland, Baltimore, MD, and the Banner Sun Health Research Institute Brain and

Body Donation Program of Sun City, AZ, and were assessed for aneuploidy. Single-cell suspensions were prepared with the samples and were processed for chromosomal analysis by fluorescence in situ hybridization (FISH) with labeled probes that recognize chromosome 12 or 21, followed by immunocytochemistry with anti-NeuN antibodies to identify neurons. Specifically, we used a dual color probe set with a chromosome 12 probe labeled with SpectrumGreen and a chromosome 21 probe labeled with Spectrum Orange (LSI TEL/AML1 ES Dual Color Translocation Probe, Abbott Molecular); for probe maps and specification, please refer to manufacturer's website at https://www.molecular.abbott. To specifically label neurons, we used an Alexa Fluor®488-conjugated anti-NeuN antibody (MAB377X, Millipore), which recognizes the neuron-specific NeuN protein that is expressed in neuronal cell nuclei. Using a Zeiss AxioObserver Z1 fluorescence microscope equipped with the following filter sets (excitation/emission, nm): DAPI (358/461), AlexaFluor®488 (460/520), and Texas Red (595/615), the cells were then analyzed for chromosomal aneuploidy using either a 40×/1.4 oil objective or a 63×/1.4 oil objective. Statistical analyses of aneuploidy levels were measured using standard programs and software, including Microsoft Excel, GraphPad Prism, and MathWorks.

2.2 Transgenic Mouse Models of AD

To assess the prevalence of chromosomal aneuploidy in animal models, we used transgenic mouse models of AD that express mutant forms of two human genes, APP and PSEN1, associated with early-onset FAD. In one set of experiments, we used transgenic mice expressing the human APP gene harboring the London (V717F) mutation (19–21 months of age) or mice lacking the APP gene (3 months of age) [3]. For analysis of PSEN1, we used transgenic mice expressing human PSEN1 harboring either the M146L or M146V mutation, transgenic mice overexpressing the wild-type human PSEN1 gene under the control of the PDGF promoter (14–19 months of age, acquired from Dr. Karen Duff, Nathan Klein Institute/New York University School of Medicine), or homozygous knock-in transgenic mice with the mutant human PSEN1 (M146V) gene replacing the endogenous mouse gene (12–15 months of age, a gift from Drs. Mark Mattson and Steven Chan of the National Institute on Aging). For comparative analyses, aneuploidy levels from both groups of transgenic mice (APP and PSEN1) were compared to their respective non-transgenic littermates. Both the brains and the spleens were harvested from these mice and either processed for FISH analysis or for further primary cell culture experiments (see Sect. 2.3). The brains were processed and fixed immediately for FISH analysis using a bacterial artificial chromosome (BAC) probe designed to hybridize to a region of mouse chromosome 16 that is syntenic to human chromosome 21 and for indirect immunofluorescence staining with the anti-NeuN antibody to identify neurons. The BAC probe (provided by Dr. Bruce Lamb at Case Western Reserve University) consists of a mouse chromosome 16-specific sequence labeled with spectrum green dUTP (Vysis) by nick translation (Roche) [58].

2.3 Primary Mouse Cell Cultures

To study mosaic aneuploidy in a defined in vitro system free from other biological factors, we used primary mouse cell cultures in our in vitro experiments. For the primary cell culture experiments, we used three different cell types: mouse brain neurons from mouse models of AD [1], neurosphere cultures and derived neuronal precursor cells (mNPCs) from non-transgenic mice [42], and splenocytes from mouse models of AD [1]. The mouse primary neuronal cultures were prepared from whole brains using a modified version of the method of Liesi et al. [59]. The mouse neurosphere cultures were established from non-transgenic prenatal mouse brains (E17–E18) following a modified protocol by Pacey et al. (2006) [60] and were maintained according to the procedure developed by Marchenko and Flanagan (2007) [61]. The spleens harvested from the transgenic mice were dissociated into single-cell suspensions and cultured in the presence of β-mercaptoethanol and Concanavilin-A (Sigma). Following incubation and growth, the cells were harvested and processed for chromosome counts by both FISH and chromosome metaphase spread karyotyping.

2.4 Human Cell Lines

In addition to primary mouse cell cultures, we have also used several different types of human cell lines in our studies of mosaic aneuploidy. They include primary fibroblast cell lines established from AD patients and NPC patients and karyotypically normal human cells lines, such as the hTERT-HME1 immortalized human mammary epithelial cell line (Clontech) [1, 3, 34, 42] and the HASM human primary aortic smooth muscle cells that were isolated from healthy human aorta and cryopreserved for secondary cell culture (ScienCell Research Laboratories) [42]. The primary fibroblast cell lines from AD patients contained FAD mutations in the APP gene and were from M. Benson (Indiana University Medical School, Indianapolis), and the age-matched control primary fibroblast cell lines were from the NIA Aging Cell Repository, Camden, NJ [23]. The fibroblast cells used for analysis of chromosome instability in association with NPC disease pathogenesis include LDL receptor-negative human skin fibroblasts with two mutations (*C240-F* and *Y160-ter*) that cause a severe form of familial hypercholesterolemia, fibroblasts with functional LDL receptor, four different fibroblast cell lines harboring NPC mutations, and age-matched controls, which were purchased from Coriell Cell Repositories [42]. The fibroblast cell lines were cultured in Minimum Essential Medium (MEM; with Eagle-Earle salts and non-essential AA) (Gibco/Invitrogen) supplemented with serum according to the manufacturer's recommended protocol [42].

The hTERT-HME1 cell line is a telomerase-immortalized primary human mammary epithelial cell line (Clontech). The stable expression of telomerase enables the cells to retain the full length of their telomeres and to divide indefinitely while retaining normal function, phenotype, and karyotype, which is important for studies of mosaic aneuploidy. The cells were cultured in Mammary Epithelium Basal Medium (MEBM, Lonza) supplemented with a growth factor MEGM kit (MEGM SingleQuot Kit Suppl. & Growth Factors, Lonza). The HASM cells were cultured in smooth muscle cell medium (SMCM; ScienCell Research Laboratories) consisting of 500 mL basal media, 2% fetal bovine serum (FBS), 5 mL smooth muscle cell supplement, and 5 mL 1× penicillin/streptomycin (PS; 5000 IU/mL each; Cellgro) during the growth process.

To generate an AD-like cell line model, the hTERT-HME1 cells were transiently transfected with plasmids for expressing mutant APP [3] or PSEN1 genes [1]. APP expression constructs used include the pcDNA3.1 vector containing the human APP gene harboring the Swedish (K595N/M596L) and London (V642I) FAD mutations, or the pAG3 vector containing the human APP gene harboring the V717I FAD mutation (provided by Chad Dickey, USF, Tampa, and Todd Golde, Mayo Clinic, Jacksonville). The PSEN1-containing plasmid constructs include the pCDNA3.1 vector (Clontech) containing either the wild-type human PSEN1 gene or the human PSEN1 gene harboring the M146L or M146V FAD mutation [1].

2.5 Aneuploidy Caused by Aβ-Mediated Inhibition of Motor Proteins

To examine the effect of beta-amyloid (Aβ) on the cell cycle, and more specifically on the function of Kinesin-5, we used several biological systems, including the hTERT-HME1 cell line and *Xenopus* egg extracts [34]. Kinesin-5 is a homotetrameric microtubule motor protein that cross-links overlapping antiparallel microtubules and slides them apart as it steps toward the plus-ends of each of the two microtubules it binds [62]. Kinesin-5 also regulates microtubule polymerization and is required for spindle formation and accurate chromosome segregation [63, 64].

The *Xenopus* egg extract we used in our experiments is a cell-free extract prepared from eggs of the South African clawed toad, *Xenopus laevis*. This extract is capable of undergoing the same key transitional steps in vitro that eukaryotic cells undergo in vivo during the cell cycle. For example, when added to the egg extract, DNA can be assembled together into a chromosome-like aggregate, which then progresses through subsequent phases of mitosis, including anaphase, whereby sister chromatids are separated [34, 65]. These distinctive features make the *Xenopus* egg extract a very unique and suitable system for studying cell cycle dynamics at a biochemical level.

3 Methods

3.1 Analysis of Chromosome Instability and Mosaic Aneuploidy in Human Fibroblast Cell Cultures and Brain Cells from Patients with AD, NPC, or FTLD

To assess aneuploidy levels in patients with AD, NPC, or FTLD, tissues samples were obtained and processed for FISH and chromosome metaphase spread karyotype analyses.

3.1.1 Fibroblasts from AD and NPC Patients

The fibroblast cells obtained from patients with AD [23] or NPC [42] were plated and grown directly on clean, non-coated glass slides. The cells were maintained in DMEM (Dulbecco's Modified MEM) supplemented with 2 mM L-glutamine and 10% fetal bovine serum (FBS); alternatively, RPMI1640 and alpha-MEM media can also be used for fibroblast cell cultures. The cells were placed in an incubator and cultured at 37 °C, 5% CO_2 for 48 h. After the growth period, the cells were trypsinized and harvested. To dislodge and fix the adherent cells from the dish, the old media was removed and the cells were washed with PBS. The cells were then treated with 5 mL of 0.05% trypsin/EDTA and incubated for 5 min at 37 °C in 5% CO_2. Following the incubation time, after all the cells had detached, an equal volume (5 mL) of complete growth media was added to the cell suspension to inactivate the trypsin and prevent cell death. Depending on the cell line, the incubation time with the trypsin can also be adjusted. The cell suspension was collected and centrifuged at 1500 rpm ($\sim 478 \times g$) for 5 min at 4 °C. After removing the supernatant by aspiration, the cells were swollen by adding 5 mL of hypotonic (75 mM) potassium chloride to the cell pellet, and the resuspended cells were incubated for 15 min in a 37 °C water bath. The cells were then removed from the water bath, and 0.5 mL of ice-cold fixative solution (methanol:acetic acid, 3:1) was added to the cell suspension drop by drop while gently flicking the tube, followed by a slow tube inversion back and forth and 1 min incubation on ice. The fixed cells were then centrifuged at 1500 rpm, 4 °C for 5 min. After removing the supernatant, the cells were resuspended in 5 mL of ice-cold fixative solution (added drop by drop) to form a single-cell suspension. The single-cell suspension was placed on ice for at least 30 min and then centrifuged for 5 min at 1500 rpm. The last two steps were repeated one more time, and the supernatant was replaced with fresh fixative solution, and the fixed cells were either used for FISH analysis, or storage at −20 °C. For subsequent FISH, the fixative amount was determined based on cell number and pellet size, and for storage ≥5 mL of fresh fix was used.

3.1.2 Cell Suspensions of Brain Tissues from Patients with Neurodegenerative Diseases

In addition to analyzing fibroblasts, brain samples from patients with AD, NPC, or FTLD ([42]; Caneus and Granic et al., unpublished observation) were also analyzed for aneuploidy by FISH. Brain samples were dissociated into a single-cell/nuclei suspension by trituration. To accomplish this, a piece of the brain tissue (~1.0–1.5 g) was cut and placed in 5 mL of ice-cold PBS in a 15 mL tube. Using 5 mL and 1 mL pipettes, the tissue was pipetted up and down multiple times (~20–30 times) until it was disassembled into smaller pieces. The mixture was further dissociated into a single-cell/nuclei suspension by pipetting up and down around 20–30 times using fire-polished Pasteur pipettes with at least two different reduced tip sizes. The solution was then placed on ice for 5 min to allow the larger pieces of the tissue and/or other cellular debris to settle down to the bottom of the tube. The single-cell suspension in the top layer was then collected and transferred to a clean 15 mL tube. Fresh ice-cold fixative (methanol:acetic acid, at least 3 mL or more, depending on the pellet size) solution was added to the cell suspension drop by drop, and the tube was placed in a 74 °C water bath for 30 min, followed by centrifugation at 1500 rpm at 4 °C, for 5 min. The supernatant was removed and discarded; the cell pellet was resuspended in fresh ice-cold fixative solution, which is ready for FISH assays or can be stored at −20 °C.

3.1.3 Fluorescence in Situ Hybridization (FISH) Assays with Human Fibroblasts and Brain Cell Suspensions

To carry out the FISH analysis, slight modifications were made to the protocol provided by the probe manufacturer (Vysis, Abbott Molecular), as described previously [23]. For freshly prepared, unfrozen cell suspensions, we used disposable plastic pipettes (Fisher Scientific) to add four drops of the cells onto wet, pre-cleaned frosted microscope glass slides (Fisher Scientific) pre-chilled in the fridge (2 h) or freezer (30 min). Cell suspensions that had been stored at −20 °C were centrifuged at 1500 rpm for 5 min, and the supernatant was discarded and replaced with fresh fixative solution (methanol:acetic acid, 3:1) before dropping the cells onto the slides. The cells/slides were left to air-dry and age overnight at room temperature. The next morning, the slides were denatured in a pre-warmed (37 °C) Coplin staining jar containing 2× SSC denaturing solution for 45 min in a 37 °C water bath. Next, the slides were dehydrated in consecutive gradients of ethanol (70, 80, and 90%) for 2 min each at room temperature. The slides were then taken out of the last ethanol solution (90%) and air-dried for 10–15 min before applying the probes. The probe solution was prepared by adding 1 μL of the probe mixture and 2 μL ddiH$_2$O to 7 μL hybridization buffer (LSI/WCP, Abbott Molecular) for each slide. The probe set (Vysis LSI ETV6 (TEL)/ RUNX1 (AML1) ES Dual Color Single Fusion Probe), which was purchased from Abbott Molecular, comes as a mixture of two probes, each labeled with a different fluorophore, that detect and hybridize to their complementary DNA sequences on either chromosome 21 (LSI/RUNXI probe for 21q22 region, labeled with SpectrumOrange) or chromosome 12 (LSI/ETV6/TEL probe for 12p13 region, labeled with SpectrumGreen). After applying 10 µL of the probe solution to the center of the slide, a coverslip was placed immediately over the slide and sealed with rubber cement. The slides were placed in a humidified Vysis HYBrite Slide Stainer for 18–20 h, with a 4 min denaturation step at 75 °C, followed by the hybridization step at 37 °C, followed by another 16–22 h incubation time at 38 °C. Alternatively, hybridization can be carried out in a humidified chamber overnight at 37 °C. Following the incubation period, the rubber cement and coverslips were removed from the slides and the slides were washed twice with 0.4× SSC/0.3% NP-40 (pre-warmed at 37 °C for ~30 min), followed by incubation in 2× SSC/0.1% NP-40 at room temperature for 3 min. Thereafter, the slides with brain cells were processed for immunostaining with an anti-NeuN-Alexa488-conjugated antibody. Prior to immunostaining, the slides were hydrated in 1× PBS for 10 min, followed by three quick washes in 1× PBS. Subsequently, the slides were blocked in 10% goat serum (1× PBS + 0.1% Triton X-100 OR Tween-20) for 1 h at room temperature, followed by three washes in 1× PBS for 10 min each. The slides were then incubated with the anti-NeuN-Alexa488-conjugated antibody (diluted 1:100 in 1× PBS containing 5% BSA and 0.5% Tween-20) for 1 h at 37 °C in a humidified shaker. The slides were then washed three times, for 5 min each, in 1× PBS at room temperature, and counterstained with DAPI (Vectashield Mounting Medium with DAPI, Vector Laboratories, Inc.). After the coverslips were mounted onto the slides and sealed with nail polish, the slides were analyzed and/or stored at 4 °C [3]. Using a Zeiss AxioObserver Z1 microscope equipped with an AxioCam MRM camera and the Zen2011 software, 40×/1.3 and 63×/1.4 oil objectives, and filter sets capable of detecting DsRed (SpectrumOrange), GFP (SpectrumGreen), and DAPI signals, the slides were examined for chromosome 12 and 21 aneuploidy in neuronal (NeuN+) cells. The chromosomes were selected and scored according to Vysis guidelines. Chromosome signals from overlapping cells were not counted: only signals from individual, non-overlapping nuclei were counted. Also, per manufacturer's recommendation, adjacent signals that were very close to each other and that appeared as double spots linked together by a thread were counted as one signal. The results of these experiments showed that fibroblasts from NPC patients (Fig. 1a, b; [42]) and AD patients (Fig. 2a; [23]), and brain cell suspensions from NPC patients (Fig. 1c, d; [42]) and FTLD patients (Granic and Caneus, et al., manuscript submitted) exhibited elevated levels of aneuploidy for chromosome 21.

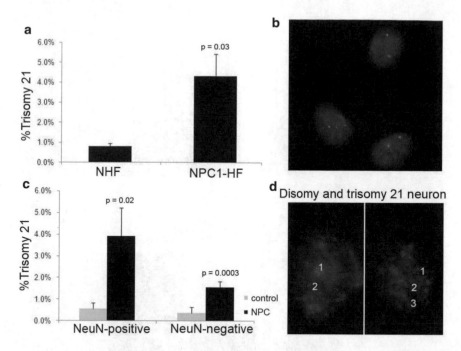

Fig. 1 Increased trisomy 21 aneuploidy in fibroblasts and in brain glia and neurons of Niemann-Pick C1 (NPC) patients. (**a, b**) FISH analysis with a DNA probe for chromosome 21 (*red*) and chromosome 12 (*green*) of fibroblasts derived from NPC patients (NPC1-HF) showed an increase in trisomy 21 cells compared to age-matched normal human fibroblasts (NHF). (**c, d**) Quantitative FISH analysis with a DNA probe for chromosome 21 (*red*) followed by staining with anti-NeuN antibody (*green*) and DAPI (*blue*) of resuspended cells from frontal cortices of control and NPC brains revealed significantly higher levels of trisomy 21 in NPC neurons and glia compared to controls. Figure adapted from [42]

3.1.4 Calculation of True Percentages of Aneuploidy with FISH Analysis

FISH analysis is one of the most popular and most used techniques to analyze and determine the numbers of specific chromosomes in cells. However, one issue with the use of FISH techniques is that although the labeled probes are designed to hybridize to their complementary sequences on a specific chromosome, sometimes incomplete hybridization occurs. In a culture of dividing cells, incomplete hybridization can result in post-S-phase cells with four copies of each chromosome that instead appear as trisomies. Therefore, in order to correct for the possibility of incomplete probe hybridization in our studies, we worked with statistician Dr. John Orav (Harvard School of Public Health) to develop a formula for calculating the true percentage of aneuploidy. Based on the assumption that each chromosome in a nucleus hybridizes as an independent event and that all/most observed monosomies are actually disomies in which one chromosome failed to hybridize, we were able to determine the actual

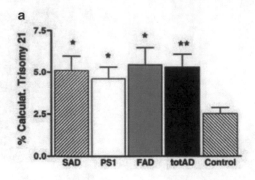

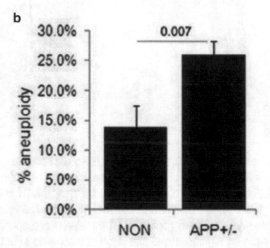

Fig. 2 Elevated levels of chromosome aneuploidy observed in AD patients and in AD mouse models. (**a**) FISH analysis of cultured fibroblasts from AD patients revealed a significant increase in trisomy 21 in sporadic AD (SAD), PSEN1 FAD (PS1), non-APP FAD (FAD), and AD (totAD). Figure adapted from [23]. (**b**) Karyotype analysis of spleen cells from transgenic AD mice expressing the human APP gene harboring the V717F FAD mutation (APP+/−) showed an increase in total aneuploidy relative to wild-type, non-transgenic mice (NON). Paired student's *t*-test was used for statistical analysis. Figure adapted from [3]

number of chromosomes in each nucleus and to calculate the true percentage of trisomy (aneuploidy) using the formula below:

$$P = \text{probability of hybridization} = \frac{m + 2d}{2(m + d)}$$

T = true percentage of trisomy

$$T = \frac{t}{\left[1 - \left(1 - P^3\right)\right]} - q\left(\frac{\dfrac{\left(1 - P^4\right)}{\left[1 - \left(1 - P^4\right)\right]}}{1 - \left(1 - P^3\right)}\right)$$

where *m* represents the percentage of monosomies observed, *d* represents the percentage of disomies observed, and *t* and *q* represent the percentages of trisomies and tetrasomies observed, respectively.

The first equation is used to estimate the probability of chromosome hybridization that occurred as equal to the ratio of the total number of observed hybridization spots in the nuclei that have either one or two such spots divided by the total number of spots such nuclei should have exhibited (2 × number of nuclei) if the hybridization were completely efficient and if all observed monosomies represented normal, but incompletely hybridized, diploid cells. The second equation uses the first of several terms in a Taylor series, together with the probability generated in the first equation, to estimate the true frequency of trisomies on a particular slide, based on all of the observed data on the number of nuclei on the same slide that exhibit various numbers (1, 2, 3, or 4) of hybridization spots. The effect of using the algorithm is to obtain a better estimate of the true number of trisomies by adjusting the observed number to take into account that some such observed trisomies may correspond to incompletely hybridized tetrasomies (post-S-phase cells), and that some observed disomies may correspond to incompletely hybridized trisomies. For each sample category we have examined, including probable sporadic AD (SAD), familial AD (FAD), PSEN1 mutation, all types of AD, and age-matched controls, both the raw trisomy measurements and the corrected levels of trisomy determined by using the algorithm were separately averaged and evaluated for statistical significance using nonparametric methods (Mann–Whitney *U* test). In practice, we found this algorithm to be most useful in situations with very low aneuploidy levels, and we have not used it routinely.

3.1.5 Metaphase Chromosome Spreads: Dividing Adherent Cells

To assess aneuploidy levels associated with AD by chromosome metaphase spread analysis, actively dividing cells (e.g., hTERT and fibroblasts) were treated with either 33 ng/mL or 100 ng/mL of colcemid for 6–10 h (depending on the genotype) to arrest cell division prior to harvesting the cells. Colcemid treatment disrupts mitotic spindles, resulting in metaphase cell cycle arrest. Following colcemid treatment, the cells were treated with 0.05% Trypsin/EDTA, harvested by trypsinization (5 mL of 0.05% trypsin/EDTA), and swollen in hypotonic (75 mM) potassium chloride in a 37 °C water bath for 30 min, followed by incubation in fixative solution for 30 min on ice. The cell suspension was centrifuged at 1500 rpm for 5 min and then resuspended in fresh fixative solution. Four drops of the cell suspension was added to a wet, cold, precleaned Fisherbrand frosted glass slide that were tilted at a 30° angle to allow the drops to spread down the slide by gravitation, and then air-dried overnight. On the next day, the slides were stained with the DNA-binding dye Giemsa (KaryoMAX® Giemsa

Stain Solution, ThermoFisher) for 2 min, washed with Gurr rinsing buffer (ThermoFisher), covered with a coverslip with Cytoseal-XYL mounting media (ThermoFisher), and incubated overnight at 50–60 °C in a hybridization oven. We used a Zeiss Imager.M1 microscope equipped with Genus 2.81 software (Applied Imaging, San Jose, CA), the Metafer 3.31 Slide Scanning System (MetaSystems, Altslussheim, Germany) with Isis 5.2 (ver. 2007; MetaSystems), and 40× and 60× oil objectives to capture and count the non-obscured and non-overlapping chromosomes.

The slides were loaded onto the automated stage of the microscope; the microscope software was programed to search for and identify the metaphase-arrested cells throughout the slides. After the microscope had scanned and recorded between 800 and 1000 metaphase-arrested cells, the total number of chromosome in each individual cell was counted. The average number of cells with more or fewer than 48 total chromosomes was computed for each sample and used for statistical analyses.

3.2 Induction of Aneuploidy by APP Overexpression or Expression of FAD Mutant Forms of APP in Mouse and Human Cells

To better understand how mosaic aneuploidy may occur in AD patients and whether increased expression of APP or expression of FAD-associated mutant forms of APP induce chromosome mis-segregation to generate aneuploidy in vivo, we used an AD transgenic mouse model expressing the human APP gene harboring the FAD-associated mutation V717F (purchased from The Jackson Laboratory) and non-transgenic controls. We also expressed the human APP gene harboring either the combined Swedish [K595N/M596L] and London [V642I] mutations or the V717I mutation in the hTERT-HME1 human cell line [3].

3.2.1 Evaluating Aneuploidy in Mouse Brain and Spleen Cells

Transgenic APP-V717F mice were sacrificed by intravenous injection of sodium barbital. Both the brain and the spleen were harvested from the mice and processed for further experiments.

3.2.2 Single-Cell Suspensions of Mouse Brain Cells

To generate single-cell suspensions from brain tissue, the brain was placed into a tube containing 5 mL of PBS and then transferred to a small (60 × 15 mm) culture dish for processing. Using a Nikon SMZ-2B stereozoom microscope equipped with two 10× eyepieces, we used a pair of forceps tips to remove the meninges, blood vessels, and cerebellum from the brain. Following the procedure previously described for human cells (Sect. 3.1.2), the remainder of the brain was placed into a 15 mL test tube containing 5 mL of 1× PBS and dissociated into a single-cell suspension using fire-polished Pasteur pipettes. The top layer containing the cell suspension was collected and centrifuged at 1500 rpm, 4 °C for 5 min. After removing the supernatant, the cell pellet was suspended in 5 mL of ice-cold fixative solution (methanol:acetic acid, 3:1), incubated on ice for at least 30 min, and either dropped onto a wet frosted glass slide for FISH or stored at −20 °C.

3.2.3 Primary Mouse Spleen Cell Cultures

To generate single-cell suspensions of primary mouse spleen cells, the spleens were washed twice with sterile 1× PBS to remove any hair/debris and placed in culture dishes containing 8 mL complete growth media (CGM: 500 mL RPMI1640 containing 10% Fetal Bovine Serum (FBS) and 5 mL Pen/Strep antibiotics). Using clean, glass slides sterilized with 70% ethanol, the spleens were ground in between the two frosted ends of the slides, followed by sedimentation for 7 min. The cell mixture was then centrifuged at 1500 rpm for 5 min. The supernatant was removed and the cell pellet was resuspended in 2 mL culture media (50 mL CGM plus 50 μL of Concanavalin-A [50 mg/mL] and 50 μL of 50 μM beta-mercaptoethanol) and then transferred to a culture flask containing an additional 8 mL of culture media. The cells were then placed into an incubator and allowed to grow at 37 °C, 5% CO_2 for 44 h. In parallel experiments, a portion of the cells was harvested and processed for FISH analysis as described below, and a portion was treated with colcemid and later harvested for karyotyping using chromosome metaphase spreads, described below. Using 0.05% trypsin/EDTA, the cells were harvested, centrifuged, and fixed in methanol:acetic acid (3:1) fixative solution.

3.2.4 FISH Analysis of Primary Mouse Spleen Cell Cultures

To evaluate chromosomal aneuploidy by FISH in mice, we used a BAC probe consisting of a DNA fragment (~300 kb) complementary to mouse chromosome 16 labeled with either SpectrumGreen or SpectrumOrange dUTPs [58]. To generate 1 μg of the labeled BAC probe, we combined 7.9 μL of nuclease-free H_2O, 20 ng/μL of extracted BAC DNA, 2.5 μL of 0.02 mM Green-dUTP, 5 μL of 0.1 mM dTTP, 10 μL of dNTP mix, 5 μL of 10× Nick Translation Buffer, and 10 μL of Nick Translation enzyme for a total volume of 50 μL in a microcentrifuge tube on ice. The mixture was briefly vortexed and centrifuged, followed by an overnight incubation (8–16 h) in a 15 °C water bath. The next day, the mixture was placed in a 70 °C water bath for 10 min and chilled on ice to stop any further reaction. To precipitate the labeled DNA (~360 ng), 60 μL of Cot-1 DNA plus 7.8 μL (1/10 of volume in tube) of 3 M sodium acetate was added to 18 μL of labeled probe in a microcentrifuge tube. In addition, 214.5 μL (2.5× of total volume in the tube) of cold 100% EtOH was added to the tube, followed by incubation in a −80 °C freezer for 20 min and centrifugation at 12,000 rpm, 4 °C for 12 min. The supernatant was removed, and the DNA pellet was washed with 500 μL of ice-cold 70% ethanol and spun for 5 min at 12,000 rpm (~ 13,362×g), 4 °C. After removing the supernatant, the DNA pellet was air-dried for 15 min at room temperature (or 5 min in a 75 °C heat block) and resuspended in 60 μL pre-warmed (45 °C) Hybrisol (Millipore), and the labeled DNA probe solution was then stored at −20 °C for future use. FISH analysis of the cells was carried out as described above using the labeled BAC probe for mouse chromosome 16.

Four drops of the cell suspension were added onto glass slides and air-dried (aged) overnight. The following day, the slides were washed with SSC wash buffer and incubated with 10 µL of pre-warmed (37 °C) labeled BAC probe. Using a Zeiss fluorescence microscope, the cells were analyzed and scored for chromosome 16 aneuploidy.

To summarize, for FISH analysis and chromosome-specific aneuploidy (e.g., trisomy 21 and 12), we used several primary cells and cell lines: human fibroblasts, mouse splenocytes, human primary aortic smooth muscle cells, mouse neural precursor cells, primary mouse neuronal culture, hTERT (human immortalized epithelial cells), and brain cell suspension (human and mouse with co-staining for neuronal population).

3.2.5 Karyotype Analysis of Primary Mouse Spleen Cell Cultures

As described, unlike FISH, in which probes bind to their complementary DNA sequence and can be detected using a fluorescence microscope, chromosome metaphase spreads are used to produce a karyotype that shows the total number of each chromosome within a metaphase-arrested cell. To arrest the cultured primary spleen cells in the metaphase stage of the cell cycle, 37 ng/mL of colcemid (Applied Imaging, San Jose, CA) was applied for 6–7 h. Alternatively, the cells can also be treated overnight with 33 ng/mL of colcemid. To dislodge and harvest the cells, 5 mL 0.05% Trypsin/EDTA solution was added to the cells in a 10 mm dish. Following the addition of 5 mL growth media, the cells were collected, spun at 1500 rpm, and fixed in fixative solution. To assess aneuploidy by chromosome metaphase spreads, the cells were dropped onto wet frosted glass slides and air-dried overnight. In the ensuing steps, the slides were stained for 2 min with the Giemsa staining dye (KaryoMAX® Giemsa Stain Solution, ThermoFisher) according to the manufacturer's protocol (with slight modifications) and rinsed immediately with Gurr rinsing buffer (ThermoFisher). The slides were air-dried, mounted with Cytoseal-XYL mounting medium (ThermoFisher), and sealed with coverslips. After the Cytoseal mounting media had completely spread under the entire coverslip, the slides were placed in a hybridization oven, and the chromosomes were stained overnight at ~50–60 °C before analyzing them. Using a Zeiss Imager.M1 microscope equipped with Genus 2.81 software (Applied Imaging, San Jose, CA) and the Metafer 3.31 Slide Scanning System (MetaSystems, Altslussheim, Germany) with Isis 5.2 (ver. 2007; MetaSystems), the chromosomes were captured and counted (see above, metaphase chromosome spread). Karyotype analyses of spleen cell suspensions from transgenic mice expressing the human APP gene harboring the V717F FAD mutation showed an increase in total aneuploidy relative to wild-type, non-transgenic control mice (Fig. 2b, [3]).

Comparing FISH with metaphase chromosome counting, it is apparent that the number of aneuploid cells for a specific chromosome is far fewer than the total number of aneuploid cells identified by karyotype analysis. This indicates that the aneuploidy that occurs in neurodegenerative diseases is likely to be rather general and not restricted to specific chromosomes. On the other hand, aneuploidy does make a cell prone to apoptosis [52], but possibly not so much for trisomy 21 alone, as people with DS can live into their 60s with proper medical care. Thus, in the brain, a selective advantage could lead to the gradual accumulation of trisomy 21 cells, as has been observed [24, 66].

3.2.6 Expression of FAD Mutant Forms of APP in Human hTERT-HME1 Cells

To investigate whether the increased levels of aneuploid cells observed in both AD patients and APP transgenic mice are a direct result of FAD-associated mutations in APP, we examined aneuploidy levels in cell cultures expressing FAD mutant APP genes [3]. We transiently transfected the hTERT-HME1 cell line with plasmid constructs for expressing APP harboring either combined Swedish [K595N/M596L] and London [V642I] FAD mutations or the V717I FAD mutation alone, or with the corresponding empty vector. The cells were seeded at a density of 1.5×10^5 cells per 2 mL in MEBM growth media (Mammary Epithelium Basal Medium, supplemented with 52 µg/mL bovine pituitary extract [BPE], 0.5 µg/mL hydrocortisone, 10 ng/mL human epidermal growth factor [hEGF], 5 µg/mL insulin, 50 µg/mL gentamicin, and 50 ng/mL amphotericin-B) in a six-well dish and allowed to proliferate for 24 h. The following day, the medium was exchanged with fresh medium and the cells were transiently transfected with one of the mutant APP plasmid constructs or with the corresponding empty vector. According to the manufacturer's recommended protocol, 1 µg of plasmid DNA was added per 3 µL of FuGene6 (Promega) in 97 µL of serum-free media (Opti-Mem 1×, ThermoFisher) and incubated at room temperature for 15–20 min. The transfection mixture was then added to the cell cultures, which were then placed in an incubator and allowed to grow at 37 °C, 5% CO_2. At 48 h post-transfection, the cells were either harvested for FISH or treated with 37 ng/mL colcemid for an additional 6–7 h (or 33 ng/mL for 10 h) for chromosome metaphase spread and karyotype analysis. The cells were washed in 1× PBS, dislodged in 0.25% Trypsin/EDTA solution, collected in a 15 mL test tube, and centrifuged at 1500 rpm for 5 min. The supernatant was discarded and the cell pellet was suspended in 5 mL of a freshly prepared hypotonic solution (75 mM potassium chloride [KCl]) and incubated at 37 °C in a water bath for 30 min. Then, 500 µL of cold fixative solution was added to the cells and mixed. After placing the mixture on ice for around 1 min, the cells were centrifuged at 1500 rpm, 4 °C for 5 min. The cell pellet was resuspended in fresh fixative solution and placed on ice for 30 min and then centrifuged at 1500 rpm, 4 °C for 5 min. After repeating the last two steps, the cells were dropped on frosted glass slides for FISH (using the dual probe set for detection of chromosomes 21 and 12) analyses, or stored at −20 °C to be used later for FISH. The slides were analyzed using a Zeiss microscope (see above). The samples expressing the mutant APP genes demonstrated significantly higher levels of aneuploidy relative to vector-alone transfected cells [3].

3.3 Chromosome Mis-Segregation Induced by Mutations in the PSEN1 Gene

In addition to mutations in the APP gene, early-onset FAD is also associated with mutations in the Presenilin genes. To determine whether FAD-associated mutations in the PSEN1 gene are also associated with chromosomal instability, we used transgenic mice and the hTERT-HME1 cell line expressing wild-type or FAD-associated mutant forms of PSEN1 to examine aneuploidy levels [1].

3.3.1 Transgenic PSEN1 Mouse Experiments

The mice were designed to express either the human wild-type or FAD mutant (M146L or M146V) PSEN1 transgene under the control of the PDGF promoter (provided by Drs. Karen Duff, Mark Mattson, and Steven Chan). Brains and spleens were harvested from the transgenic mice and were either fixed immediately for in situ hybridization and immunocytochemistry after generating single-cell suspensions or were processed to generate primary cell cultures. To prepare primary neuronal cultures from whole mouse brains, we removed the meninges together with the cerebellum, followed by trituration (see procedures above) of the brains in serum-free Modified Eagle Medium (MEM) for ~20–30 times. The dissociated cell mixture was left on ice for approximately 10–20 min, and the layer containing the single-cell suspension on top was transferred to Neurobasal Medium (NBM, ThermoFisher) supplemented with B-27 (ThermoFisher) and 0.5 mM L-glutamine in a chamber glass slide (Lab-Tek). The cells were briefly incubated for 1 h at 37 °C, 5% CO_2. Later, the original media together with the remaining unattached cells (glia) were aspirated, and 2 mL of fresh NBM-B27 media was added to the cells on the chamber slides. The cells were cultured for an additional 12–16 h followed by trypsinization with 0.05% Trypsin/EDTA. The cells were collected in a 15 mL tube, spun at 1500 rpm, and fixed in methanol:acetic acid fixative solution for FISH, as described above (Sect. 3.2.4), using the mouse chromosome 16 BAC probe. Primary mouse spleen cell cultures were prepared as described above (Sect. 3.2.3) for FISH analysis (Sect. 3.2.4) and chromosome metaphase spread karyotype experiments (Sect. 3.2.5).

3.3.2 Expression of FAD Mutant Forms of PSEN1 in Human hTERT-HME1 Cells

To further explore the link between FAD-associated mutations in PSEN1 and chromosomal instability, we transfected hTERT-HME1 cells with plasmids for expression of wild-type or FAD-associated mutant (M146L or M146V) PSEN1 genes. Low passage number (4–6) hTERT-HME1 cells were plated at a density of 1.5×10^5 cells/2 mL in six-well plates and incubated overnight at 37 °C, 5% CO_2 in MEBM medium supplemented with growth factors and antibiotics (MEGM SingleQuot Kit Suppl. & Growth Factors, Lonza). The following day, 100 μL of transfection solution, prepared according to the manufacturer's instructions (1 μg of plasmid DNA plus 3 μL of FuGene-6 was added to 97 μL of reduced serum media [Opti-MEM 1×] and incubated for 15 min at room temperature) was added to the cells in the plates. Prior to transfecting the cells, the medium was removed and replaced with fresh medium. At 48 h post-transfection, the cells were either collected and fixed immediately for FISH analysis (Sect. 3.1.3) or treated with 33 ng/mL colcemid for 10 h for chromosome metaphase spread karyotype (Sect. 3.1.5). Using the microscope described earlier (Sect. 3.1), the cells were analyzed and scored for aneuploidy.

As shown in Fig. 3 (adapted from [1]), we observed elevated levels of chromosomal aneuploidy in both the transgenic mice (Fig. 3a, b) and the hTERT-HME1 cell cultures (Fig. 3c, d) expressing FAD mutant PSEN1 or WT-PSEN1 genes relative to age-matched wild-type littermates and control vector-alone transfected cell cultures, respectively. In parallel experiments, transfected hTERT-HME1 cells were grown on chamber glass slides and processed for immunostaining with anti-α-tubulin monoclonal antibodies and DAPI for mitotic spindle assessments. In contrast to the untransfected cells or cells expressing the pcDNA3 vector alone (Invitrogen), which showed normal mitotic spindle formation and chromosome segregation, the hTERT-HME1 cells expressing FAD-associated mutant (M146L or M146V) PSEN1 genes exhibited a significantly higher percentage of cells with abnormal mitotic spindles and lagging chromosomes (Fig. 3e, f).

3.4 Treatment with Aβ Peptides Induces Aneuploidy in hTERT-HME1 Cells

Based on our observation that expression of FAD-associated mutant forms of APP or PSEN1 and overexpression of wild-type PSEN1 promote chromosome mis-segregation and aneuploidy, and that these same mutations also lead to increased production of the neurotoxic form of the Aβ peptide (Aβ-42), we investigated whether treatment with Aβ peptides is also capable of disrupting the cell cycle. In parallel experiments, low passage number hTERT-HME1 cells were plated at a density of 1.5×10^5 cells/2 mL in a six-well dish. The cells were maintained in MEBM growth media supplemented with growth factors and incubated overnight at 37 °C, 5% CO_2. The following day, the media was removed and replaced, and the cells were then treated with 0.5 or 1.0 μM of synthetic Aβ1-40, Aβ1-42, control reversed Aβ42-1 (American Peptide Company, Sunnyvale, CA; BioSource International, Camarillo, CA), scrambled (negative control) Aβ peptides designed by random sequencing of the Aβ1-42 peptide (NH2-ADFVGSVI

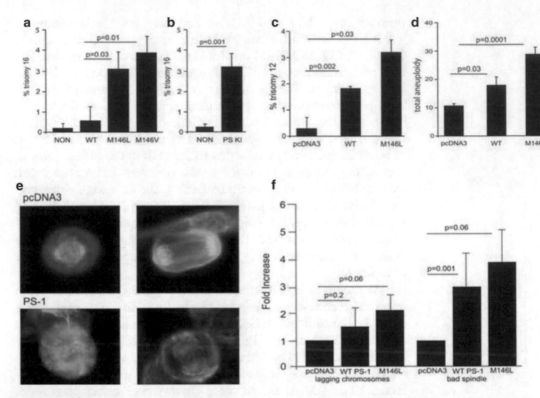

Fig. 3 Expression of the human PSEN1 gene harboring FAD mutations results in elevated levels of chromosomal aneuploidy in both transgenic mouse neurons and hTERT-HME1 human cell cultures. (**a**) Quantitative FISH of neurons from transgenic mice (14–19 months of age) expressing wild-type human PSEN1 (WT), FAD mutant M146L PSEN1 (M146L), or FAD mutant M146V PSEN1 (M146V), and non-transgenic mice (NON) revealed significantly higher levels of trisomy 16 only in the neurons from transgenic mice expressing FAD mutant forms of PSEN1. (**b**) Quantitative FISH of cultured neurons from FAD mutant (M146V) PSEN1 Knock-In mice (PS KI) and non-transgenic mice revealed significantly higher levels of trisomy 16 in the (M146V) PS1 KI neurons. PS KI mice were 10–15 months of age and the non-transgenic mice were 17 months of age. (**c**) FISH analysis of hTERT-HME1 cells transiently transfected with a vector for expression of wild-type PSEN1 (WT) or FAD mutant PSEN1(M146L) revealed significantly higher levels of trisomy 12 relative to cells transfected with the empty vector (pcDNA3). (**d**) Karyotype analysis of hTERT-HME1 cells transiently transfected with a vector for expression of wild-type PSEN1 (WT) or FAD mutant PSEN1(M146L) revealed significantly higher levels of total aneuploidy relative to cells transfected with the empty vector (pcDNA3). Up to 30% of cells expressing wild-type or FAD mutant PSEN1 were aneuploid at 48 h post-transfection, and all chromosomes were affected, regardless of their size. (**e**) Microtubule and DAPI staining of hTERT-HME1 cells at 48 h post-transfection with a vector for expression of wild-type PSEN1 (WT), FAD mutant PSEN1(M146L), or the empty vector (pcDNA3) showed that cells expressing wild-type or FAD mutant PSEN1 had significantly more abnormal mitotic spindles. (**f**) Quantification of abnormal mitotic spindles in cells from (**e**). Figure adapted from [1]

NIGKLELKMVGQVGVHGIAEVHFDYSFADHEARG-OH), or the Aβ12-28 peptide (NH2-VHHQKLVFFAEDVGSNK-OH) from Bio-Synthesis, Lewisville, TX, and Sigma Genosys, St. Louis, MO. At 48 h post-treatment, the cells were either harvested with 0.25% Trypsin/EDTA solution and fixed for FISH analysis of chromosome 21 and 12 aneuploidy (Sect. 3.1.3) or treated with 33 ng/mL colcemid for ~10 h prior to harvesting for chromosome

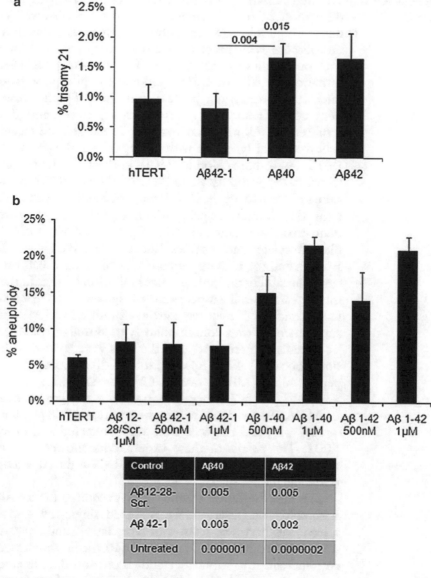

Fig. 4 Aβ treatment induces aneuploidy in hTERT-HME1 cells. hTERT-HME1 cells treated with Aβ1-40 or Aβ1-42 for 48 h show a significant increase in trisomy 21 by FISH analysis (**a**) and a significant dose-dependent increase in the percentage of total aneuploidy by chromosome metaphase spread (**b**) relative to untreated cells (hTERT) or cells treated with scrambled Aβ42-1. Figure adapted from [3]

metaphase spread (Sect. 3.1.5). As shown in Fig. 4, treatment with either the Aβ1-40 or Aβ1-42 peptide induced significantly increased levels of chromosomal aneuploidy relative to treatment with the scrambled Aβ, Aβ12-28, or reversed Aβ42-1 peptides [3]. These results provide evidence that expression of FAD-associated mutant forms of APP and PSEN1 are likely to generate chromosomal aneuploidy via their ability to induce the production of increased levels of the neurotoxic Aβ peptides.

3.5 Tau is Required for Aβ-Induced Aneuploidy

We then carried out experiments to test whether the Aβ-mediated disruption of mitotic spindle assembly we observed occurs by direct inhibition of microtubule structure/function or by interference with other downstream pathways and/or proteins. Previous studies had shown that Aβ treatment or FAD-associated mutations in APP or PSEN1 induces the phosphorylation of the microtubule-associated protein Tau (MAPT), the main component of AD-associated neurofibrillary tangles, and that Tau is required for Aβ-mediated toxicity. Therefore, we treated spleen cells from wild-type mice with Aβ peptides in the presence of either BAPTA or lithium chloride (LiCl), both of which had been shown to prevent Aβ toxicity via indirect inhibition of Tau phosphorylation and/or toxicity. BAPTA is an extracellular calcium (Ca^{2+}) chelator that inhibits calpain, which cleaves Tau and produces a neurotoxic Tau fragment [67, 68]. LiCl inhibits GSK-3β, which phosphorylates and activates Tau [69–71]. We hypothesized that BAPTA and/or LiCl may inhibit Aβ-mediated effects on chromosomal instability through indirect inhibition of Tau. In these experiments, single-cell suspensions of spleen cells from wild-type, tau$^{+/-}$, and tau$^{-/-}$ mice were grown for 44 h in RPMI 1640 media supplemented with concanavalin A to stimulate cell division [3]. The media was removed, and the cells were treated with Aβ peptide alone (Aβ1-40 or Aβ1-42) for 44–48 h, were treated with Aβ peptide alone (Aβ1-40 or Aβ1-42) for ~41 h with the addition of 2.5 μM LiCl at 7 h before harvesting, or were pretreated with 1 μM BAPTA for 3 min before the addition of Aβ peptide (Aβ1-40 or Aβ1-42), and the cells were then collected and processed for FISH. The results of these experiments showed that treatment with BAPTA or LiCl inhibited Aβ1-42-induced aneuploidy in wild-type mouse spleen cells.

To directly address whether Tau is required for the Aβ-induced aneuploidy we observed, we prepared single-cell suspensions of spleen cells harvested from wild-type, tau$^{+/-}$, and tau$^{-/-}$ mice that were cultured for 44 h in RPMI 1640 media supplemented with concanavalin A to stimulate cell division, and then treated with Aβ peptide alone (Aβ1-40 or Aβ1-42) in fresh medium for ~48 h [3]. The cells were then collected and processed for chromosome 16 FISH. The results of these experiments showed that the loss of just one copy of Tau, and even more effectively of both copies of Tau, resulted in increased aneuploidy even without Aβ treatment [3]. Notably, Aβ treatment failed to induce a significant increase in aneuploidy in the Tau-deficient cells, suggesting that Aβ requires and disrupts Tau-stabilized microtubules to induce its aneugenic effects.

3.6 Aβ-Mediated Inhibition of Kinesin-5 and Other Microtubule Motor Proteins

To determine how Aβ disrupts mitotic spindle function, we used two different systems to examine mitotic spindles following Aβ treatment: hTERT-HME1 cells and cell-free *Xenopus* egg extracts. To investigate whether Aβ induces chromosome mis-segregation by disrupting mitotic spindle formation, hTERT-HME1 cells were seeded on pre-coated (poly-L-lysine, 2 µg/mL, Sigma) single chamber glass slides (BD Falcon) at a density of 1.5×10^5 cells/2 mL in supplemented MEBM growth media. The cells were cultured at 37 °C, 5% CO_2 overnight, and then treated with two different concentrations (0.5 µM or 1 µM) of different Aβ peptide fragments (Aβ1-40, Aβ1-42, Aβ42-1, Aβ1-11, Aβ1-28, Aβ1-33, and scrambled Aβ1-42, from AnaSpec, American Peptide Company, BioSource International and Sigma Genosys). After 48 h of incubation, the cells were washed and fixed for indirect immunofluorescence microscopy with anti-α-tubulin antibody (Sigma) and DAPI (4,6-dimidino-2-phenylidone). These experiments showed that treatment with Aβ peptides, and particularly Aβ1-42, resulted in abnormal spindle formation and maintenance (Fig. 5a) [34].

Based on the spindle defects we observed when hTERT-HME1 cells were treated with Aβ1-42, we hypothesized that Aβ may inhibit microtubule-based mitotic motor proteins [34]. To test our hypothesis, we used cell-free Xenopus egg extracts, which require microtubule motors for proper mitotic spindle formation and spindle microtubule dynamics. To study spindle assembly, Xenopus egg extracts were treated with 0, 0.5, 1.5, or 2.0 µM Aβ1-40 or Aβ1-42, or with 2.0 µM scrambled Aβ peptide, incubated on ice for 1 h, and then incubated at room temperature for 60 min, to allow us to analyze spindle assembly. To examine effects on spindle stability, extracts were incubated for 75 min at room temperature to allow spindle assembly, followed by Aβ1-42 or Aβ1-42 treatment for 1 h at room temperature. We found that samples treated with either Aβ1-40 or Aβ1-42 had a significantly higher percentage of abnormal mitotic spindles compared to the untreated samples or the samples treated with the scrambled Aβ peptide [34]. In subsequent experiments, we found that the mitotic spindle defects could be rescued by the addition of purified recombinant motor domains of Kinesin-5 (Eg5), KIF4A, or KIF2C (MCAK) to the cell-free Xenopus egg extracts [34]. In those experiments, 0.4 µM Kinesin-5 or KIF4A or 0.1 µM MCAK was added to the extracts, followed by the addition of 1 µM Aβ1-42 peptide or buffer and incubated for 60 min at room temperature, followed by mitotic spindle analysis. Pretreatment with the purified recombinant motor domains prior to Aβ treatment resulted in a significant reduction in mitotic spindles abnormalities; addition of polymerized microtubules or of the microtubule stabilizing protein Tau (MAPT) also reduced the effects of Aβ treatment on spindle formation.

To confirm that Aβ-mediated disruption of the microtubule network and its ability to induce aneuploidy occurs at least in part through its inhibition of Kinesin-5/Eg5, hTERT-HME1 cells were plated in 10 cm dishes, and treated 1 day later with 10 µM monastrol, a specific inhibitor of Kinesin-5, for 48 h. The cells were collected, processed, and fixed for FISH analysis with labeled

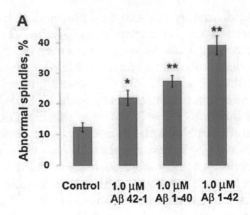

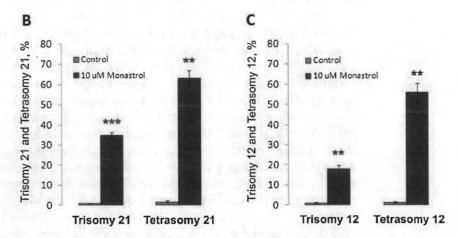

Fig. 5 Aβ-mediated inhibition of Kinesin-5 leads to abnormal spindle formation and aneuploidy. (**a**) hTERT-HME1 cells treated with Aβ1-40, Aβ1-42, or Aβ42-1 for 48 h show a significantly higher percentage of abnormal mitotic spindles relative to untreated control cells. In addition, significant increases in aneuploidy (i.e., trisomy and tetrasomy) for chromosome 21 (**b**) and for chromosome 12 (**c**) were observed in hTERT-HME1 cells treated with the Kinesin-5 inhibitor monastrol relative to untreated control cells. At least 500 cells were scored per sample. Error bars indicate statistical significant at: *$p \leq 0.05$, **$p \leq 0.005$, ***$p \leq 0.0005$. Figure adapted from [34]

3.7   Comparison of FISH and Single-Cell Sequencing for Detecting Mosaic Aneuploidy in Brain Cells during Aging and in Neurodegenerative Diseases

probes for detection of human chromosomes 21 and 12. The cells were examined and scored for aneuploidy. The results showed elevated levels of chromosome 21 and 12 trisomy and tetrasomy in the monastrol-treated samples relative to untreated control samples (Fig. 5b), providing evidence that Aβ-mediated inhibition of Kinesin-5 alone is sufficient to disrupt mitotic spindle formation and stability, which is likely to result in chromosome mis-segregation and aneuploidy in AD.

As many chapters and their references in this book attest, mosaic aneuploidy in neurons and nonneuronal cells in the brain and other tissues can be observed to arise during aging and neurodegenerative diseases in humans and in animal models thereof, as assessed by chromosome-specific FISH and other methods (see especially Chaps. 1–5, 13–15, and 17). Furthermore, we have shown by karyotype analysis and FISH that neurodegeneration-causing mutant genes or their protein product (i.e., Aβ peptide) induce massive chromosome mis-segregation and mosaic aneuploidy, affecting over 20% of cells after only two generations in culture [1, 3]. In contrast, several studies using whole genome single-cell sequencing (wgSCS) of human brain neurons failed to replicate this finding in either aging or neurodegenerative disease [72, 73]. How can these two apparently incompatible findings be reconciled? Our recent analyses of brain cells from FTLD patients with mutations in the MAPT gene and of transfected cells expressing an FTLD-causing mutant form of MAPT may offer an explanation. Specifically, we have found that the vast majority (80%) of apoptotic (TUNEL+) cells in FTLD brains or in mutant MAPT-transfected cells are also aneuploid, whereas less than 20% of aneuploid cells are apoptotic (Granic and Caneus et al., manuscript submitted). These results suggest that aneuploidy induces apoptosis, and that apoptosis does not induce aneuploidy. By definition and based on TUNEL measurements, chromosomal DNA in apoptotic cells is damaged. Therefore, aneuploid cells that become apoptotic are likely to show an aneuploid metaphase chromosome spread for karyotype analysis and their aneuploidy also would be detectable by FISH. However, the nicks and gaps, and, even worse, double-strand breaks that are characteristic of apoptotic cells will naturally be exposed during the standard procedures of wgSCS. During the first step of the wgSCS protocol, gaps in the chromosomal DNA will be clipped by micrococcal nuclease and may then be unable to fully accept the end linkers in the second step, and any nicks or gaps would most certainly block subsequent PCR amplification. The consequence would be that damaged (aneuploid/apoptotic) cells would be less likely to contribute to the final collection of sequenced data. Indeed, 40–50% of the brain neurons analyzed by wgSCS usually fail quality control (see, for example, [72, 73]), a strong indication of damaged DNA. Thus, aneuploid cells, which our studies have shown to be prone to subsequent apoptosis, would be severely undercounted in standard wgSCS experiments.

One approach to solving the potential inability of the wgSCS method to detect genomic aberrations, such as aneuploidy, in the midst of associated apoptosis, would be to sequentially pretreat the isolated nuclei prepared for wgSCS (or for FISH) with T4 DNA ligase plus ATP to seal single-strand nicks in the DNA and to potentially reseal double-strand breaks that are still adjacent to each other in the still-intact chromatin, and to then treat with T4 DNA polymerase plus nucleotide triphosphate DNA precursors together with additional ligase and ATP to fill in and seal single-strand gaps.

Only after the DNA has been thus repaired should the single-cell sequencing procedure be continued. By assuring that even aneuploid/apoptotic cell nuclei are also assessed by wgSCS, this preparative procedure should bring the results of wgSCS and FISH experiments into alignment by assuring that aneuploid cells are equally counted using both methods.

4 Conclusions

Neurodegenerative disease encompasses several different disorders that affect the CNS. While each one of these conditions arises under different circumstances and shows distinct clinical symptoms, many of them are associated with cognitive deficits (dementia) at some time point during the disease course. The current view is that dementia develops as a result of neuronal dysfunction and cell loss in the brain. To date, many factors or abnormal molecular processes have been identified with the capacity to promote neuronal dysfunction, including chromosome instability and aneuploidy. Aneuploidy arises from chromosome mis-segregation during the cell cycle. Several lines of evidence have shown cell cycle abnormalities, including aberrant cell cycle reentry and mis-expression of cell cycle proteins, in the brains of AD patients, which may lead to chromosome mis-segregation and the generation of aneuploid cells. Aneuploid cells, which may arise from adult neurogenesis [74, 75] or by neuronal reentry into the cell cycle or even transdifferentiation from astrocytes (for discussion see [76]) are highly unstable, and may in turn undergo apoptosis resulting in cell death, brain atrophy, and the development of dementia. Induction of aneuploidy in neurodegenerative diseases is significant because, in addition to studies conducted by Arendt et al. (2010), who have demonstrated that the loss of aneuploid cells in the brain of AD patients is positively correlated with the development of dementia [14], recent studies from our lab have also revealed a positive correlation between aneuploidy and cell death in FTLD (Granic and Caneus, et al., manuscript submitted).

To investigate whether aneuploidy may serve as a common mechanistic pathway responsible for cognitive deficits associated with different neurodegenerative diseases, brain tissues from patients with different neurodegenerative conditions (AD, NPC, and FTLD) were prepared and analyzed using FISH and chromosome karyotyping techniques. Using labeled chromosome probes that can hybridize to their complementary sequences on human chromosomes 12 and 21, in combination with the anti-NeuN antibody for detection of neurons, we were able to assess the levels of aneuploidy in both neuronal and nonneuronal cells by FISH in patients with these different neurological disorders. In addition to using FISH to count the individual signals for chromosomes 21 and 12, we have often also scored the total number of chromosomes in each dividing cell in culture that was expressing disease-causing mutant genes, using chromosome metaphase spreads for karyotype analysis. Taken together, we have observed significant increases in chromosomal aneuploidy levels in the patients with neurodegenerative disorders compared to age-matched control subjects. Furthermore, statistical analyses of data collected from in vivo and in vitro experiments using AD mouse models and human cell lines have indicated that FAD-associated mutations in genes that code for APP and PSEN1 may play a direct role in the disruption of chromosome segregation during the cell cycle and in promoting aneuploidy. Both APP mutations and abnormal APP processing by γ-secretase result in overproduction of the toxic forms of the Aβ peptides. Additionally, using Aβ peptides in vitro, we were able to demonstrate that Aβ inhibits microtubule motor proteins to disrupt the mitotic spindle machinery and induce aneuploidy. Taken together, these findings suggest that aneuploidy is not only associated with AD, NPC, and FTLD pathologies, but may also play a pivotal role in the generation and progression of dementia in other neurodegenerative diseases.

Based on the data discussed herein, we identified/postulated two possible pathways or targets through which Aβ might be acting to induce aneuploidy and generate cognitive decline in AD patients. In addition to us showing Aβ-mediated inhibition of the motor protein Kinesin-5, Aβ is also known to promote phosphorylation of the microtubule-associated protein Tau (MAPT). Since both Kinesin-5 and Tau are microtubule-based proteins that are needed for microtubule stability and polymerization, inhibition of either by Aβ will also disrupt mitotic spindle assembly, resulting in chromosome mis-segregation and aneuploidy. Subsequently, due to chromosome imbalance and genomic instability, aneuploid cells will be more prone to undergo apoptosis, leading to neurodegeneration.

References

1. Boeras DI, Granic A, Padmanabhan J, Crespo NC, Rojiani AM, Potter H (2008) Alzheimer's presenilin 1 causes chromosome missegregation and aneuploidy. Neurobiol Aging 29(3):319–328. doi:10.1016/j.neurobiolaging.2006.10.027

2. Gerrish A, Russo G, Richards A, Moskvina V, Ivanov D, Harold D, Sims R, Abraham R, Hollingworth P, Chapman J, Hamshere M, Pahwa JS, Dowzell K, Williams A, Jones N, Thomas C, Stretton A, Morgan AR, Lovestone S, Powell J, Proitsi P, Lupton MK, Brayne C, Rubinsztein DC, Gill M, Lawlor B, Lynch A, Morgan K, Brown KS, Passmore PA, Craig D, McGuinness B, Todd S, Johnston JA, Holmes C, Mann D, Smith AD, Love S, Kehoe PG, Hardy J, Mead S, Fox N, Rossor M, Collinge J, Maier W, Jessen F, Kolsch H, Heun R, Schurmann B, van den Bussche H, Heuser I, Kornhuber J, Wiltfang J, Dichgans M, Frolich L, Hampel H, Hull M, Rujescu D, Goate AM, Kauwe JS, Cruchaga C, Nowotny P, Morris JC, Mayo K, Livingston G, Bass NJ, Gurling H, McQuillin A, Gwilliam R, Deloukas P, Davies G, Harris SE, Starr JM, Deary IJ, Al-Chalabi A, Shaw CE, Tsolaki M, Singleton AB, Guerreiro R, Muhleisen TW, Nothen MM, Moebus S, Jockel KH, Klopp N,

Wichmann HE, Carrasquillo MM, Pankratz VS, Younkin SG, Jones L, Holmans PA, O'Donovan MC, Owen MJ, Williams J (2012) The role of variation at AbetaPP, PSEN1, PSEN2, and MAPT in late onset Alzheimer's disease. J Alzheimers Dis 28(2):377–387. doi:10.3233/JAD-2011-110824

3. Granic A, Padmanabhan J, Norden M, Potter H (2010) Alzheimer Abeta peptide induces chromosome mis-segregation and aneuploidy, including trisomy 21: requirement for tau and APP. Mol Biol Cell 21(4):511–520. doi:10.1091/mbc.E09-10-0850

4. Selkoe DJ, Hardy J (2016) The amyloid hypothesis of Alzheimer's disease at 25 years. EMBO Mol Med 8(6):595–608. doi:10.15252/emmm.201606210

5. van der Kant R, Goldstein LS (2015) Cellular functions of the amyloid precursor protein from development to dementia. Dev Cell 32(4):502–515. doi:10.1016/j.devcel.2015.01.022

6. Goate A, Hardy J (2012) Twenty years of Alzheimer's disease-causing mutations. J Neurochem 120(Suppl 1):3–8. doi:10.1111/j.1471-4159.2011.07575.x

7. Cai Y, An SS, Kim S (2015) Mutations in presenilin 2 and its implications in Alzheimer's disease and other dementia-associated disorders. Clin Interv Aging 10:1163–1172. doi:10.2147/CIA.S85808

8. Selkoe DJ (1998) The cell biology of beta-amyloid precursor protein and presenilin in Alzheimer's disease. Trends Cell Biol 8(11):447–453

9. Wolfe MS (2007) When loss is gain: reduced presenilin proteolytic function leads to increased Abeta42/Abeta40. Talking point on the role of presenilin mutations in Alzheimer disease. EMBO Rep 8(2):136–140. doi:10.1038/sj.embor.7400896

10. Xia D, Watanabe H, Wu B, Lee SH, Li Y, Tsvetkov E, Bolshakov VY, Shen J, Kelleher RJ 3rd (2015) Presenilin-1 knockin mice reveal loss-of-function mechanism for familial Alzheimer's disease. Neuron 85(5):967–981. doi:10.1016/j.neuron.2015.02.010

11. Xia W (2000) Role of presenilin in gamma-secretase cleavage of amyloid precursor protein. Exp Gerontol 35(4):453–460

12. Palop JJ, Mucke L (2010) Amyloid-beta-induced neuronal dysfunction in Alzheimer's disease: from synapses toward neural networks. Nat Neurosci 13(7):812–818. doi:10.1038/nn.2583

13. Tu S, Okamoto S, Lipton SA, Xu H (2014) Oligomeric Abeta-induced synaptic dysfunction in Alzheimer's disease. Mol Neurodegener 9:48. doi:10.1186/1750-1326-9-48

14. Arendt T, Bruckner MK, Mosch B, Losche A (2010) Selective cell death of hyperploid neurons in Alzheimer's disease. Am J Pathol 177(1):15–20. doi:10.2353/ajpath.2010.090955

15. Yang Y, Mufson EJ, Herrup K (2003) Neuronal cell death is preceded by cell cycle events at all stages of Alzheimer's disease. J Neurosci 23(7):2557–2563

16. Lee VM, Trojanowski JQ (2006) Progress from Alzheimer's tangles to pathological tau points towards more effective therapies now. J Alzheimers Dis 9(3 Suppl):257–262

17. Gong CX, Iqbal K (2008) Hyperphosphorylation of microtubule-associated protein tau: a promising therapeutic target for Alzheimer disease. Curr Med Chem 15(23):2321–2328

18. Lee G, Leugers CJ (2012) Tau and tauopathies. Prog Mol Biol Transl Sci 107:263–293. doi:10.1016/B978-0-12-385883-2.00004-7

19. Duesberg P, Rasnick D (2000) Aneuploidy, the somatic mutation that makes cancer a species of its own. Cell Motil Cytoskeleton 47(2):81–107. doi:10.1002/1097-0169(200010)47:2<81::AID-CM1>3.0.CO;2-#

20. Jefford CE, Irminger-Finger I (2006) Mechanisms of chromosome instability in cancers. Crit Rev Oncol Hematol 59(1):1–14. doi:10.1016/j.critrevonc.2006.02.005

21. Ried T (2009) Homage to Theodor Boveri (1862–1915): Boveri's theory of cancer as a disease of the chromosomes, and the landscape of genomic imbalances in human carcinomas. Environ Mol Mutagen 50(8):593–601. doi:10.1002/em.20526

22. Potter H (1991) Review and hypothesis: Alzheimer disease and down syndrome—chromosome 21 nondisjunction may underlie both disorders. Am J Hum Genet 48(6):1192–1200

23. Geller LN, Potter H (1999) Chromosome missegregation and trisomy 21 mosaicism in Alzheimer's disease. Neurobiol Dis 6(3):167–179. doi:10.1006/nbdi.1999.0236

24. Iourov IY, Vorsanova SG, Yurov YB (2011) Genomic landscape of the Alzheimer's disease brain: chromosome instability—aneuploidy, but not tetraploidy—mediates neurodegeneration. Neurodegener Dis 8(1–2):35–37. doi:10.1159/000315398; discussion 38-40

25. Kingsbury MA, Yung YC, Peterson SE, Westra JW, Chun J (2006) Aneuploidy in the normal and diseased brain. Cell Mol Life Sci 63(22):2626–2641. doi:10.1007/s00018-006-6169-5

26. Migliore L, Botto N, Scarpato R, Petrozzi L, Cipriani G, Bonuccelli U (1999) Preferential occurrence of chromosome 21 malsegregation in peripheral blood lymphocytes of Alzheimer disease patients. Cytogenet Cell Genet 87(1–2):41–46

27. Mosch B, Morawski M, Mittag A, Lenz D, Tarnok A, Arendt T (2007) Aneuploidy and DNA replication in the normal human brain and Alzheimer's disease. J Neurosci 27(26):6859–6867. doi:10.1523/JNEUROSCI.0379-07.2007

28. Ringman JM, Rao PN, PH L, Cederbaum S (2008) Mosaicism for trisomy 21 in a patient with young-onset dementia: a case report and brief literature review. Arch Neurol 65(3):412–415. doi:10.1001/archneur.65.3.412

29. Thomas P, Fenech M (2008) Chromosome 17 and 21 aneuploidy in buccal cells is increased with ageing and in Alzheimer's disease. Mutagenesis 23(1):57–65. doi:10.1093/mutage/gem044

30. Trippi F, Botto N, Scarpato R, Petrozzi L, Bonuccelli U, Latorraca S, Sorbi S, Migliore L (2001) Spontaneous and induced chromosome damage in somatic cells of sporadic and familial Alzheimer's disease patients. Mutagenesis 16(4):323–327

31. Westra JW, Barral S, Chun J (2009) A reevaluation of tetraploidy in the Alzheimer's disease brain. Neurodegener Dis 6(5–6):221–229. doi:10.1159/000236901

32. Frade JM, Lopez-Sanchez N (2010) A novel hypothesis for Alzheimer disease based on neuronal tetraploidy induced by p75 (NTR). Cell Cycle 9(10):1934–1941. doi:10.4161/cc.9.10.11582

33. Lopez-Sanchez N, Frade JM (2013) Genetic evidence for p75NTR-dependent tetraploidy in cortical projection neurons from adult mice. J Neurosci 33(17):7488–7500. doi:10.1523/JNEUROSCI.3849-12.2013

34. Borysov SI, Granic A, Padmanabhan J, Walczak CE, Potter H (2011) Alzheimer Abeta disrupts the mitotic spindle and directly inhibits mitotic microtubule motors. Cell Cycle 10(9):1397–1410

35. Falnikar A, Tole S, Baas PW (2011) Kinesin-5, a mitotic microtubule-associated motor protein, modulates neuronal migration. Mol Biol Cell 22(9):1561–1574. doi:10.1091/mbc.E10-11-0905

36. Ferenz NP, Gable A, Wadsworth P (2010) Mitotic functions of kinesin-5. Semin Cell Dev Biol 21(3):255–259. doi:10.1016/j.semcdb.2010.01.019

37. Scholey JE, Nithianantham S, Scholey JM, Al-Bassam J (2014) Structural basis for the assembly of the mitotic motor Kinesin-5 into bipolar tetramers. elife 3:e02217. doi:10.7554/eLife.02217

38. Chen Y, Hancock WO (2015) Kinesin-5 is a microtubule polymerase. Nat Commun 6:8160. doi:10.1038/ncomms9160

39. Potter H (2005) Cell cycle and chromosome segregation defects in Alzheimer's disease. In: ACAF N (ed) Cell cycle mechanisms and neuronal cell death. Landes Bioscience/Eurekah.com/Kluwer Academic/Plenum Publishers, Georgetown, TX/New York, NY, pp 55–78

40. Yang Y, Geldmacher DS, Herrup K (2001) DNA replication precedes neuronal cell death in Alzheimer's disease. J Neurosci 21(8):2661–2668

41. Iourov IY, Vorsanova SG, Liehr T, Kolotii AD, Yurov YB (2009) Increased chromosome instability dramatically disrupts neural genome integrity and mediates cerebellar degeneration in the ataxia-telangiectasia brain. Hum Mol Genet 18(14):2656–2669. doi:10.1093/hmg/ddp207

42. Granic A, Potter H (2013) Mitotic spindle defects and chromosome mis-segregation induced by LDL/cholesterol-implications for Niemann-Pick C1, Alzheimer's disease, and atherosclerosis. PLoS One 8(4):e60718. doi:10.1371/journal.pone.0060718

43. Yang Y, Shepherd C, Halliday G (2015) Aneuploidy in Lewy body diseases. Neurobiol Aging 36(3):1253–1260. doi:10.1016/j.neurobiolaging.2014.12.016

44. Rossi G, Conconi D, Panzeri E, Paoletta L, Piccoli E, Ferretti MG, Mangieri M, Ruggerone M, Dalpra L, Tagliavini F (2014) Mutations in MAPT give rise to aneuploidy in animal models of tauopathy. Neurogenetics 15(1):31–40. doi:10.1007/s10048-013-0380-y

45. Rossi G, Conconi D, Panzeri E, Redaelli S, Piccoli E, Paoletta L, Dalpra L, Tagliavini F (2013) Mutations in MAPT gene cause chromosome instability and introduce copy number variations widely in the genome. J Alzheimers Dis 33(4):969–982. doi:10.3233/JAD-2012-121633

46. Bouge AL, Parmentier ML (2016) Tau excess impairs mitosis and kinesin-5 function, leading to aneuploidy and cell death. Dis Model Mech 9(3):307–319. doi:10.1242/dmm.022558

47. Mandelkow E, Mandelkow EM (1995) Microtubules and microtubule-associated proteins. Curr Opin Cell Biol 7(1):72–81

48. Freund RK, Gibson ES, Potter H, Dell'Acqua ML (2016) Inhibition of the motor protein Eg5/Kinesin-5 in amyloid beta-mediated impairment of hippocampal long-term potentiation and dendritic spine loss. Mol Pharmacol 89(5):552–559. doi:10.1124/mol.115.103085

49. Kopeikina KJ, Hyman BT, Spires-Jones TL (2012) Soluble forms of tau are toxic in Alzheimer's disease. Transl Neurosci 3(3):223–233. doi:10.2478/s13380-012-0032-y

50. Takashima A (2008) Hyperphosphorylated tau is a cause of neuronal dysfunction in tauopathy. J Alzheimers Dis 14(4):371–375

51. Zheng WH, Bastianetto S, Mennicken F, Ma W, Kar S (2002) Amyloid beta peptide induces tau phosphorylation and loss of cholinergic neurons in rat primary septal cultures. Neuroscience 115(1):201–211

52. Oromendia AB, Amon A (2014) Aneuploidy: implications for protein homeostasis and disease. Dis Model Mech 7(1):15–20. doi:10.1242/dmm.013391

53. Dalton WB, Yu B, Yang VW (2010) p53 suppresses structural chromosome instability after mitotic arrest in human cells. Oncogene 29(13):1929–1940. doi:10.1038/onc.2009.477

54. Ganem NJ, Pellman D (2007) Limiting the proliferation of polyploid cells. Cell 131(3):437–440. doi:10.1016/j.cell.2007.10.024

55. Jeganathan K, Malureanu L, Baker DJ, Abraham SC, van Deursen JM (2007) Bub1 mediates cell death in response to chromosome missegregation and acts to suppress spontaneous tumorigenesis. J Cell Biol 179(2):255–267. doi:10.1083/jcb.200706015

56. Kuffer C, Kuznetsova AY, Storchova Z (2013) Abnormal mitosis triggers p53-dependent cell cycle arrest in human tetraploid cells. Chromosoma 122(4):305–318. doi:10.1007/s00412-013-0414-0

57. Storchova Z, Kuffer C (2008) The consequences of tetraploidy and aneuploidy. J Cell Sci 121(Pt 23):3859–3866. doi:10.1242/jcs.039537

58. Kulnane LS, Lehman EJ, Hock BJ, Tsuchiya KD, Lamb BT (2002) Rapid and efficient detection of transgene homozygosity by FISH of mouse fibroblasts. Mamm Genome 13(4):223–226. doi:10.1007/s00335-001-2128-5

59. Liesi P, Fried G, Stewart RR (2001) Neurons and glial cells of the embryonic human brain and spinal cord express multiple and distinct isoforms of laminin. J Neurosci Res 64(2):144–167. doi:10.1002/jnr.1061

60. Pacey L, Stead S, Gleave JA, Tomczyk K, Doering L (2006) Neural stem cell culture: neurosphere generation, microscopical analysis and cryopreservation. Protoc Exchan. doi:10.1038/nprot.2006.21

61. Marchenko S, Flanagan L (2007) Passing human neuronal stem cells. J Vis Exp 7

62. Waitzman JS, Rice SE (2014) Mechanism and regulation of kinesin-5, an essential motor for the mitotic spindle. Biol Cell 106(1):1–12. doi:10.1111/boc.201300054

63. Walczak CE, Heald R (2008) Mechanisms of mitotic spindle assembly and function. Int Rev Cytol 265:111–158. doi:10.1016/S0074-7696(07)65003-7

64. Wittmann T, Hyman A, Desai A (2001) The spindle: a dynamic assembly of microtubules and motors. Nat Cell Biol 3(1):E28–E34. doi:10.1038/35050669

65. Gillespie PJ, Gambus A, Blow JJ (2012) Preparation and use of Xenopus egg extracts to study DNA replication and chromatin associated proteins. Methods 57(2):203–213. doi:10.1016/j.ymeth.2012.03.029

66. Iourov IY, Vorsanova SG, Liehr T, Yurov YB (2009) Aneuploidy in the normal, Alzheimer's disease and ataxia-telangiectasia brain: differential expression and pathological meaning. Neurobiol Dis 34(2):212–220. doi:10.1016/j.nbd.2009.01.003

67. Khorchid A, Ikura M (2002) How calpain is activated by calcium. Nat Struct Biol 9(4):239–241. doi:10.1038/nsb0402-239

68. Wei Z, Song MS, MacTavish D, Jhamandas JH, Kar S (2008) Role of calpain and caspase in beta-amyloid-induced cell death in rat primary septal cultured neurons. Neuropharmacology 54(4):721–733. doi:10.1016/j.neuropharm.2007.12.006

69. Alvarez G, Munoz-Montano JR, Satrustegui J, Avila J, Bogonez E, Diaz-Nido J (1999) Lithium protects cultured neurons against beta-amyloid-induced neurodegeneration. FEBS Lett 453(3):260–264

70. Bertsch S, Lang CH, Vary TC (2011) Inhibition of glycogen synthase kinase 3[beta] activity with lithium in vitro attenuates sepsis-induced changes in muscle protein turnover. Shock 35(3):266–274. doi:10.1097/SHK.0b013e3181fd068c

71. Takashima A, Honda T, Yasutake K, Michel G, Murayama O, Murayama M, Ishiguro K, Yamaguchi H (1998) Activation of tau protein kinase I/glycogen synthase kinase-3beta by amyloid beta peptide (25-35) enhances phosphorylation of tau in hippocampal neurons. Neurosci Res 31(4):317–323

72. Knouse KA, Wu J, Whittaker CA, Amon A (2014) Single cell sequencing reveals low levels of aneuploidy across mammalian tissues. Proc Natl Acad Sci U S A 111(37):13409–13414. doi:10.1073/pnas.1415287111

73. van den Bos H, Spierings DC, Taudt AS, Bakker B, Porubsky D, Falconer E, Novoa C,

Halsema N, Kazemier HG, Hoekstra-Wakker K, Guryev V, den Dunnen WF, Foijer F, Tatche MC, Boddeke HW, Lansdorp PM (2016) Single-cell whole genome sequencing reveals no evidence for common aneuploidy in normal and Alzheimer's disease neurons. Genome Biol 17(1):116. doi:10.1186/s13059-016-0976-2

74. Kuhn HG, Dickinson-Anson H, Gage FH (1996) Neurogenesis in the dentate gyrus of the adult rat: age-related decrease of neuronal progenitor proliferation. J Neurosci 16(6): 2027–2033

75. Mu Y, Gage FH (2011) Adult hippocampal neurogenesis and its role in Alzheimer's disease. Mol Neurodegener 6:85. doi:10.1186/1750-1326-6-85

76. Potter H, Granic A, Caneus J (2016) Role of trisomy 21 mosaicism in sporadic and familial Alzheimer's disease. Curr Alzheimer Res 13(1): 7–17

Chapter 17

Identification of Low Allele Frequency Mosaic Mutations in Alzheimer Disease

Carlo Sala Frigerio, Mark Fiers, Thierry Voet, and Bart De Strooper

Abstract

Germline mutations of *APP*, *PSEN1*, and *PSEN2* genes cause autosomal dominant Alzheimer disease (AD). Somatic variants of the same genes may underlie pathogenesis in sporadic AD, which is the most prevalent form of the disease. Importantly, such somatic variants may be present at very low allelic frequency, confined to the brain, and are thus very difficult or impossible to detect in blood-derived DNA. Ever-refined methodologies to identify mutations present in a fraction of the DNA of the original tissue are rapidly transforming our understanding of DNA mutation and their role in complex pathologies such as tumors. These methods stand poised to test to what extend somatic variants may play a role in AD and other neurodegenerative diseases.

Key words Single- cell sequencing, Mosaicism, Somatic variant, Alzheimer's disease, Parkinson's disease

1 Introduction

Many neurodegenerative diseases, such as Alzheimer disease (AD) and Parkinson disease (PD), have an important genetic component, and may present as familial or sporadic forms. In AD, familial forms (FAD) show dominant autosomal inheritance and are associated with point mutations in three genes (*APP*, *PSEN1*, and *PSEN2*) [1], although cases of duplication of the *APP* locus are also known [2]. The causes of sporadic AD (SAD) have not yet been identified; however, the clinical, histopathological, and biochemical similarities between SAD and FAD cases suggest a common pathological cascade.

Given the enormous genomic heterogeneity found in cells comprising the human brain [3–7], it is fair to hypothesize that during brain organogenesis some individuals may acquire somatic mutations in genes known to be causative in FAD forms. These individuals would then have patches of brain tissue bearing pathogenic mutations, which could start the same pathological cascade

José María Frade and Fred H. Gage (eds.), *Genomic Mosaicism in Neurons and Other Cell Types*, Neuromethods, vol. 131, DOI 10.1007/978-1-4939-7280-7_17, © Springer Science+Business Media LLC 2017

of events as seen in FAD patients, causing sporadic AD. Indeed, patches of neurons bearing somatic mutations of FAD genes would produce Aβ or tau aggregates, which could then spread in the brain parenchyma seeding further aggregation of amyloid or tau, respectively, in a process known as template-seeded aggregation [8, 9] (Fig. 1).

Single nucleotide variants (SNVs) are the most abundant class of mutations responsible for genomic variation between humans at a population level [10], and hence a main cause of cellular genomic heterogeneity within an individual. Indeed, it is estimated that per cell division approximately 10^{-10} errors per base pair accumulate [11], suggesting that a large majority of cells within a body may be genetically dissimilar. Whole genome sequencing of single human neurons suggests the presence of approximately 1500 somatically acquired SNVs per neuron [3]. Interestingly, transcriptionally active genes were enriched for somatic SNVs and the latter were often private to single neurons, indicating that also mechanisms other than DNA replication are causal for somatic DNA mutations [3] (Fig. 1).

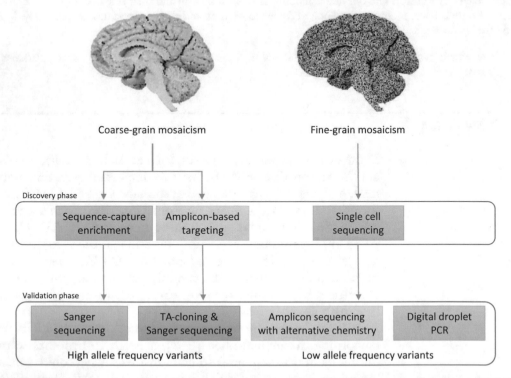

Fig. 1 Different degrees of mosaicism and recommended workflows. Depending on the developmental timing of appearance of a mutation, it can appear in multiple cells (exemplified by the red patch in the "coarse-grain mosaicism") or it can be private to a single cell (cell-private mutations are exemplified by the colored dots in "fine-grain mosaicism," where each dot is a different mutation). Depending on the prevalence of a mutation in a tissue, different sequencing approaches can be undertaken in the discovery phase. Methodologies for the validation phase are mainly dictated by the allelic frequency of the candidate variants

Analysis of somatic mutations is a rapidly evolving field that has seen a fast improvement in recent years, thanks to the development of high throughput/next generation sequencing (NGS) and more refined data analysis algorithms. There is not yet a unique gold standard method for somatic mutation detection; moreover, different methods are required for investigating fine-grain mosaicism (due to private somatic mutations present in very small numbers of cells) when compared to coarse-grain mosaicism (due to somatic variants present in a sizeable percentage of brain cells) (Fig. 1). In this chapter we will present methodologies available for investigating somatic variants, with a particular focus on those we have employed in our laboratory. Of importance, the same methodologies could be applied to the study of other neurodegenerative diseases for which a similar somatic mutation-based hypothesis can be envisaged, e.g., Parkinson disease (by targeting *SNCA*, *LRRK2*, *VPS35*, and possibly *PINK1*, *DJ-1*, and *PARK2* genes) and prion diseases such as Creutzfeldt–Jakob (by targeting the *PRNP* gene) [12].

2 Methodologies

The search for somatic variants present in a sizeable fraction of brain cells can be performed by analyzing bulk DNA extracted from a frozen postmortem brain tissue sample. The technical challenge of this approach lies in the fact that when only a few clonal cells contribute the signal for the mutant allele, thousands or millions of other cells will deliver a wild-type signal. Classic Sanger DNA sequencing is not very well suited to detect somatic mutations with low allele frequency, having a sensitivity threshold that approximates 20% mutant allele frequency (ALT) [13, 14]. The development of NGS technologies has allowed more profound analysis of bulk DNA samples: by independently sequencing hundreds to thousands of alleles from the starting DNA sample, wild-type and mutant signals at a particular genomic locus can be efficiently detected thus allowing to reach sensitivities of 5% ALT and lower. Although powerful, NGS sample preparation often involves several steps of PCR amplification of the original DNA, leading to polymerase errors. Moreover, the sequence-by-synthesis chemistry of Illumina (a widely used sequencing technology today) is also error-prone. Effectively, all in vitro polymerase errors limit the sensitivity of detection of genuine low-frequency mutations. To overcome such limitations of NGS, several methods for sample preparation and data analysis have been developed.

To achieve the high sequencing depth that is required for the detection of somatic mutations with low ALT cost efficiently, it is recommended to focus the analysis only on specific target regions of the genome that are already suspected to be involved in the disease studied. In the case of AD, it makes sense to selectively sequence the *APP, PSEN1,* and *PSEN2* genes, as these are the only genes known to cause FAD when mutated, foregoing the analysis of the rest of the genome/exome. This targeted approach can be attained with two different methodologies: (1) targeted enrichment or capture using a custom probeset on a genome sequencing library, or (2) target amplicon generation using a custom multiplexed PCR. Several vendors offer custom-designed sequence-enrichment and/or target-amplicon panels, e.g., Agilent SureSelect Target Enrichment system, Illumina Nextera Rapid Capture Custom Enrichment kit and TruSeq Custom Amplicon kit, Qiagen GeneRead DNAseq custom panels, Raindance ThuderStorm and ThunderBolt systems, and Roche NimbleGen's SeqCap. It is recommended to test the sensitivity and accuracy of the targeting approach chosen at the beginning of a research project. This can be done by setting up a preliminary experiment analyzing a series of "synthetic mosaic" samples, which can be obtained by serially diluting a bulk DNA sample containing known heterozygous mutations in the region of interest with bulk *wild-type* DNA. The expected allelic frequency of the "synthetic mosaic" variants can then be compared to the observed values and to the overall sequencing noise. This provides information about the lowest allelic frequency reliably detectable and about which parameters in sample preparation/data analysis would need optimization.

The study of fine-grained mosaicism requires to sequence the DNA of single cells: by querying each cell on its own, the contribution of each cell to the overall genomic heterogeneity of the brain can be determined. In addition to the limitations inherent to NGS (see above), single-cell DNA sequencing is confounded by the minute amounts of starting material (6.6 pg of DNA for a diploid cell) that has to be amplified prior to sequencing. Such whole-genome amplification (WGA) procedures can lead to false positives—e.g., due to DNA polymerase errors in early rounds of amplification—as well as to false negatives—e.g., due to locus or allelic dropout. Several WGA methods have been developed, the choice for a specific method is primarily guided on the desired classes of genetic variation to be detected genome wide [15, 16].

2.1 Considerations on the Tissue Samples to be Investigated

Depending on the developmental timing of its appearance, a somatic mutation will be spread more or less throughout the body; while early events could lead to a mutation being present in a fraction of both blood and neuronal cells, it is possible to have brain-private somatic mutations if they appeared after gastrulation. Therefore, in order not to miss somatic mutations present only in the brain, we have analyzed brain tissue samples from deceased AD patients instead of blood-derived DNA.

Somatic mutation analysis in neurodevelopmental diseases and cancer is facilitated by the fact that diseased tissue can be clearly identified, thanks to specific histological features. DNA isolated from the diseased tissue will be enriched for the mutant signal, while parallel sequencing of healthy tissue provides a background reference to exclude germline and de-novo mutations and to control for sequencing errors. Unfortunately, in AD it is not possible to clearly discern a brain area which is more likely to harbor a somatic variant. Reports of a patterning in the perceived spread of tau aggregates suggest that the entorhinal cortex could be one of the earlier areas affected [17]; however, the mechanistic implications of the "prionoid spread" of amyloid seeds suggest that any brain area could be the source of the first amyloid seeds. Therefore, the search for somatic mutations in sporadic AD should be directed towards several different brain areas.

2.2 Bulk DNA Sequencing: Sequence-Enrichment Approach

We have previously employed a custom Roche NimbeGen SeqCap panel to enrich sequencing libraries for the loci of *APP*, *PSEN1*, *PSEN2*, and *MAPT* [18]. The *MAPT* gene was included in the analysis even though germline *MAPT* mutations do not cause AD for the reason that tau, the product of the *MAPT* gene, is a primary player in the biochemical pathological cascade of AD; in a somatic mutation scenario it can be thus hypothesized that somatic *MAPT* mutations could lead to AD in a "two-hit" mechanism [19]. A more conservative approach would be of course to only consider the three known FAD-causative genes (*APP*, *PSEN1*, and *PSEN2*).

We chose to target the entire loci of our genes of interest, so that we could leverage the sequencing data to simultaneously analyze both somatic SNVs and CNVs, exploiting disturbances in B-allele fractions of germline heterozygous SNPs for the detection of subclonal CNVs. Both types of somatic mutations could be relevant for the development of AD. We also included 10 kbp pad regions upstream and downstream of each locus to avoid drastic drops in sequencing coverage at both ends of the loci, which would otherwise complicate the analysis. The regions targeted were (based on the Human Genome release hg19): *APP* (chr21:27,242,859–27,553,138), *PSEN1* (chr14:73,593,141–73,700,399), *PSEN2* (chr1:227,048,271–227,093,804), and *MAPT* (chr17:43,961,646–44,115,799). As the hg19 release also foresees an alternative assembly for chromosome 17, we also included the *MAPT* regions specific for the alternate assembly (chr17_ctg5_hap1:762,280–895,830). The actual probes were designed by NimbleGen according to our desired target areas, manufactured, and shipped in solution.

The experimental workflow begins with the isolation of high quality gDNA from tissue samples. To isolate bulk gDNA, frozen brain tissue is chopped with a scalpel and incubated overnight with

Protease K at 50 °C with mild agitation. RNA is then degraded during a 15 min incubation at 37 °C with RNase A (Qiagen, Venlo, The Netherlands). DNA is isolated with phenol:choloroform:isoamyl alcohol, washed twice with choloroform:isoamyl alcohol, and precipitated with 100% ethanol. The DNA pellet is further washed with 70% ethanol, dried, and finally resuspended in Tris–EDTA buffer. Sample preparation must avoid excessive vortexing or heating, as this would fragment or denature gDNA and render subsequent steps impossible. The concentration of the DNA is determined with a QuBit fluorimeter (Life Technologies, Gent, Belgium), which specifically detects double-stranded DNA, and quality is determined with a NanoDrop spectrophotometer (NanoDrop, Wilmington, DE) to exclude residual contaminants.

High quality gDNA is sheared with a Covaris sonicator (Covaris, Woodingdean Brighton, UK) to produce 300 bp fragments on average and indexed libraries are then prepared with the TrueSeq DNA kit from Illumina (Illumina, San Diego, CA). Individual libraries can then be pooled prior to sequence enrichment, the number of samples that can be pooled is function of the total number of bases captured, the intended sequencing depth for every base, and the sequence output of the instrument that will be used to analyze the samples. For this, the projected output of the instrument (according to the manufacturer) can be divided by the size of the target region and by the desired sequencing depth. We have obtained high sequencing depth (>2000× per position, on average) by pooling ten samples enriched for a ~600 kb target region and sequencing with an Illumina HiSeq 2500 in rapid mode. Demultiplexing the sequencing data amongst the pooled samples is a critical step to avoid the wrong assignment of reads to a particular sample which may result in false signals. Sample-specific indices are preferably at least three nucleotides different, allowing maximum one mismatch in the index sequence during demultiplexing. After demultiplexing, sequencing data are usually encoded in FASTQ files.

Sample-individual FASTQ files are aligned to the human reference genome using BWA [20] and converted to a BAM file format for downstream analysis [21]. Next, since indels can cause misalignment of the reads, the alignment should be refined by local realignment around indels using the GATK IndelRealigner tool [22], and base qualities should be recalibrated using the GATK BaseRecalibrator tool [22] to correct systematic technical errors in base quality calling by the sequencing instrument. These steps of data preprocessing will yield a BAM file which can then be used to call variants. There is a great variety of variant calling algorithms (see Table 1) based on different statistical algorithms. We have efficiently used Samtools mPileup function [23] together with VarScan 2.0 [24] to generate a list of candidate somatic variants. A useful approach for variant calling involves the generation of

Table 1
Bioinformatics analysis software for DNA sequencing and variant calling analysis

Name	Purpose	Link
BWA	Alignment of raw reads to reference genome	bio-bwa.sourceforge.net/
SAMtools	Handling of aligned reads, pileup of aligned reads	samtools.sourceforge.net/
GATK	Handling of aligned reads, error correction, variant calling	https://software.broadinstitute.org/gatk/
VarScan	Variant calling	http://dkoboldt.github.io/varscan/
SNVer	Variant calling	snver.sourceforge.net/
LoFreq	Variant calling	http://csb5.github.io/lofreq/
UMI-tools	Handling of unique molecular identifiers	https://github.com/CGATOxford/UMI-tools
SnpEff	Variant annotation	snpeff.sourceforge.net/
Annovar	Variant annotation	www.openbioinformatics.org/annovar/
Monovar	Variant calling in single-cell sequencing data	https://bitbucket.org/hamimzafar/monovar
Single Cell Genotyper	Variant calling in single-cell sequencing data	https://bitbucket.org/aroth85/scg/wiki/Home
R	Statistical analysis	http://www.r-project.org/

In the table we provide a description and links for software mentioned in the methodologies, along with similar software that can be used for DNA sequencing analysis and somatic variant discovery. Links are updated and valid as of December 2016

a conservative list of candidate variants called by different algorithms. Indeed, each variant calling software may identify different sets of variants, due to the unique properties of each algorithm. Variants called by multiple algorithms may be considered as high confidence candidates.

Of importance, some somatic variant calling algorithms expect that a test sample is compared with a "matched normal" sample (e.g., a tumor sample compared to a healthy tissue sample) to efficiently rule out false positive calls and germline variants. In the case of AD tissue, it is not possible to perform such comparison, since, differently from tumor studies, it is not obvious which tissue should have a somatic mutation and which should be devoid of it. Hence, we performed variant calling on each sample on its own.

Candidate variants are then annotated, i.e., information on the genomic region, presence in databases and potential functional consequences (synonymous, nonsynonymous, nonsense,...) is retrieved, finally yielding a VCF (variant call format) file. Various tools exist also for variant annotation, we have used SnpEff [25] and Annovar [26]. Further analysis of variants and sequencing data can be carried out efficiently using R (http://www.r-project.org/). Annotation of variants can be useful to prioritize a long list of candidate variants for further validation and is instrumental in gaining insight on the biological consequences of true somatic variants.

It is important to notice that all the abovementioned software is constantly updated, hence it is recommended to use the latest versions, to consistently use one version for the analysis of all samples in a project, and in any case to correctly report the version number of each software used.

The bioinformatics analysis can be computationally intense, in particular the alignment step, the GATK-based steps, and the steps in R if dealing with a big target region or a high number of candidate variants. Analysis can be efficiently tackled by computing clusters or by the use of dedicated servers.

2.3 Bulk DNA Sequencing: Targeted Approach with Tagged Reads

Preparation of sequencing libraries involves several PCR steps, which introduce errors which may complicate somatic variant analysis. The main errors due to PCR are: (1) the incorporation of wrong nucleotides and (2) the skew in amplification of mutated alleles and wild-type alleles (see Fig. 2) which could result in false negative (in case the mutant allele is under-represented) or false positive (in case a PCR error is over-represented) calls. The

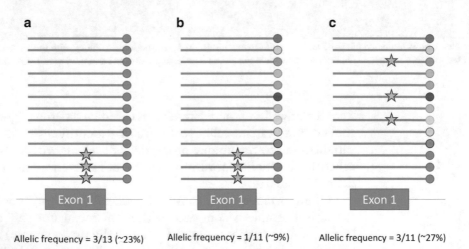

Allelic frequency = 3/13 (~23%)   Allelic frequency = 1/11 (~9%)   Allelic frequency = 3/11 (~27%)

Fig. 2 Principles of amplicon tagging. A low-frequency mutation (indicated by a star) is present only in a fraction of the reads (represented by a *blue line ending with a circle*) aligning on a genomic target of interest. When no read tagging is used (e.g., after sequence-enrichment of sheared gDNA), the allelic frequency of a somatic mutation cannot be corrected for PCR duplication artifacts (**a**). When reads are tagged prior to PCR amplification, duplicates can be easily spotted and the allelic frequency of the mutation can be corrected (**b**, tags indicated by the *colored circles* at the end of each read). Deduplication of PCR artifacts leads to correct assessment of the allelic frequency of somatic variant also in the case that the somatic variant is not present in the duplicated reads (**c**)

skewed amplification behavior and the introduction of incorrect nucleotides are inherent to PCR and DNA polymerases [27, 28], however several smart workarounds have been developed to prevent and counter them.

Two approaches (Duplex Sequencing [29] and UMI-TSCA [30]) share the general principle to add barcodes to the original DNA molecules in order to track the daughter molecules produced during PCR amplification. This allows (1) to correctly remove PCR duplicates (deduplication) by counting unique barcodes instead of raw reads, and (2) to generate consensus reads by pooling all reads sharing the same barcodes (Fig. 2).

In the Duplex Sequencing approach [29, 31], gDNA is sheared and fragments are ligated to duplex sequencing adapters, resulting in DNA fragments bearing 12-nucleotide long barcode sequences at both ends. Fragments barcoded at each end are PCR amplified and sequenced, then reads are grouped by the barcodes and consensus sequences are derived. This approach also allows to double-check a variant by identifying its presence in both families of sequencing reads that derive from the complementary Watson and Crick strands of the original gDNA molecule. However, this method still requires the development of a sequence-enrichment panel to prepare a targeted sequencing library.

An alternative approach [30] applies a modification of the TruSeq Custom Amplicon (TSCA) kit from Illumina to introduce a Unique Molecule Identifer (UMI) in place of the P5 sample index. The TSCA kit is a custom-designed panel of probes recognizing a list of user-defined genomic targets; for each target, two probes are designed, one upstream and one downstream. After hybridization to the gDNA targets, the upstream probe is extended by a DNA polymerase onto the downstream probe, producing a copy of the original gDNA region. Next, a unique barcode is added to each copy and the product is PCR amplified to produce the sequencing library: during this step the barcodes are copied together with the genomic target, thus keeping the original labelling. This approach is more straightforward than the former, as it directly generates a targeted amplicon library; however, it lacks the possibility of copying both strands of the original gDNA target.

A third method, CirSeq [32], dispenses from using barcodes altogether: instead of PCR amplifying the intended target, the gDNA is sheared, fragments are circularized and using random primers and a high processivity DNA polymerase concatenated copies of the original template are made. The strength of the method lies in the fact that mutations are not propagated by PCR, as each new copy comes from the original DNA sequence. However, it also requires an extra step to select the regions of interest when dealing with a targeted approach, which would require additional PCR reactions.

We have applied the UMI-modified TSCA approach to analyze somatic mutations in AD, our choice was based on the fact that this approach is the most straightforward for a targeted sequencing analysis, and built around a commercially available kit already aimed at the analysis of large genomes (such as the human genome). In order to maximize the sequencing coverage of areas of interest, we targeted the exons of *APP, PSEN1, PSEN2,* and *MAPT* known to harbor pathogenic mutations. The TSCA panel was designed by Illumina, to accommodate 250 bp-long amplicons, the total area covered is ~7 kb long thus allowing a very high coverage when sequencing 12 pooled samples on an Illumina MiSeq 2×300 sequencing run.

To prepare sequencing libraries with the modified TSCA method, first the gDNA is denatured and slowly cooled in the presence of the probeset, allowing specific annealing of the probes with their cognate sites on the gDNA. Next, a DNA polymerase extends the upstream probe and a DNA ligase joins the newly synthesized copy of the gDNA to the downstream probe. The ligation-extension products are purified using a filter plate (provided with the kit, per manufacturer's recommendations). Next, in place of the canonical direct PCR amplification of the extension-ligation products, we performed one cycle of PCR with the Illumina P7 primers (bearing a sample-specific index) and a modified P5 primer (P5′) which contains a random 12-nucleotide sequence (which constitutes the UMI) in place of the second sample-specific index. Given the high number of possible UMIs ($4^{12} = 16,777,216$), it is highly likely that each copy of a specific genomic target in the original sample will get a different UMI. Next, PCR products are purified with Ampure XP magnetic beads and a second round of PCR is carried out using the same sample-specific P7 primer as before and a P5″ primer that anneals downstream of the UMI and carries the Illumina-specific P5 sequence handle for correct loading on an Illumina flowcell. The final PCR product is again purified using Ampure XP magnetic beads, quantified using the KAPA library quantification kit for Illumina libraries (Kapa Biosystems, Wilmington, MA), and equimolar sample-specific libraries are pooled. Since only the P7 sample index can be used to label different biological samples, the maximum amount of samples that can be pooled on a single sequencing run is 12, as Illumina provides only 12 different P7 indices, restricting high sample throughput.

The Illumina MiSeq sample sheet (which instructs the sequencing instrument on the run parameters) has to be modified to account for a longer (12 nucleotides instead of 8) i5 index read (which covers the UMI). In order to keep the UMI tied to each specific R1 and R2 read pair (which cover the actual amplicon), we have appended the UMI sequence to the header of each R1 and R2 read. To analyze the data, we first aligned the reads to the reference genome using BWA-MEM, and performed local realignment around indels and base quality recalibration with the dedicated GATK software.

Next, UMIs are leveraged to correct for PCR artifacts and to generate consensus nucleotide calls at each position assessed. The overall approach is to group reads aligned to one genomic locus that share the same UMI (called "UMI family"). Subsequently, analysis can focus on a read-by-read basis or on a position-by-position basis. Following the latter approach, we have developed an algorithm (called Scotoplanes, Fig. 3) which generates a pileup of the nucleotides aligned at each position of our target region, while taking the UMI into account. Deduplication of reads nucleotides with the same UMI depends on the level of UMI duplication and whether or not all the reads with the same UMI support the same nucleotide: (1) a UMI appears only once: the associated nucleotide is counted (e.g., UMI1, Fig. 3); (2) a UMI is observed more than once, but all instances support the same nucleotide: the nucleotide

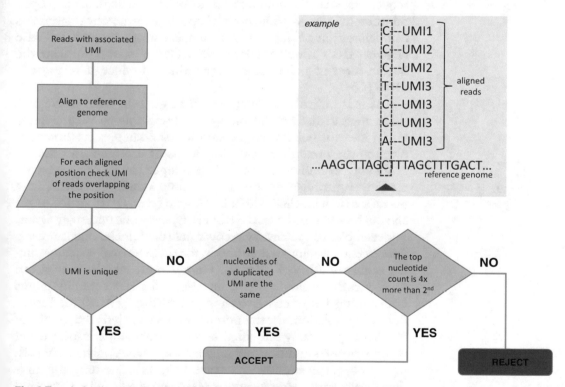

Fig. 3 Tag-deduplication algorithm. The algorithm represented has been implemented in a software developed in the lab to process UMI-labelled reads. The *green box* represents different possible scenarios of duplicated UMIs and connected reads. After alignment to the genome, reads and their associated UMIs are analyzed position by position. Reads are grouped based on their UMIs into UMI families (e.g., UMI1, UMI2, UMI3). Depending on the duplication levels and on the supporting nucleotides found for different duplicate molecules, a read is either retained (accept) or discarded (reject). In the example (*green box*) UMI1 is unique and thus retained; UMI2 is duplicated but all nucleotides are the same and thus UMI2 is retained; UMI3 is duplicated and the most represented nucleotide is less than fourfold more abundant than the second, thus UMI3 is rejected.

is retained but counted only once (e.g., UMI2, Fig. 3); (3) a UMI is observed more than once, and associates with a number of different nucleotides: the most abundantly present nucleotide is retained (and counted once), but only if it is at least four times more abundant than the second most frequent nucleotide associated with the same UMI at that position (e.g., UMI3, Fig. 3). After such position-by-position UMI deconvolution, allele frequencies are calculated at each position, and major and minor alleles are called. Candidate variants can then be annotated using SnpEff and Annovar as seen above to prioritize variants for further validation.

2.4 Single-Cell DNA Sequencing

Single-cell DNA sequencing has the capacity to detect somatic variants down to the biological unit of organs. Although powerful, the method is still technically highly challenging, which has to be taken into account when designing a project [16]. In particular, a cell's gDNA has to be amplified to obtain enough material for sequencing analysis, and methods for WGA are still in need of optimization to improve uniformity of coverage—whereby unevenness in amplification can lead to locus or allelic drop outs and thus false negative SNVs as well as false positive CNVs—and to mitigate the introduction of polymerase errors which produce false positive SNVs [16].

As it is difficult to isolate intact single cells from a complex tissue such as the human brain, it is preferable to isolate single nuclei instead. For nuclear isolation, tissue samples are homogenized in the presence of very low levels of detergent in order to avoid compromising the nuclear envelope. Nuclei can be recovered by centrifugation on an Optiprep density gradient. The nuclei are then labelled with a DNA stain (e.g., DAPI or DRAQ5) and can be additionally marked for cell type-specific nuclear antigens. For example, to specifically recover neuronal nuclei one can use a fluorescently conjugated antibody for the neuronal-specific splicing regulator Rbfox3 (NeuN). Fluorescently labelled nuclei can subsequently be single-sorted into 96 well plates containing lysis buffer using fluorescent activated cell sorting (FACS) platforms. Following isolation, for the purpose of SNV analysis the gDNA of the cell is preferably amplified by isothermal amplification using random primers and the Φ29 polymerase, a DNA polymerase with high processivity and low error rates [15, 33]. The resulting multiple displacement amplification (MDA) product can be converted in a sequencing library using conventional methods, or can be used in conjunction with a sequence-enrichment approach. As for bulk DNA approaches, it is best to include a sequence-enrichment step to focus on genomic regions which, if mutated, could drive AD pathogenesis.

Although variant calling algorithms developed for bulk DNA have also been used for calling variants in single-cell DNA sequencing data, this latter kind of data has peculiar aspects which warrants the use of bespoke algorithms. Specifically, the technical artifacts introduced during WGA need to be taken into account, to prevent calling false variants. Due to the error rate of WGA polymerases—e.g., the per-cycle per-base error rate of MDA has been estimated to approximate 3.2×10^{-6} [34]—it is currently impossible to call SNVs that are private to a cell with absolute confidence [15]. Instead, the candidate SNV has to be reported by at least a few cells to increase reliability. Algorithms specifically designed for analyzing SNVs in single-cell DNA sequencing data, as Monovar [35] and Single Cell Genotyper [36], are emerging.

Single-cell DNA sequencing has been used widely to study tumor evolution, and recently it has been used for the identification of somatic mutational signatures in the human brain [3]. A possible future application to investigate somatic genetic variation underlying the cause of AD would be to couple single nucleus sorting and DNA library preparation with a sequence-enrichment panel targeted for the *APP, PSEN1,* and *PSEN2* loci. Alternatively, a direct amplicon-based assay targeting the most pathogenic mutant sites for AD may be directly applied on single-cell DNA at scale without upfront WGA [37]. Although single-cell sequencing allows SNV analysis at relatively low coverage, many different cells may need to be pooled and sequenced for the discovery of low-frequency somatic SNVs in the diseased tissue.

3   Confirmation of Candidate Somatic Variants

Although bioinformatics analysis of sequencing data is based on continuously improving algorithms, the stochastic nature of errors introduced at any step of sample preparation and sequencing means that the analysis can at best give a list of candidate variants with differing degrees of confidence. Confirmation of called somatic variants by an alternative approach remains therefore important to validate the results and to develop increasingly reliable somatic variant callers. There are several ways to confirm candidate variants, depending on the approach used at first pass (bulk tissue or single-cell analyses), the allelic frequency of the candidate variant and the number of candidates to be tested (for budgeting reasons).

3.1 Sanger Sequencing

The resolution of Sanger sequencing for detecting somatic variants is limited to variants with an allelic frequency of 20% or more [13, 14]. In this procedure, the candidate mutant locus is amplified from bulk DNA with specific PCR primers and then directly sequenced. However, it requires that the sequencing runs have very low levels of noise to be able to reliably identify the somatic variant. Parameters that have to be carefully controlled are in particular the clean-up and resuspension of the sample after the PCR with fluorescent dideoxy nucleotides. In practice, it is more efficient to use alternative methods even for high allele frequency somatic variants.

3.2 TA-Subcloning and Sanger Sequencing

The principle of this approach is to exploit bacteria to partition the bulk DNA signal, so that Sanger sequencing can be efficiently used to validate a candidate somatic variant.

The candidate mutation is amplified from the original bulk gDNA sample using a pair of specific PCR primers. The PCR amplicons are then cloned into a plasmid vector, ideally one which allows quick and efficient subcloning without the need for restriction digest of the amplicon such as those provided by the TOPO TA cloning kit from ThermoFisher (ThermoFisher, Waltham, Massachusetts). After bacterial transformation and plating, single independent colonies are isolated and sequenced. Each colony will bear only one allele, hence Sanger sequencing will yield a yes/no answer for the presence of the candidate mutation. By counting the ratio of mutant colonies, we can infer the allelic frequency of the target mutation in the original DNA sample.

This approach works best with mutations having a relatively high allelic frequency, to avoid having to sequence many colonies, but could in theory be up-scaled to detect low-frequency variants.

3.3 Digital Droplet PCR

Following the same principle of partitioning the original bulk gDNA signal, digital droplet PCR (ddPCR) allows to achieve very good sensitivity, down to 0.1% ALT or lower (Fig. 4) [38]. In a ddPCR assay, a PCR reaction is mixed with oil in a microfluidics chip, leading to emulsification of the PCR reaction into thousands of nanoliter droplets containing either zero, one, or more than one template DNA molecules. After completion of the PCR reaction, droplets are read one by one, measuring the presence of fluorescently labelled allele-specific probes. Statistical analysis of the count data, based on Poisson statistics, allows calculating confidence levels for the allelic frequency determined for each sample. The extremely high partition of the template DNA and the discrete nature of droplet counting allows to detect very low levels of mutant allele molecules.

For each candidate variant a specific PCR primer set is designed together with two hydrolysis probes (e.g., Taqman probes), one recognizing the wild-type allele and one recognizing the mutant allele, labelled with fluorophores with different spectral wavelengths (e.g., FAM and HEX). The use of two differentially labelled probes allows the simultaneous detection of the two alleles in the same reaction, thus offering a better quantification than running mutant and wild-type assays in different tubes. For each newly developed assay, several parameters need to be optimized, including correct annealing temperature, starting gDNA quantity, and the numbers of wells/droplets required.

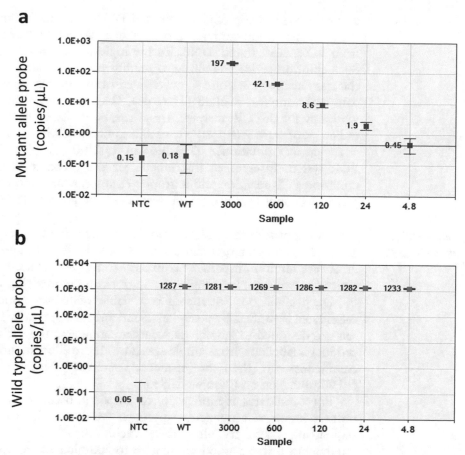

Fig. 4 Titration of digital droplet PCR assay for a candidate variant. A ddPCR assay for a candidate variant is first tested for sensitivity by running appropriate negative and positive controls. Negative controls include a non-template control (NTC) sample and a wild-type DNA sample, to check background probe fluorescence and nonspecific hybridization, respectively. As positive controls we used a series of wild-type DNA samples containing a spiked-in gBLOCK construct containing the candidate variant. We spiked in 3000, 600, 120, 24, and 4.8 molecules of mutant gBLOCK construct per reaction, which corresponds to 10, 2, 0.4, 0.08, and 0.016% of the number of alleles present in the template DNA in each reaction (100 ng of DNA, ~30,000 haploid genome copies). Data were analyzed with BioRad QuantaSoft, and we report the calculated number of copies per μL of mutant (**a**) and wild-type allele (**b**); error bars are the 95% Poisson confidence intervals. The red line highlights the lowest sensitivity attainable, i.e., the upper boundary of the mutant allele count in the wild-type sample. Hence, for this specific assay a variant can be called if its lower Poisson confidence interval is above 0.42 copies/μL

Each assay should be run with appropriate negative and positive control samples. As negative control samples, assays should be run without template gDNA to assess the levels of inherent background fluorescence of the probes, and should be run with gDNA devoid of the target somatic variant (wild-type control) to assess the signal coming from nonspecific PCR products. As positive controls, a dilution series of a construct containing the target somatic variant spiked to wild-type DNA could be analyzed. For this purpose, a gBLOCK synthetic double-stranded DNA molecule mimicking the genomic region containing the somatic variant can be ordered from IDT (Integrated DNA Technologies, Leuven, Belgium). Such dilution series will allow to readily assess the sensitivity of the assay against the noise signal derived from the analysis of pure wild-type samples. BioRad (Hercules, CA) provides a complete workflow for ddPCR analysis. Assays can be designed using their online software (https://www.bio-rad.com/digital-assays/#/assays-create/mutation). The BioRad QX200 Droplet Digital PCR system foresees an instrument for the preparation of the emulsion PCR reaction and a droplet counter. Moreover, BioRad provides a software (QuantaSoft) to analyze droplet counts (Fig. 4).

3.4 Resequencing of Candidate Variants

Another approach to validate candidate variants is to re-sequence at very high depth target amplicons: as most PCR and sequencing errors are random, it would be unlikely to retrieve a false positive variant in two separate sample preparations and sequencing runs.

Deep amplicon sequencing is a viable validation option for candidates identified by a sequence-enrichment approach. Primer sets are designed to amplify each target candidate, and each PCR product is labelled with a sample-specific index to allow for pooling. Amplicons can then be sequenced at very high coverage (>100,000×) on an Illumina MiSeq.

For candidates identified in an amplicon-based approach a more relevant re-sequencing approach would entail change of sequencing chemistry altogether. Pacific Biosciences (PacBio) sequencing offers a good alternative to Illumina sequencing for this kind of validation, as the sequencing chemistry, and therefore the error pattern, are radically different between the two. For PacBio sequencing, after preparing indexed amplicons, library preparation foresees the ligation of bubble adaptors on both ends of each DNA molecule, enabling recursively sequencing of the Watson and Crick DNA strands of that molecule on an RSII or Sequel platform (Pacific Biosciences, Menlo Park, CA). Following sequencing, all subreads from a DNA molecule are piled and a consensus sequence is derived. As PacBio polymerase errors are stochastic, the consensus sequence generation is able to correct most sequencing errors, thus providing a high confidence validation.

4 Conclusions

Somatic variant analysis is a rapidly developing field, with continuously improved sample preparation and sequencing methodologies and continuously refined statistical algorithms for variant detection. Interest is that somatic variant detection goes across multiple fields of biology, e.g., developmental biology, tumor biology, neurobiology, toxicology, and has even forensics applications.

In the field of AD research, where the primary cause of sporadic AD is still unknown, the possibility of somatic mutations being pathogenic drivers is a long-standing question [39, 40]. We foresee that continuous optimization of the methodologies will finally clarify the role of somatic mutations in AD. Moreover, the methodologies illustrated in this chapter can be extended to other neurodegenerative diseases, in particular for those in which a template-seeded aggregation mechanism is involved, e.g., Parkinson disease [41, 42] and prion diseases.

References

1. Cruts M, Theuns J, Van Broeckhoven C (2012) Locus-specific mutation databases for neurodegenerative brain diseases. Hum Mutat 33:1340–1344. doi:10.1002/humu.22117

2. Rovelet-Lecrux A, Hannequin D, Raux G et al (2006) APP locus duplication causes autosomal dominant early-onset Alzheimer disease with cerebral amyloid angiopathy. Nat Genet 38:24–26. doi:10.1038/ng1718

3. Lodato MA, Woodworth MB, Lee S et al (2015) Somatic mutation in single human neurons tracks developmental and transcriptional history. Science 350:94–98. doi:10.1126/science.aab1785

4. Upton KR, Gerhardt DJ, Jesuadian JS et al (2015) Ubiquitous L1 mosaicism in hippocampal neurons. Cell 161:228–239. doi:10.1016/j.cell.2015.03.026

5. Evrony GD, Lee E, Park PJ, Walsh CA (2016) Resolving rates of mutation in the brain using single-neuron genomics. elife. doi:10.7554/eLife.12966

6. McConnell MJ, Lindberg MR, Brennand KJ et al (2013) Mosaic copy number variation in human neurons. Science 342:632–637. doi:10.1126/science.1243472

7. Evrony GD, Cai X, Lee E et al (2012) Single-neuron sequencing analysis of L1 retrotransposition and somatic mutation in the human brain. Cell 151:483–496. doi:10.1016/j.cell.2012.09.035

8. Aguzzi A, Lakkaraju AK (2015) Cell biology of prions and prionoids: a status report. Trends Cell Biol 26(1):40–51. doi:10.1016/j.tcb.2015.08.007.

9. Brettschneider J, Del Tredici K, Lee VM, Trojanowski JQ (2015) Spreading of pathology in neurodegenerative diseases: a focus on human studies. Nat Rev Neurosci 16:109–120. doi:10.1038/nrn3887

10. Auton A, Brooks LD, Durbin RM et al (2015) A global reference for human genetic variation. Nature 526:68–74. doi:10.1038/nature15393

11. Nussbaum R, McInnes RR, Willard HF (2007) Thompson & Thompson genetics in medicine, 7th edn. Saunders, Philadelphia

12. Alzualde A, Moreno F, Martinez-Lage P et al (2010) Somatic mosaicism in a case of apparently sporadic Creutzfeldt-Jakob disease carrying a de novo D178N mutation in the PRNP gene. Am J Med Genet B Neuropsychiatr Genet 153B:1283–1291. doi:10.1002/ajmg.b.31099

13. Tsiatis AC, Norris-Kirby A, Rich RG et al (2010) Comparison of Sanger sequencing, pyrosequencing, and melting curve analysis for the detection of KRAS mutations: diagnostic and clinical implications. J Mol Diagn 12:425–432. doi:10.2353/jmoldx.2010.090188

14. Jamuar SS, Lam AT, Kircher M et al (2014) Somatic mutations in cerebral cortical malformations. N Engl J Med 371:733–743. doi:10.1056/NEJMoa1314432

15. Macaulay IC, Voet T (2014) Single cell genomics: advances and future perspectives. PLoS Genet 10(1):e1004126. doi:10.1371/journal.pgen.1004126

16. Gawad C, Koh W, Quake SR (2016) Single-cell genome sequencing: current state of the science. Nat Rev Genet 17:175–188. doi:10.1038/nrg.2015.16

17. Braak H, Braak E (1991) Neuropathological stageing of Alzheimer-related changes. Acta Neuropathol 82:239–259

18. Sala Frigerio C, Lau P, Troakes C et al (2015) On the identification of low allele frequency mosaic mutations in the brains of Alzheimer's disease patients. Alzheimers Dement 11:1265–1276. doi:10.1016/j.jalz.2015.02.007

19. Small SA, Duff K (2008) Linking Abeta and tau in late-onset Alzheimer's disease: a dual pathway hypothesis. Neuron 60:534–542. doi:10.1016/j.neuron.2008.11.007

20. Li H, Durbin R (2010) Fast and accurate long-read alignment with burrows-wheeler transform. Bioinformatics 26:589–595. doi:10.1093/bioinformatics/btp698

21. Li H, Handsaker B, Wysoker A et al (2009) The sequence alignment/map format and SAMtools. Bioinformatics 25:2078–2079. doi:10.1093/bioinformatics/btp352

22. McKenna A, Hanna M, Banks E et al (2010) The genome analysis toolkit: a MapReduce framework for analyzing next-generation DNA sequencing data. Genome Res 20:1297–1303. doi:10.1101/gr.107524.110

23. Li H (2011) A statistical framework for SNP calling, mutation discovery, association mapping and population genetical parameter estimation from sequencing data. Bioinformatics 27:2987–2993. doi:10.1093/bioinformatics/btr509

24. Koboldt DC, Zhang Q, Larson DE et al (2012) VarScan 2: somatic mutation and copy number alteration discovery in cancer by exome sequencing. Genome Res 22:568–576. doi:10.1101/gr.129684.111

25. Cingolani P, Platts A, Wang le L et al (2012) A program for annotating and predicting the effects of single nucleotide polymorphisms, SnpEff: SNPs in the genome of Drosophila melanogaster strain w1118; iso-2; iso-3. Flying 6:80–92. doi:10.4161/fly.19695

26. Wang K, Li M, Hakonarson H (2010) ANNOVAR: functional annotation of genetic variants from high-throughput sequencing data. Nucleic Acids Res 38:e164. doi:10.1093/nar/gkq603

27. Kanagawa T (2003) Bias and artifacts in multi-template polymerase chain reactions (PCR). J Biosci Bioeng 96:317–323. doi:10.1016/S1389-1723(03)90130-7

28. Gundry M, Vijg J (2011) Direct mutation analysis by high-throughput sequencing: from germline to low-abundant, somatic variants. Mutat Res 729:1–15. doi:10.1016/j.mrfmmm.2011.10.001

29. Schmitt MW, Kennedy SR, Salk JJ et al (2012) Detection of ultra-rare mutations by next-generation sequencing. Proc Natl Acad Sci U S A 109:14508–14513. doi:10.1073/pnas.1208715109

30. Smith EN, Jepsen K, Khosroheidari M et al (2014) Biased estimates of clonal evolution and subclonal heterogeneity can arise from PCR duplicates in deep sequencing experiments. Genome Biol 15:420. doi:10.1186/s13059-014-0420-4

31. Kennedy SR, Schmitt MW, Fox EJ et al (2014) Detecting ultralow-frequency mutations by duplex sequencing. Nat Protoc 9:2586–2606. doi:10.1038/nprot.2014.170

32. Lou DI, Hussmann JA, McBee RM et al (2013) High-throughput DNA sequencing errors are reduced by orders of magnitude using circle sequencing. Proc Natl Acad Sci U S A 110: 19872–19877. doi:10.1073/pnas.1319590110

33. Dean FB, Hosono S, Fang L et al (2002) Comprehensive human genome amplification using multiple displacement amplification. Proc Natl Acad Sci U S A 99:5261–5266. doi:10.1073/pnas.082089999/8/5261

34. de Bourcy CF, De Vlaminck I, Kanbar JN et al (2014) A quantitative comparison of single-cell whole genome amplification methods. PLoS One 9:e105585. doi:10.1371/journal.pone.0105585PONE-D-14-24544

35. Zafar H, Wang Y, Nakhleh L, et al (2016) Monovar: single-nucleotide variant detection in single cells. doi: 10.1038/pj.2016.37

36. Roth A, McPherson A, Laks E et al (2016) Clonal genotype and population structure inference from single-cell tumor sequencing. Nat Methods 13:573–576. doi:10.1038/nmeth.3867

37. Eirew P, Steif A, Khattra J et al (2014) Dynamics of genomic clones in breast cancer patient xenografts at single-cell resolution. Nature 518: 422–426. doi:10.1038/nature13952

38. Hindson BJ, Ness KD, Masquelier DA et al (2011) High-throughput droplet digital PCR system for absolute quantitation of DNA copy number. Anal Chem 83:8604–8610. doi:10.1021/ac202028g

39. Geller LN, Potter H (1999) Chromosome missegregation and trisomy 21 mosaicism in Alzheimer's disease. Neurobiol Dis 6:167–179. doi:10.1006/nbdi.1999.0236

40. Beck JA, Poulter M, Campbell TA et al (2004) Somatic and germline mosaicism in sporadic early-onset Alzheimer's disease. Hum Mol Genet 13:1219–1224. doi:10.1093/hmg/ddh134ddh134

41. Proukakis C, Houlden H, Schapira AH (2013) Somatic alpha-synuclein mutations in Parkinson's disease: hypothesis and preliminary data. Mov Disord 28:705–712. doi:10.1002/mds.25502

42. Proukakis C, Shoaee M, Morris J et al (2014) Analysis of Parkinson's disease brain-derived DNA for alpha-synuclein coding somatic mutations. Mov Disord 29:1060–1064. doi:10.1002/mds.25883

INDEX

A

Aging..28, 170, 271–285, 287–295,
    331, 333, 334, 352–354
Aicardi–Goutières syndrome (AGS) 198
Allele frequency (AF)..............................6–11, 13, 14, 17–20,
    179, 301, 306–308, 361–376
Allelic dropout events (ADO)...265
Alu element..190, 220
Alzheimer's disease (AD) 28, 34, 38, 44,
    83, 90, 92, 95, 97–103, 156, 329–342, 345, 352,
    354, 355, 361–376
Amplicon-seq...................15, 16, 19, 307, 311, 316, 318, 376
Amyloid precursor protein (APP) 329–331,
    333–335, 340, 342–347, 350, 355
Annovar program 368, 372
Apoptosis....................................44, 272, 332, 345, 352–355
Ataxia-telangiectasia...28, 210
Ataxia telangiectasia mutated (ATM) 197, 198
Autism..28
Autism spectrum disorders (ASD) 197

B

BAM files..125, 366
BED files...111, 126, 127
Bedtools...125, 126, 177
Beta-amyloid (Aβ)......................................83, 330, 332, 335,
    347–351, 353, 355, 362
Bioinformatic statistics 110
Bipolar disorder 209, 210
Bowtie aligner..239, 311
Brain..3–11, 13–20, 27–37,
    81–83, 87, 92–98, 163, 169, 170, 180, 182, 220,
    221, 226, 227, 238, 240, 242, 253, 254, 266,
    271–285, 287–295, 299–306, 308, 310, 311,
    314–316, 318, 320, 322, 330–334, 336–342,
    344–346, 352, 354, 361, 363–365, 372, 373
Burrows-Wheeler Aligner (BWA-ALN).................311, 313
Burrows-Wheeler Aligner-Maximal Exact Matches
    algorithm (BWA-MEM)........................ 264, 310,
    311, 313, 322, 370

C

Cell cycle 4, 27, 44, 58, 61,
    164, 175, 182, 265, 331, 335, 341, 344,
    347, 354, 355

Cell nuclei isolation..61
CellRaft technology 116–118
Charcot-Marie-Tooth neuropathy type 1 (CMT1)..........144
Chick forebrain 59
Chicken erythrocyte nuclei (CEN) 45, 46, 50, 52
Chromogenic in situ hybridization (CISH) 60, 82,
    83, 87, 90–93, 95–97
Chromosomal mosaicism 27–37
Chromosome instability 329–352, 354, 355
Chromosome mis-segregation........................ 330–332, 342,
    346, 347, 350, 352, 354, 355
Circular binary segmentation (CBS) algorithm...............126
Clinical Dementia Rating (CDR)..............................83, 99
Clonal expansion 6, 10, 11, 17
CNV coldspots 128
CNV detection 109–121, 123–128,
    144, 146, 152, 154
CNV hotspots 128
Coefficient of variation (CV) 151, 160
Combined Annotation-Dependent Depletion (CADD)
    Value 314
Comparative genome hybridization array
    (aCGH) 5, 12, 13
Competitive PCR.. 143–152, 155,
    156, 158–160
ContEst tool...314
CpG methylation 65, 70–75
CRISPR technology.................................... 318, 322, 323
C value..58
Cystic fibrosis 143, 144, 146, 158

D

Degenerate oligonucleotide-primed PCR
    (DOP-PCR).........................9, 110, 119, 255–257
4',6'-Diamidino-2-phenylindole (DAPI) 29, 35,
    45, 47, 48, 53, 60, 62, 68, 69, 77, 116, 258,
    276–278, 284, 287, 288, 293, 333, 338, 339,
    347, 348, 351, 372
Digital droplet PCR (ddPCR) 16, 18–20, 374–376
Digital PCR 5, 145, 160, 376
DNA content variation (DCV)........................... 44, 82, 83,
    89, 96–102
DNA fragment capture and sequencing 14, 15
DNA libraries...110, 114, 124, 373
Φ29 DNA polymerase...256, 257
DNA probes for FISH 28, 32

José María Frade and Fred H. Gage (eds.), *Genomic Mosaicism in Neurons and Other Cell Types*, Neuromethods, vol. 131,
DOI 10.1007/978-1-4939-7280-7, © Springer Science+Business Media LLC 2017

DNA sequencing 3, 6, 10, 13–16, 18, 37, 44, 50, 71, 73, 112, 143, 189, 220, 221, 255, 265, 315, 338, 344, 369

DNA sequence variations
  single nucleotide variant (SNV) 3, 4, 8, 9, 13, 14, 17–19, 82, 138, 166, 176–178, 254, 264, 265, 302, 303, 306–308, 310–312, 314, 316–318, 322, 362, 365, 372, 373
  small insertion and deletion (indel) 3, 4, 13–15, 17–19, 166, 176–178, 366, 371

DNA-sequencing 363–373

DNA transposons 189

Down syndrome (DS) 144, 145, 156, 157, 330, 331, 345

DRAQ5 .... 45, 372

Duchenne muscular dystrophy (DMD) 144

F

False-negative (FN) errors 265

False-positive (FP) errors 265

Familial AD (FAD) 329, 330, 333–335, 340–350, 355, 361, 362, 364, 365

FastQC 113, 124, 125, 303, 308, 309

FASTQ file 309, 366

FASTQ file 124, 125, 175, 176

FASTX-Toolkit 125

Flanking genomic PCR assay 265

Flow cytometric analysis (FCM) 44–50, 52–54, 57–65, 67, 69–72, 74–76

Flow cytometry 12, 44–50, 52–54, 57–62, 64, 66–69, 145, 211, 213

Fluorescence-activated cell sorting (FACS) 44, 46, 47, 54, 61, 62, 65, 69, 116, 117, 168, 180, 199, 203, 255, 259–261, 372

Fluorescence-activated nuclear sorting (FANS) 44, 46, 116

Fluorescence in situ hybridization (FISH) 5, 8, 11, 12, 27–37, 44–50, 52–54, 144, 271–285, 287–295, 329–352, 354, 355

Focal cortical dysplasia (FCD) 300, 301, 312, 317, 323

Focal cortical malformation (FCM) 300–302, 305, 306, 310, 311, 317, 318, 320, 322, 323

Four-color interphase FISH 273, 277–279, 282–289, 295

Fragment size analyzer 306

Frontotemporal lobar degeneration (FTLD) 331–333, 336–342, 352, 354, 355

Functional prediction score 311, 314

G

GATK BaseRecalibrator tool 366

GATK IndelRealigner tool 366

Gel2Gel 110, 113, 121, 122

Genome analysis toolkit (GATK) 178, 368, 371

Genome structural variations
  aneuploidy 3, 4, 8, 12, 14, 27, 28, 44, 82, 138, 141
  copy number alterations (CNAs) 3, 4, 8, 9, 12, 14, 15, 18
  copy number variation/variants (CNVs) 7, 44, 82, 109–121, 123–128, 136, 137, 139, 143–145, 149, 150, 152, 154, 155, 159, 177, 178, 254, 256, 264
  inversions 3, 4, 31, 32, 242, 248
  losses of heterozygosity (LOH) 3, 4
  mobile element insertion (MEI) 3, 4, 9, 173, 178, 254, 264
  multiploidy 3
  translocations 3, 4, 143, 333

Genomic DNA extraction 69, 148, 238, 242, 317

Genomic evolutionary rate profiling (GERP) 311

Genomic imprinting 65, 70–76

Genomic instability (GIN) 295

Genomic mosaicism 27, 44, 54

Genomic quality number (GQN) 302, 305, 306

Germinal polyploidy 57

Ginkgo 111, 124, 125, 127, 128

H

Haploid 10, 28, 57, 59, 138

Hemimegalencephaly (HME) 300

High-resolution melting analysis (HRMA) 145, 147, 151

High-throughput DNA sequencing 221

Human embryonic stem cells (hESCs) 192, 194–197, 200, 201, 203

Hyperploid 61, 101

I

Illumina platform 110, 120, 124

Image cytometry (IC) 60, 82, 86–91, 95, 96, 98, 101

Induced pluripotent stem cells (iPSCs) 128, 192, 196–198, 200, 201

Interphase FISH 30, 32, 44–50, 52–54, 61, 273, 279, 282–287, 295

Interphase high-resolution chromosome-specific multicolor banding (ICS-MCB) 33, 34, 82

Interphase molecular cytogenetics 27, 30, 33, 37

Intron-flanking PCR 203–205

In utero electroporation (IUE) 304, 319, 320, 322, 323

Inverse PCR (iPCR) 200, 204

L

Laser capture microdissection (LCM) 301, 303, 311, 316–318, 322

LCGreen Plus 159

L1-EGFP retrotransposition assay 193

L1 element 201, 220, 240, 241, 248
L1-enrichment by PCR amplification..........................15–16
Lewy body diseases..28
LINE-1 capture library 257
LINE-1 copy number..209–217
LINE-1 insertion 219–243, 245–247,
        249, 265, 266
LINE-1 retrotransposition 190–204, 210, 257
L1 insertions.. 16, 190–193, 195,
        196, 200, 203, 222, 226, 227, 238
L1 mobilization...220, 222
L1 profiling 227
L1 retrotransposition................................. 16, 190–204, 226
L1 retrotransposition assay...192
Long interspersed nuclear element-1 (L1/LINE-1) 200
Long terminal repeat (LTR)..190
LTR retrotransposon..190

M

Major depressive disorder (MDD) 195
Malformations of cortical development (MCD) 300
Massively parallel sequencing (MPS)........................144–146
McCune–Albright syndrome 300
Median absolute deviation (MAD) 121, 122, 127
Melting curve 147, 149, 158
Melting temperature (Tm) 146, 149, 159, 248
Metaphase chromosome spread................ 341, 342, 344, 352
Microtubule-associated protein tau (MAPT) 330–332,
        350–352
Microwell displacement amplification system
        (MIDAS)..10
Mobile element scanning (ME-Scan) 227
Mobile genetic elements..219
Molecular identifier (MID)..316
Mosaic aneuploidy...................272, 273, 329–352, 354, 355
Mosaic variant discovery
        bulk analysis...5–7
        single cell analysis 5, 7, 8, 14
Multiple annealing and looping-based amplification cycles
        (MALBAC).................................. 119, 134–138,
        222, 226, 256, 257, 266
Multiple displacement amplification
        (MDA) 9, 110, 119, 134,
        136–139, 222, 224–226, 232–238, 256,
        266, 372, 373
Multiplex ligation-dependent probe amplification
        (MLPA)...144, 145
Multiplex PCR.. 148, 149, 158
Multiplex PCR efficiency...159
Mutant allele frequency (ALT)................................363, 374
MuTect algorithm 308, 312

N

Neural progenitor cells (NPC) 189–205,
        210, 274, 300, 319, 320, 322, 331–334, 336–339,
        354, 355

Neurodegenerative disease................................... 37, 82, 182,
        197, 329–352, 354, 355, 361, 363, 377
Neurofibromatosis type I (NF1)..144
Neurogenesis 128, 254, 354
Neuronal nuclei (NeuN)..............................5, 44, 47, 50–52,
        67–70, 116, 117, 203, 211, 231, 258, 261, 303,
        317, 321, 333, 338
Neurons... 11, 27, 37, 44, 57–65, 67,
        69–72, 74–76, 81–84, 86–93, 95–97, 99, 101,
        102, 109–121, 123–128, 144, 155, 163–182, 193,
        195–201, 203, 205, 209–217, 219–243, 245–247,
        249, 253–266, 271–285, 287–295, 300, 302, 317,
        320, 321, 323, 330–334, 338, 339, 344, 346, 348,
        352, 354, 362, 364, 372
Next-generation sequencing (NGS)......................... 82, 119,
        121, 134, 135, 242, 256, 257, 264, 302, 303,
        305–307, 311, 314, 316, 318, 363, 364
Niemann-Pick C1 (NPC)...................................331–334,
        336–342, 354, 355
Noncoding DNA..189
Non-LTR retrotransposon 190
Nucleofection 199–202, 205
N value 58

O

Open reading frames (ORFs)........................... 190, 191, 201

P

Parkinson's disease (PD)................................... 361, 363, 377
PCR based targeted high-depth sequencing 5
PhastCons score 311
Phred score 177, 309
PicoPLEX 10, 110, 112, 118–121, 125
PolyPhen score 311
Polyploidy............................. 12, 28, 57, 58, 60, 273, 288, 295
Post-traumatic stress disorder (PTSD)............................195
Presenilin (PSEN).. 330, 333, 335,
        340, 341, 346–350, 355
Propidium iodide (PI) 29, 45, 47,
        50, 60–63, 67–70
Proteus syndrome 300

Q

Quantitative FISH (QFISH) 28, 35, 36
Quantitative PCR (qPCR)......................83, 93–96, 144, 145
Quantitative PCR of alu repeats............... 60, 83, 93–95, 100

R

Read-depth...14, 15, 111, 125–128,
        176–178, 306, 312, 313, 316
Restricting dNTPs 143–152, 155, 156, 158–160
Retroelement......................................15, 190–192, 195, 198
Retrotransposition 16, 189–204, 210,
        220–222, 226, 248, 257, 264, 266
Retrotransposition-competent L1 (RC-L1)....................190

Retrotransposon 4, 15, 110, 189, 190, 195, 198, 219, 220, 225, 226, 238, 239, 254

Retrotransposon capture sequencing (RC-seq) 195, 196, 222, 225–227, 237–239, 247, 248, 257, 258

Rett syndrome (RTT) 197, 210

Reverse transcriptase (RT) 145, 189–191, 198, 210, 213, 254

Ribonucleoprotein particle (RNP) 191

S

SAI file..125

Samtools...125, 366

Sanger-sequencing.........................5, 17, 18, 63, 71–73, 75, 178, 204, 239, 242, 265, 307, 311, 317, 373, 374

Schizophrenia...28, 209, 210

SCNT-ES cell 165–167, 172–179, 181

Senescence-associated secretory phenotype (SASP) 272

Sequence capture...15, 164

Sequencing, identification and mapping of primed L1 elements (SIMPLE) 227

SINE-VNTR-Alu element (SVA) 190, 220

Single allele base extension reaction assay (SABER) 301, 316–319

Single cell clonal expansion 10

Single cell CNV 109–121, 123–128

Single-cell mosaicism 264–265

Single cell RC-seq............................222, 225–227, 237–239, 247–249, 257, 258

Single cell sequencing (SCS) 164

Single-cell SLAV-Seq data.....................................255

Single-cell WGA 10, 17, 257–259, 261, 262, 265

Single nuclei isolation........................ 117, 222–224, 227–232

Single nucleotide polymorphism (SNP) arrays...............5, 7, 13, 144

snpEFF program 303

SOAP2 Aligner 311

Somatic cell nuclear transfer (SCNT) 165–168, 171–179, 181

Somatic hypermutation 220

Somatic L1-associated variant (SLAV)............................196

Somatic L1 retrotransposition 192–198, 226

Somatic LINE-associated variant sequencing (SLAV-seq)..................................... 227, 254, 255, 257–259, 262–266

Somatic mosaicism 3–11, 13–20, 27, 109, 110, 128, 220, 253, 254, 266, 300

Somatic mutation 15, 28, 163–167, 173–176, 178, 179, 181, 182, 209, 210, 299–306, 308, 310, 311, 314–316, 318, 320, 322, 361, 363–365, 367, 368, 370, 373, 377

Somatic polyploidy..58, 59

Somatic tetraploidy..59

Somatic variant......................................4, 173, 176, 221, 319, 363, 365–368, 372–376

Somatic variant discovery 16, 174, 175, 312

Somatic variant validation 16, 306, 315, 368, 373

Sorting intolerant from tolerant (SIFT)...........................311

S-phase...58, 59, 166, 339, 341

Spinal muscular atrophy (SMA).............. 143, 144, 146, 155

Sporadic Alzheimer's disease (SAD) 361, 365, 377

Sporadic disease...300, 329, 331, 340, 341, 361

Structural variant mutation......................................170, 173

Structural variants (SVs)...................................... 3, 4, 12, 14, 16–19, 170, 173, 176–179, 254, 264

Structural variation breakpoint detection 177

Sturge–Weber syndrome 300

T

Target primed reverse transcription (TPRT)........... 191, 192, 220, 221, 239, 242, 248

Target site duplication (TSD) 191, 220, 221, 239–241, 248

Tetraploidy.................................. 59–62, 64–75, 273

Transposable elements (TEs) 189, 190

Trout erythrocyte nuclei (TEN)..................................45, 50

TruePrime 10

Turner's Syndrome 330

U

Untranslated region (UTR).............................. 190, 194, 198

V

Variant allele frequency (VAF).........................166, 177, 178

Variant annotation..368

Variant call format (VCF) file 176, 367

Variant evaluation...................................... 302, 303, 310

V(D)J recombination..220

Vecuum...303, 314

Virmid algorithm 311

W

Whole chromosome instability (W-CIN).......................272

Whole exome sequencing (WES) 302, 303, 306, 310, 312, 316, 318

Whole genome amplification (WGA) 5, 6, 9, 10, 17, 110, 114, 118–121, 125, 133–135, 138, 226, 235, 237–240, 243, 245–247, 254, 255, 257–259, 261, 262, 265, 305, 364, 372, 373

Whole genome sequencing (WGS) 3, 5, 6, 14, 16, 20, 254

Z

Zebrafish 65–69, 75

Printed in the United States
By Bookmasters